Thermodynamics FOR DUMMIES®

by Michael Pauken, PhD

John Wiley & Sons, Inc.

Thermodynamics For Dummies®

Published by
John Wiley & Sons, Inc.
111 River St.
Hoboken, NJ 07030-5774
www.wiley.com

Copyright © 2011 by John Wiley & Sons, Inc., Hoboken, New Jersey

Published simultaneously in Canada

WILEY

About the Author

Michael Pauken is a senior mechanical engineer at NASA's Jet Propulsion Laboratory (JPL), an operating division of the California Institute of Technology (CalTech) in Pasadena, California. He develops spacecraft thermal control systems and balloon systems for exploring other planets in the solar system. He enjoys flying model rockets, but they never seem to make it into orbit. He is also a lecturer in Mechanical Engineering (ME) at CalTech, teaching courses in thermodynamics and heat transfer.

Before joining JPL, he was an assistant professor in ME at Washington University in St. Louis, Missouri.

He received his PhD in ME from the Georgia Institute of Technology in Atlanta. Vanderbilt University in Nashville, Tennessee, gave him a bachelor of science in ME in exchange for a large sum of cash (and passing a lot of classes).

He has been a member of ASHRAE, ASME, and AIAA professional engineering societies and is a registered professional engineer.

Dedication

To Frances and Victor, my little heat engines.

Author's Acknowledgments

Although writing itself is a solitary effort, it takes a team of players to make it presentable. I thank the folks at Wiley who made this book possible: Erin Mooney, my acquisitions editor; Chad Sievers and Chrissy Guthrie, my project editors; Christy Pingleton, my copy editor; and James Mayhew and Michael Moorhead of Rose-Hulman Institute of Technology, my technical editors. Traci Cumbay deserves special thanks for her advice, questions, and way with words.

I could not have written this book without the teachings and discussions on thermodynamics from my mentors and colleagues from long ago: Dr. Sheldon Jeter, Dr. Said Abdel-Khalik, Dr. Jim Hartley, and Dr. Theo Alexander.

I owe a million thanks to my students who made numerous contributions to this book by working examples, writing out equations, looking up information, and even writing a few parts of chapters — Keir Gonyea, Teri Juarez, Grace Li, and Lin Li. You are the next generation of engineers.

Publisher's Acknowledgments

We're proud of this book; please send us your comments at http://dummies.custhelp.com. For other comments, please contact our Customer Care Department within the U.S. at 877-762-2974, outside the U.S. at 317-572-3993, or fax 317-572-4002.

Some of the people who helped bring this book to market include the following:

Acquisitions, Editorial, and Media Development

Senior Project Editors: Christina Guthrie, Chad Sievers

Acquisitions Editor: Erin Calligan Mooney

Copy Editor: Christine Pingleton

Assistant Editor: David Lutton

Editorial Program Coordinator: Joe Niesen

Technical Editors: James E. Mayhew, Michael S. Moorhead, PhD

Editorial Manager: Christine Meloy Beck

Editorial Assistants: Rachelle Amick, Alexa Koschier

Cover Photos: © iStockphoto.com / Ovidiu Iordachi

Cartoons: Rich Tennant (www.the5thwave.com)

Composition Services

Project Coordinator: Nikki Gee

Layout and Graphics: Carl Byers, Carrie A. Cesavice, Mark Pinto, Corrie Socolovitch

Proofreader: Henry Lazarek

Indexer: Potomac Indexing, LLC

Special Help

Jennifer Tebbe

Publishing and Editorial for Consumer Dummies

　　Kathleen Nebenhaus, Vice President and Executive Publisher

　　David Palmer, Associate Publisher

　　Kristin Ferguson-Wagstaffe, Product Development Director, Consumer Dummies

Publishing for Technology Dummies

　　Andy Cummings, Vice President and Publisher

Composition Services

　　Debbie Stailey, Director of Composition Services

Contents at a Glance

Contents at a Glance

Table of Contents

Introduction

*T*oo often, thermodynamics is one of those "gateway" classes in engineering schools. Either you pass through the gate and study engineering, or you don't pass and start thinking about studying something else, like political science or psychology.

I've met countless people who find out my career choice is in engineering and say, "I started out in engineering. . . ." Then they struggled and moved on to something else. Usually, they stopped at the beginning, right around the time they studied statics, dynamics, and thermodynamics.

A good teacher and a good book can make all the difference in the world in determining whether you enjoy and understand a subject. Even then, people click in different ways. Some students loathe "Professor A," while others think Professor A is the best they've ever had. Some students understand the textbook the professor chooses to use in the class; others don't.

It's too bad when people quit something they struggle against. Many things aren't as hard as they may seem. You just need the light to click on in your brain for a tough subject to become understandable. Think of that childhood story *The Little Engine That Could*. You *can* understand thermodynamics. This book can help you through that gate into engineering.

About This Book

Thermodynamics For Dummies can help you understand how energy is used in the things that make our lives comfortable and convenient, such as automobiles, airplanes, and electric power plants. This book is written with you in mind. I use examples of things — real things that you can identify with — to help you understand a concept. For example, when I show you how to figure out how much heat goes into something, I give you an example of boiling an egg — something you're familiar with — not dropping a block of iron into a vat of water. Have you ever dropped a block of iron in water? Unless you're a blacksmith, I don't think you have.

In engineering schools, thermodynamics is often taught in both mechanical engineering and chemical engineering curriculums. This book comes at thermodynamics from a mechanical engineering perspective. I focus on how energy is used in planes, trains, and automobiles (and refrigerators, air conditioners, and power plants). Although I cover nonreacting gas mixtures and combustion reactions, I don't go into chemical equilibrium.

I also make the examples in the book connect with each other in different areas. For example, when you find out about the thermodynamics of a gasoline engine, you discover that the energy source for the engine comes from combustion. Then when you read about combustion reactions, I show you how much energy the reaction provides for an automobile engine.

Conventions Used in This Book

Every subject has its own language, and thermodynamics is no different. I use the following conventions in this book:

- ✔ Whenever I introduce a technical term, I use *italics* so you can quickly see it and look for an explanation.

- ✔ I also use *italics* to indicate variables in mathematical equations.

- ✔ I work all the examples in the metric system, because it's less confusing than the system of feet, inches, pounds, and so on that we use in the U.S.

- ✔ I use **boldface** for velocity (**V**) and *italics* for total volume (*V*) to distinguish between these two variables.

- ✔ I also use **boldface** to denote the action parts of numbered steps and to highlight key words or phrases in bulleted lists.

What You're Not to Read

If you want to read this book cover-to-cover, that's up to you. But if you just want to get an explanation of something you're stuck on, you can skip the sidebars (they appear in gray-shaded boxes).

Sidebars are tidbits of information that have interesting information related to a topic. You can grasp the fundamentals without reading them, but they do enhance your overall understanding (and possibly enjoyment) of a topic.

Foolish Assumptions

I assume that you've taken an introductory physics class. If so, you may have seen a little bit of thermo already. But if you haven't had physics, don't worry; you can probably grasp the concepts in this book anyway.

I also assume you've had some calculus. In some parts of thermodynamics, you have to understand how to use an integral (Chapters 8 and 9). You don't have to be an expert in calculus to follow along because these parts of thermodynamics involve the simplest kinds of equations. Even if you don't know a thing about calculus, you can still solve almost all the problems in this book — that is, assuming you can do basic algebra.

How This Book Is Organized

I've organized this book along the lines of most undergraduate thermodynamic textbooks, which start with the basics and progress to more difficult subjects. Four parts of the book deal directly with thermodynamics, plus one part gives you a quick peek at some interesting people and processes in thermodynamics. You can follow this book from beginning to end along with your own thermodynamics textbook (assuming you're taking the class), or you can just dip into any section and chapter to get help with something you may be stuck on.

Part 1: Covering the Basics in Thermodynamics

I start off Part I by giving you some examples of both natural and man-made thermodynamic systems so that you can recognize and relate to thermodynamics. After you're comfortable with these examples, I explain how energy can be used to perform work and how work can be used to move energy. I show you that a thermodynamic system is made up of several processes, and each process has a starting point and an endpoint. These two points are described by the properties of the materials used in the process. Some of the basic properties I discuss include temperature, pressure, internal energy, enthalpy, entropy, and specific heat. Oh! And I know you're eager to learn about the laws of thermodynamics, so I introduce them to you here.

Part II: Employing the Laws of Thermodynamics

In this part, I discuss the fundamental concepts of the conservation of mass and the conservation of energy. This jargon means that neither mass nor energy can be created or destroyed, but both mass and energy can change form. In a nutshell, the amount of mass and energy at the start of any process equals the amount of mass and energy at the end of the process. While this idea seems simple enough, it can get confusing when mass changes from, say, a liquid to a gas and energy changes from heat to work during a process. I break down the confusion by showing you how to apply the first law of thermodynamics to different kinds of systems or processes.

The second law of thermodynamics is the most complicated part of thermodynamics, although that doesn't mean you can't understand it. What the second law is really about is the concept that energy, like a river, flows in a certain direction: downhill. The basic idea is that a thermodynamic process starts with energy flowing from a reservoir at a high elevation, does something useful — like spin a motor — and ends with the remaining energy settling into another reservoir at a lower elevation. Sometimes the energy in the lower reservoir can still be used for other desirable things; sometimes it can't. The abstract concept of entropy describes this phenomenon and lets you know whether a process is possible or impossible. I hope you don't get depressed when I tell you that the universe is going to end one day because of the "increase in entropy" principle.

Part III: Planes, Trains, and Automobiles: Making Heat Work for You

Part III definitely covers the most interesting part of thermodynamics. If you get bogged down somewhere as you begin to study thermodynamics, I recommend you spend a few minutes looking over this part to see the wonderful things you can do once you grasp the fundamentals. Think of it as window shopping. You'll be motivated to persevere so you can get to the fun part of thermodynamics. I describe how to use the first and second laws of thermodynamics on things like gasoline and diesel engines, jet engines, electric power plants, refrigerators, and air conditioners. You find out how to determine the efficiency of these kinds of machines and how to improve them. Your mouth is watering already, isn't it?

Part IV: Handling Thermodynamic Relationships, Reactions, and Mixtures

In Part IV, I cover how gases behave and relate to one another in different situations. Many gases obey a special relationship law called the ideal-gas law; others don't behave that way and are called real gases. Some gas mixtures react with each other, like the combustion of gasoline vapor in air, and form carbon dioxide and water vapor. Combustion reactions are especially important because they're the energy source for many kinds of thermodynamic engines. Other gas mixtures don't react with each other at all, like air and water vapor. The presence of moisture in the air is very important in understanding applications related to heating, ventilating, and air conditioning. I help you sort out these thorny relationship issues.

Part V: The Part of Tens

In this book, I cover a lot of ground and throw in a bunch of names along the way, like Celsius, Watt, Fahrenheit, and Diesel. Who were these people, and how did they get into a thermodynamics book? In Part V, I give you a thumbnail sketch of ten early pioneers in thermodynamics. I also talk about ten new or less common ways of producing work from energy — in things like automobiles, jets, and power plants — that you may be interested in learning more about.

Finally, because solving problems in thermodynamics relies on material property data for substances, I've provided an appendix that includes abridged versions of thermodynamic property tables. You can use these tables to follow along with examples presented throughout the book. Although these tables aren't as extensive as ones you find in textbooks, they provide all the information you need to grasp the fundamental concepts.

Icons Used in This Book

You'll find some icons in the margins of this book. These icons are flags that point out different things. Here's what the icons stand for.

This icon tells you that you should either remember a certain fact for future reference or recall this fact from an explanation that appears earlier in the book.

I use this icon when I give you a bit of extra information to help you understand a topic or a suggestion for a shortcut to working a problem.

When you see this icon, it means pay attention! I'm giving you important information to keep you from making a common mistake.

Where to Go from Here

Each chapter in this book is written with the idea that you may want to jump around and read about individual topics. For example, if you're stuck on entropy, you can turn to Chapter 8 to get a grasp on the fundamentals. You don't need to read the first seven chapters. If you need to understand certain basic concepts before you start reading a particular chapter, I act as your traffic cop and direct you to where those concepts are explained more fully.

Part I
Covering the Basics in Thermodynamics

In this part . . .

Thermodynamics is part of natural law — it governs the use of energy in everything from the weather to your diet. I introduce you to basic concepts of energy, describing how it changes form in both natural and man-made systems. With just four simple laws, a table of material properties, and a calculator, you can figure out how much energy it takes to boil an egg or operate a power plant. Soon you'll be calculating all kinds of interesting facts related to energy.

Chapter 1

Thermodynamics in Everyday Life

Thermodynamics is as old as the universe itself, and the universe is simply the largest known thermodynamic system. When the universe ends in a whimper and the total energy of the universe dissipates to nothingness, so will thermodynamics end.

Broadly speaking, thermodynamics is all about energy: how it gets used and how it changes from one form to another. In many cases, thermodynamics involves using heat to provide work, as in the case of your automobile engine, or doing work to move heat, as in your refrigerator. With thermodynamics, you can find out how efficient things are at using energy for useful purposes, such as moving an airplane, generating electricity, or even riding a bicycle.

The word *thermodynamics* has a Greek heritage. The first part, *thermo,* conveys the idea that heat is somehow involved, and the second part, *dynamics,* makes you think of things that move. Keep these two ideas in mind as you look at your world in terms of the basic laws of thermodynamics. This book is written to help you understand that thermodynamics is about turning heat into power, a concept that really isn't so complicated after all.

Grasping Thermodynamics

Many thermodynamic systems are at work in the natural world. That sun you see in the sky is the ultimate energy source for the earth, warming the air, the ground, and the oceans. Huge masses of air move over the earth's surface. Giant currents of water swirl in the oceans. This movement and swirling happens because of the transformation of heat into work.

Energy takes many different forms — it can't be created or destroyed, but it can change form. This statement is one of the fundamental laws of thermodynamics. Consider how energy changes form in storm clouds:

- Storm clouds have motion within them.
- Motion between moisture droplets in clouds rubbing against each other creates friction.
- Friction causes a buildup of static charge.
- When the charge becomes high enough, the clouds produce lightning.
- This electrical surge of energy can then start a fire on the ground, and before you know it, you have a combustion problem on your hands.

Not only does energy change form, but *matter* (that is, a material or substance) also changes form in many thermodynamic systems. Storm clouds are formed by water evaporating into the air. As the water vapor reaches the colder parts of the atmosphere, it condenses to form clouds. Eventually, the amount of moisture the clouds contain becomes great enough to collect into droplets and form liquid water again, so it rains.

One thing people have observed about energy is that it flows in a preferred direction. This observation is another fundamental law of thermodynamics. Heat flows from a hot object to a cold object. Wind blows from a region of high pressure to a region of low pressure. Some forms of energy are developed by forces of nature. Air bubbles move upwards in water against gravity because buoyancy forces them to rise. Water droplets fall in the atmosphere because the force of gravity pulls them toward the ground.

Another brilliant observation about energy is that if you have absolutely no energy at all, you have no temperature. The concept of absolute zero temperature is a fundamental law of thermodynamics.

I cover the changing forms of energy and matter and the fundamental laws that govern how these changes work in Part I.

Examining Energy's Changing Forms

Many clever people have observed the fundamental laws of thermodynamics in natural systems and applied them to create some wonderful ways of doing work by harnessing energy. Heat is used to generate steam or heat up air that moves a piston in a cylinder or spins a turbine. This movement is used to turn a shaft that can operate a lawn mower; move a car, a truck, a locomotive, or a ship; turn an electric generator; or propel an airplane.

Other clever people have used thermodynamic principles to use work to move heat from one place to another. Refrigerators and heat pumps remove heat from one location to produce a desirable cooling or heating effect. The work required for this cooling shows up on your electric bill every month.

In Part II, I show you how the fundamental laws of thermodynamics can tell you how much heat you need to provide to produce work that can be used to move a car, fly an airplane, or turn an electric generator. You can also use the laws of thermodynamics to find out how efficient something is at using energy.

Energy is the basis of every thermodynamic process. When you use energy to do something, it changes form along the way. When you start your car, the battery causes the starter to turn. The battery is a big, heavy box of chemical energy. The battery's job is to change chemical energy into electrical energy. An electric motor rotates (a form of kinetic energy) the engine, and the spark plugs fire. These sparks ignite fuel via a combustion process wherein the chemical energy from gasoline is turned into a form of thermal energy called internal energy. In the few seconds it takes to start your car, energy changes from chemical to electrical to kinetic to thermal or internal energy.

Kinetic energy

A car battery provides electricity to operate your starter. As the motor turns, the electrical energy is converted into a form of mechanical energy called *kinetic energy*. Kinetic energy involves moving a mass so that it has velocity. The mass doesn't have to be very large to have kinetic energy — even electrons have kinetic energy — but the mass has to be moving. Before you start the car, nothing in the engine is moving so it has no kinetic energy. After the engine is started, it has kinetic energy because of its moving pistons and rotating shafts. If the car is parked while the engine is running, the car as a "system" has no kinetic energy until the engine makes the car move.

Potential energy

If you drive your car up a hill and park it there, you change the kinetic energy of the car into another form of energy called *potential energy*. Potential energy is only available with gravity. You must have a mass located at an elevation above some ground state. Potential energy gets its name from its potential to be converted into kinetic energy. You see this conversion process when you park on a hill and forget to apply the parking brake. Potential energy changes back into kinetic energy as your car rolls down the hill.

Internal energy

When you apply the brakes to stop your car, you make energy change form again. You know the car has kinetic energy because it's moving. Stopping the car changes all this kinetic energy into heat. Brake pads squeeze onto steel disks or steel drums, creating friction. Friction generates heat — sometimes a lot of heat. When materials heat up, another form of energy called *internal energy* increases. Have you ever smelled a burning odor while driving down long hills? That odor indicates that someone used their brakes to slow down, and the brakes overheated. Do your brakes a favor: Shift into a lower gear and allow the engine to do the braking for you. When the engine is used as a brake, the kinetic energy of the moving car compresses the air in the cylinders, and the energy changes into internal energy because the air heats up from compression. All that internal energy just goes out the tailpipe.

Watching Energy and Work in Action

Until the invention of the steam engine, man had to slug it out against nature with nature. Horses pulled coaches, mules pulled plows, sails moved ships, windmills ground grain, and water wheels pressed apples into cider that fermented and made man feel happy for all his labors. The steam engine was able to replace these natural work sources and move coaches, plows, and ships, among many other things. For the first time, fire was harnessed to provide something more than just heat — it was used to do work. This use of heat to accomplish work is what Part III is all about. Over time, many different kinds of work machines were developed, theories were made, and experiments were done until a rational system of analyzing heat and work was developed into the field of thermodynamics.

Engines: Letting energy do work

A *heat engine* is a machine that can take some source of heat — burning gasoline, coal, natural gas, or even the sun — and make it do work, usually in the form of turning a shaft. With a rotating shaft, you can make things move — think of elevators or race cars. Every heat engine uses four basic processes that interact with the surroundings to accomplish the engine's job. These processes are heat input, heat rejection, work input, and work output.

Take your automobile engine as an example of a heat engine. Here are the four basic processes that go on under the hood:

1. **Work input**

   Air is compressed in the cylinders. This compression requires work from the engine itself. Initially, this work comes from the starter. As you can imagine, this process takes a lot of work, which is why they don't have those crank handles on the front of cars any more.

2. **Heat input**

   Fuel is burned in the cylinder, where the heat is added to the engine. The heated air in the cylinder naturally wants to increase in pressure and expand. The pressure and expansion move the piston down the cylinder.

3. Work output

As the expanding gas in the cylinder pushes the piston, work is output by the engine. Some of this work compresses the air in adjacent cylinders.

4. Heat rejection

The last process removes heat with the exhaust from the engine.

Refrigeration: Letting work move heat

When Willis Carrier made air conditioners a popular home appliance, he did more than make people comfortable and give electric utilities a reason for growth and expansion. He brought thermodynamics into the home. Thermodynamics has been there all along, and you never realized it. Refrigerators, freezers, air conditioners, and heat pumps are all the same in thermodynamics. Only three basic processes involve energy interacting with the surroundings in what is known as the *refrigeration cycle:*

1. Heat input

Heat is absorbed from the cold space to keep it cold.

2. Work input

Work is added to the system to pump the heat absorbed from the cold space out to the hot space.

3. Heat rejection

Heat is rejected to the hot space.

Actually, a fourth process takes place in most refrigeration cycles, but it doesn't involve a change in energy. Instead of having a work-output process in the cycle like heat engines do, refrigerators simply utilize a pressure-reducing device in the system. Energy doesn't change form in such a device.

Getting into Real Gases, Gas Mixtures, and Combustion Reactions

Using energy to generate electric power, cool your house, fly a jet, or race cars around the Indianapolis Motor Speedway is the glamorous side of thermodynamics. But behind the movie stars are a supporting cast and crew of *thermodynamic relationships* (this is jargon for "mathematical equations") for real gases, gas mixtures, and combustion reactions that make it all happen.

In Part IV, you discover the difference between a real gas and an ideal gas. There you see that real gases behave a bit differently than ideal gases. You also figure out the thermodynamic properties of a mixture of gases, such as water vapor and air for heating, air conditioning, and ventilating purposes. Lastly, you calculate how much energy you can get out of fuel in a combustion reaction to power your jet, your race car, or your lawn mower.

If you want to sell jet engines to an aircraft manufacturer, you have to show that your engine burns fuel efficiently. To build a jet engine, you need to know how much energy a combustion reaction adds to an engine and how much the air in the engine heats up as a result of the combustion. To figure out the latter, you use thermodynamic relationships of real gases to calculate properties such as temperature, pressure, and energy.

Discovering Old Names and New Ways of Saving Energy

As you learn about thermodynamics, you'll run across a number of names. Some of the names may be familiar; others may be new to you. For example, when you get your electric bill, it tells you how many watt-hours of electricity you used last month. If you reheat yesterday's leftover pizza, you set your oven to 350 degrees Fahrenheit. (Or, if you live outside the U.S., you set your oven to some temperature in degrees Celsius.) That big rig that's riding your bumper on the highway burns diesel fuel.

How did these terms — watt, Fahrenheit, Celsius, and Diesel — become part of our language? In Part V, you discover that these words (and six more) are actually the last names of characters bent on figuring out what energy is and how to harness it for the benefit of mankind (and maybe to line their pockets with some folding money).

Pioneers in thermodynamics didn't just work in the good old days; there are modern-day pioneers as well. The world's demand for energy steadily increases while energy resources dwindle. Part V shows you ten ways innovative thinkers have improved energy consumption for automobiles, air conditioners, refrigerators, and electric power plants. Making a better future for all has motivated many people to think of better ways to use energy.

Even Albert Einstein got a patent for making a better air-conditioning system (see Chapter 18). Maybe you'll be inspired to create your own innovation and make a name for yourself in thermodynamics.

Chapter 2

Laying the Foundation of Thermodynamics

- -

In This Chapter

▶ Getting comfortable with thermodynamic properties

▶ Discovering how to use thermodynamic processes

▶ Understanding the laws of thermodynamics

- -

*E*very builder knows the importance of a good foundation. Imagine building a house without making sure the footing is secure. Making everything go together just right and having a house stand the test of time would be difficult. The same can be said for studying a new subject. Every subject has its own lingo, and this chapter is where you find out about some key concepts that are used throughout thermodynamics. Many of the terms are already familiar to you, like "temperature," "pressure," and "density." Others may be words you've heard before, but you may not know exactly what they mean — for example, "specific heat capacity" and "latent heat." Here, I go beyond merely defining words and concepts; I tell you the basic ways in which these ideas come together in thermodynamics.

In thermodynamics, processes are the means by which objectives — heating your house, for example — are accomplished. One house may use a boiler to make hot water that circulates in radiators throughout the house for heating. Another house may use a furnace to heat the air and a fan to circulate the hot air in the house. Both processes accomplish the same objective but take different paths to do it. Just as you may sometimes take a different route to get from Point A to Point B, the same thing can happen in thermodynamics.

In this chapter, I describe the concept of thermodynamic processes and the paths taken during those processes to go from Point A to Point B. You may have heard that thermodynamics has two laws. Actually, it has four laws; the two lesser known ones deal with temperature and aren't really used in thermodynamic problems. I discuss these laws in this chapter so you'll know what all the fuss is about.

Defining Important Thermodynamic Properties

Whether they're solids, liquids, or gases, all materials have properties that tell you two things:

✔ Some properties, such as specific heat capacity, tell you how a material behaves during a thermodynamic process.

✔ Other properties, such as temperature or pressure, tell you what condition or state a material is in at any point in a thermodynamic process.

Suppose, for example, that you use a hot water bottle to warm up your bed before you get into it. The water is initially very warm. You can use the thermodynamic properties of mass and temperature to describe its condition or state when you first fill it up and its condition after it has warmed up your bed. The mass of water in the bottle remains constant, but the temperature of the water bottle decreases during the process.

The thermodynamic property of specific heat capacity describes how quickly the water cools when you put the water bottle in your bed. In this way, the specific heat capacity tells you how the hot water bottle behaves while it warms up your bed. You find out more about several important material properties in the upcoming sections.

Eyeing general measurement basics

Before I delve into the properties themselves, you first need a basic understanding of how properties are measured. If you can measure something, it has a *dimension*. Some dimensions are described as *primary* or *fundamental dimensions* such as length, mass, temperature, and time. When you combine dimensions to describe properties like volume, pressure, or energy, you have *secondary dimensions* or *derived dimensions*. Dimensions have units associated with them, such as Celsius or Fahrenheit for temperature, inches or meters for length, and kilograms or pounds for mass. Units quantify the size of a dimension.

Some property measurements don't have dimensions per se and are known as dimensionless properties. Often dimensionless properties are ratios or fractions where the dimensions cancel each other out. One very common dimensionless property is relative humidity, which is reported as a percentage.

The world is divided into two different unit systems:

✔ **English:** Most common in the United States, the English system is formally called the United States Customary System, but that's a big mouthful, so *English* is the term you hear most often and the one I use in this book.

✔ **Système Internationale (SI):** Everywhere except the United States, the Système Internationale is the system of choice. Often the SI system is called the *metric system.* SI units are now the preferred unit system in thermodynamics textbooks, so they're used throughout this book.

Furthermore, thermodynamic properties come in two different types; *extensive properties* depend on mass and *intensive properties* don't.

✔ **Intensive:** An intensive property doesn't depend on mass. Temperature and pressure are examples of intensive properties. No matter how much of a material you have, the temperature and pressure remain the same. They don't change simply because you have more or less of the material on hand.

✔ **Extensive:** An extensive property does depend on mass; it depends on how much material you have. Examples of extensive properties include mass, volume, and energy.

Not everything is black and white. Some intensive properties are represented by extensive properties that are divided by a unit mass. These properties are often referred to as *specific properties,* such as specific volume or specific internal energy. Intensive properties are usually represented by a lowercase symbol. Exceptions include temperature (T) and pressure (P). Extensive properties are usually represented by uppercase symbols except for mass (m), which uses the lowercase. You gotta love exceptions to rules.

Mass

Mass is a property in thermodynamics that describes the amount of material used in a system or process. Many people think mass is the same thing as weight. But it's not. Weight is actually a force exerted on an object by gravity. The weight (W) of an object is calculated by multiplying its mass (m) by the acceleration of gravity (g): $W = m \cdot g$. On earth, the acceleration of gravity is 9.81 meters per second squared (m/s^2) or 32.2 feet per second squared (ft/s^2).

The SI unit for mass is the kilogram (kg), and the English unit is pound mass (lbm). The lbm abbreviation is derived from the Roman word "libra" plus an "m" for mass. The SI unit for weight is the newton (N), and in English units it's pounds-force (lbf).

The calculation of weight demonstrates that the dimensions of force (F) are derived from a product of mass (m) times acceleration (a): $F = m \cdot a$. The newton is therefore defined as $1 \text{ N} = 1 \text{ kg} \cdot \text{m/s}^2$.

When mass (m) appears as a variable in an equation, it's italicized in this book. When "m" appears in units, as it does in the preceding equation, it stands for "meters" and is not italicized in this book.

Pressure

If you blow up a balloon, there is pressure inside the balloon. Pressure is the result of molecules inside the balloon moving around and colliding into the walls of the balloon. Because molecules are also colliding with the outside of the balloon, there's pressure on the outside as well. These molecular collisions create a normal force that acts on the surfaces of the balloon. Figure 2-1 shows that a normal force is perpendicular to a surface. *Pressure* is defined as a normal force acting over an area and is a thermodynamic property in liquids and gases. (When acting on a solid, the same concept is called *stress*).

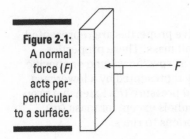

Figure 2-1:
A normal
force (*F*)
acts per-
pendicular
to a surface.

In honor of Blaise Pascal's experiments in hydraulics, the SI system defines the unit of pressure as the Pascal (Pa). The unit of pressure is related to the units of force acting over an area by this equation:

$$1 \text{ Pa} = 1 \text{ N/m}^2$$

Because 1 Pascal isn't very much pressure, the SI system prefers to use the kilopascal (kPa) for most practical engineering calculations:

$$1 \text{ kPa} = 1{,}000 \text{ Pa}$$

To put this into perspective, the pressure of the atmosphere at sea level is about 101 kilopascals.

In the English system, units of pressure are defined as pounds-force per square inch (psi). You can convert English to SI units using this formula:

1 psi = 6.895 kPa

Pressure as a thermodynamic property has a few different scales. In the absolute pressure scale, the zero pressure point is defined by the perfect vacuum. The standard atmospheric pressure is defined as 101.325 kilopascals in SI units or 14.7 pounds-force per square inch absolute (psia) in the English unit system. The *a* at the end of psi indicates it is the *absolute* pressure scale.

A tire pressure gauge doesn't read 14.7 pounds-force per square inch when it's just sitting around in the garage. It reads 0 pounds-force per square inch gauge (psig). The *g* at the end of psi means that the *gauge* pressure scale is being used. When you measure your tire pressure and the gauge reads 25 pounds-force per square inch, it means that the tire pressure is 25 pounds-force per square inch above the atmospheric pressure. You can convert between gauge pressure and absolute pressure with the following formula. In this formula, P_{gage} is the gauge pressure, P_{abs} is the absolute pressure, and P_{atm} is the atmospheric pressure:

$$P_{gage} = P_{abs} - P_{atm}$$

Vacuum pressure (P_{vac}) refers to a situation in which the absolute pressure of a gas or liquid is less than the atmospheric pressure. You measure vacuum pressure with a vacuum gauge. You can convert between vacuum pressure and atmospheric pressure with this formula:

$$P_{vac} = P_{atm} - P_{abs}$$

Pressure is always expressed with positive numbers (unlike temperature when the Celsius or Fahrenheit scale is used and negative temperatures are possible). When you use the thermodynamic property tables in the appendix for your thermodynamic calculations, the entries for pressure are in absolute pressure values.

Temperature

Temperature describes how hot or cold something is. You're probably familiar with this thermodynamic property. On a cold and windy winter day, you may listen to the weather report and hear that the air temperature is 25 degrees Fahrenheit (°F), but with the wind chill it feels like 15 degrees. Or you may feel very comfortable when it's 75 degrees outside, but if you go swimming in 75-degree water, you feel very cold. The human body feels temperature but isn't really a good thermometer.

When you feel something that's hot or cold, what you're really feeling is heat moving to or from your skin. When air is still, it doesn't remove heat from your skin very effectively, but wind removes heat more easily. That's why the air "feels" colder when the wind blows. Water removes heat more effectively than air, so it feels colder to you than air. A metal surface feels cold compared to a cotton cloth because the metal surface conducts heat away from you more quickly

Temperature is really a concept used to describe the direction in which heat transfer takes place. Heat is a form of energy that naturally flows from hot to cold temperatures. You can feel heat transferring to or from your body when you are in an environment warmer or colder than you. At *absolute zero temperature,* you can't remove any more heat from a material because it has no energy. I discuss energy in detail later in the "Energy" section.

Four different measurement scales are commonly used to quantify temperature:

- **Celsius (°C):** Developed by Anders Celsius, this scale defines 100 equally spaced points or degrees between the boiling and freezing points of water at 1 atmosphere (atm) pressure.

- **Fahrenheit (°F):** Daniel Gabriel Fahrenheit chose the temperature of an ice bath melting in a solution of salt water as the zero point on his scale. He then chose body temperature (96 degrees, in his estimation) as the second point on his scale. After he made up his scale, it was discovered that water freezes around 32 degrees and boils around 212 degrees. It never occurred to him that temperature scales would one day be related to the freezing point and boiling point of pure water.

- **Kelvin (K):** Known as the 1st Baron Kelvin of Largs, Scotland, William Thompson suggested the need for an absolute temperature scale defining zero temperature as the point where there is zero energy. He based his scale on the Celsius scale and calculated that the freezing point of water is 273.15 Kelvin (usually rounded to 273 Kelvin in calculations).

- **Rankine (R):** An absolute temperature scale that uses the Fahrenheit scale, William Rankine's system declares the freezing point of water to be 459.67 Rankine. You usually round this number up to 460 Rankine when using it in calculations.

Note that the Kelvin and Rankine scales don't use the degree symbol. The degree symbol was dropped from the Kelvin scale in 1967, and some textbooks have followed suit with Rankine.

You use the absolute temperature scales of Kelvin or Rankine exclusively when using the ideal-gas law (I introduce you to the ideal-gas law in Chapter 3). The *ideal-gas law* depends on ratios between pressure, temperature, and volume. These ratios are based on absolute zero as a starting point to make the pressure and volume scale properly with temperature.

You can use the Celsius or Fahrenheit scales when a calculation involves a temperature difference. You cannot use Celsius or Fahrenheit scales when a calculation involves a temperature ratio. If you're ever in doubt as to which scale to use, you can't go wrong with the Kelvin or Rankine scales.

You only need three formulas to convert from one temperature scale to another. You can rearrange these formulas to solve for any other conversion, like Fahrenheit to Celsius or Rankine to Fahrenheit.

- ✔ **Celsius to Fahrenheit:** °F = 1.8 · °C + 32.

- ✔ **Celsius to Kelvin:** °C = K – 273.15. Remember, you can just round off the 273.15 to 273 if you want. Doing so doesn't really change the answer that much.

- ✔ **Fahrenheit to Rankine:** °F = R – 459.67. Rounding 459.67 to 460 requires fewer keystrokes on the calculator without losing much accuracy.

When you need to calculate a temperature difference, the change in degrees Celsius equals the change in Kelvin. That is: Δ°C = ΔK. The same is true in English units; one degree change in Fahrenheit equals one degree change in Rankine. That is: Δ°F = ΔR. Don't make the mistake of adding 273 to the change in Celsius to find the change in Kelvin or adding 460 to the change in Fahrenheit to get the change in Rankine.

Density

Density refers to the amount of mass of a material per unit volume. Density (ρ) is determined by using the mass (m) of a material and dividing it by the volume (V) it occupies. In the SI system, the units are kilograms per cubic meter (kg/m^3), and in the English system, they're pound-mass per cubic foot (lbm/ft^3). You calculate density of a uniform material with this equation: $\rho = m/V$.

The densities of liquids are often considered incompressible because they don't change a lot with temperature and pressure. However, densities of gases vary quite a bit with both temperature and pressure. You see this effect in relation to the ideal-gas law in Chapter 3.

When you use the reciprocal of density, you get the *specific volume* (v). The specific volume is used in the ideal-gas law equations I describe in Chapter 3. Specific volume is an intensive property. This is the volume per unit mass. The units are cubic meters per kilogram (m^3/kg) in SI units and cubic feet per pound-mass (ft^3/lbm) in the English system. You can calculate specific volume with this equation: $v = V/m = 1/\rho$.

Sometimes the density of a material in a thermodynamic property table is given as the *specific gravity,* meaning that the density of the material is divided by the density of water at 4 degrees Celsius, which is 1,000 kilograms per cubic meter. You calculate the specific gravity for a material with this equation:

$$SG = \frac{\rho}{\rho_{H_2O}}$$

Because the specific gravity is a fractional ratio of two densities, it's a dimensionless property. When you have a fraction made of two identical properties, the units cancel each other out so it becomes dimensionless.

The specific gravity of water at room temperature is around 1.0, but because density can change with temperature and pressure, even for liquids, the specific gravity of water can vary a little bit depending on the conditions. For example, when water is at 100 degrees Celsius, its density is 958 kilograms per cubic meter. So at 100 degrees Celsius, the specific gravity of water is 0.958. If a material or substance has a specific gravity less than 1.0, it will float in water. When the specific gravity is greater than 1.0, a material will sink in water.

Another property related to density that involves liquids and solids is called the *specific weight* (γ) of a material. You may see this when determining the potential energy of a column of liquid, for example. To get the specific weight, you multiply the density of the material (ρ) by the acceleration of gravity (g): $\gamma = \rho \cdot g$. In SI, the units are newtons per cubic meter (N/m^3), and in English, the units are pounds-force per cubic foot (lbf/ft^3).

In this case, the use of "specific" is an exception to the rule of "per unit mass." This is a "per unit volume" property, even though you may think it's related to the mass of a material.

Energy

Energy is a familiar word — you probably know intuitively what it is and what it does — but how can you define it? You may think of it as the capacity of a physical system to make a change on another physical system. By physical system, I mean anything of any size, from subatomic particles to the universe itself — so everything has some form of energy associated with it.

In thermodynamics, energy can transfer from one system to another system through work, heat transfer, and mass transfer. You (as one physical system) use energy to lift a bag of groceries (another physical system) out of the car — this is work. You turn on the oven (one physical system), and energy bakes a batch of cookies (another physical system) using heat transfer. A river (one physical system) flows into a reservoir (another physical system) and adds energy to the reservoir by mass transfer.

Energy comes in many different forms, and the quality and quantity of energy of a physical system depend on the physical state of the system relative to a reference frame. I discuss some of these energy forms and their reference frames in the following sections.

The quality of energy determines its ability to do work or be useful in some way. For example, the energy in a combustion process has a high quality (relative to the environment), and it can be used to power an automobile engine. The quality of the energy in the exhaust is relatively lower than the combustion process and is no longer useful for the engine. However, the exhaust is warmer than the environment, so it has a higher energy quality than the environment. In theory, the energy in the exhaust could be made to do something useful. The reference frame for the quality of energy is often the ambient environment.

From a broad thermodynamic perspective, energy can be grouped as either macroscopic or microscopic. The *total energy* (E) of a system is the sum of the energy from both groups.

- ✔ **Macroscopic:** A system has *macroscopic energy* if it can move relative to a reference frame in response to an external force such as gravity, magnetic fields, or electric fields. Typical forms of macroscopic energy are kinetic energy and potential energy. For example, a bulldozer has macroscopic energy when it pushes a pile of dirt. The good thing about macroscopic energy is that it can do work. In thermodynamics, work is defined by moving a force over a distance. I discuss work in detail in Chapter 4.

- ✔ **Microscopic:** *Microscopic energy* in a system comes in many forms and works on a molecular scale. Individual molecules have kinetic energy by rotation, translation, and vibration. The combination of all forms of molecular energy is called *internal energy*. Internal energy depends on the degree of molecular activity within the system, and it can be harnessed or changed in form to be useful on a macroscopic scale. For example, steam can be heated at high pressure so it can flow through a turbine and produce work. Microscopic energy can transfer heat from one system to another system as long as one system is warmer than the other. I discuss heat in detail in Chapter 4. Lastly, microscopic energy does not depend on any external frame of reference.

How much energy is in a donut?

Another commonly used unit of energy is the *calorie*. Originally, a calorie was the amount of energy required to raise one gram of water by one degree Celsius. But because the amount of energy it takes to increase the temperature of water varies with temperature, it's now defined as the specific heat of water at 15 degrees Celsius. (I discuss specific heat in the following section.) The kilocalorie (1,000 calories) is the *calorie* used by nutritionists for the energy content of food. Your average 320-calorie donut has enough energy to increase the temperature of 32 kilograms of water by about 10 degrees Celsius. You may want to think twice about having that second donut.

In the SI system, the unit of energy is joules (J). One joule is equal to the amount of energy required to apply a force of 1 newton over a distance of 1 meter. In electricity, a joule is also the amount of work you get with 1 watt for 1 second. In the English system, the unit for energy is the British thermal unit (Btu). A Btu is the amount of energy required to increase the temperature of 1 pound of water by 1 degree Fahrenheit.

The following sections identify specific types of energy you encounter when working with thermodynamics and how to calculate the total energy (E) of a system.

Kinetic energy

Kinetic energy is the energy associated with a mass moving relative to a reference frame. In many cases, the ground is the frame of reference because it remains stationary. But for an airplane, a bird, or Superman, the air may be a more appropriate frame of reference. You calculate the kinetic energy (KE) of an object with mass (m) moving at velocity (V) relative to the reference frame, using the following equation:

$$KE = \frac{mV^2}{2}$$

So, for example, if you throw a standard 5-ounce (0.14-kilogram) baseball at 80 miles per hour (35.8 meters/second), the kinetic energy of the ball is as follows:

$$KE = \frac{1}{2}(0.14 \text{ kg})(35.8 \text{ m/s})^2 = 89.7 \text{ J}$$

When you calculate kinetic energy, the units in the equation are kg $\cdot$ m^2/s^2, which is equivalent to a joule.

The following energy units are equivalent to each other:

1,000 J = 1 kJ

$$1 \text{ kJ} = 1,000 \text{ kg} \cdot \text{m}^2/\text{s}^2 = 1,000 \text{ N} \cdot \text{m} = 1 \text{ kPa} \cdot \text{m}^3$$

If you need to work with energy using English units, the conversion from SI to English units is 1 kJ = 0.948 Btu.

Potential energy

Potential energy is the form of energy associated with a mass at an elevation above a reference frame. The mass has the potential to do work by moving downward in a gravitational field. Often the reference frame for potential energy is the ground, but it can be between any two vertical elevations.

You can calculate the potential energy (*PE*) of an object with mass (*m*) at elevation (*z*) above the reference frame with the gravitational acceleration (*g*), using this equation:

$$PE = m \cdot g \cdot z$$

Suppose you're winding up your grandfather clock. The weights are 2 kilograms each, and you raise them up by 30 centimeters. You can figure out how much potential energy the weights have to operate your clock:

$$PE = (2 \text{ kg})(9.81 \text{ m/s}^2)(0.30 \text{ m}) = 5.9 \text{ J}$$

You can verify the units are correct in this equation by the equivalent energy unit equations I show earlier.

Internal energy

Internal energy is associated with both the motion and structure of molecules in a substance. Molecular or atomic movement by rotation, translation, and vibration in a substance increases as energy is added to the system. A change in kinetic energy of molecules often results in a temperature change of a substance. This form of internal energy is called *sensible energy* because it can be "sensed" by a thermometer. The symbol used for internal energy is *U*.

The molecular structure of a substance can change when energy is added to the system — for example, by melting or evaporating. Various forces bind molecules and atoms together in a substance. These *binding forces* are strongest in solids, which is why solids retain their shape. The binding forces are weaker in liquids, which stay together but change shape and fill containers. In gases, the binding forces are weakest; a gas freely expands until it fills a closed container. The following forms of internal energy interact with different kinds of binding forces:

- **Latent energy:** When a substance changes phase (such as a solid melting into a liquid), some of the molecular binding force energy is overcome by thermal energy. The amount of energy required to break these bonds is called *latent energy* because thermal energy is added to the substance and its temperature doesn't change during the phase-change process. A thermometer can't sense the energy added during a phase change, so the energy is "hidden" or latent. See Chapter 3 for more details on phase changes. Latent energy is given different names for different kinds of phase changes:

  - When a liquid freezes (or melts), the latent energy is called the *latent heat of fusion* (h_{sf}). The subscript "s" stands for the solid; the subscript "f" stands for the liquid.

  - When a liquid evaporates (or condenses), it's called the *latent heat of vaporization* (h_{fg}). The subscript "f" stands for the liquid; the subscript "g" stands for the gas.

  - When a solid evaporates directly to a gas (like dry ice does), it's called the *latent heat of sublimation* (h_{sg}). The subscript "s" stands for the solid; the subscript "g" stands for the gas.

- **Chemical energy:** *Chemical energy* is used when fuels are burned in combustion reactions. When a fuel is burned, some chemical bonds are destroyed while new ones are formed. This can release energy stored in the molecular bonds of a combustible material. Chemical energy is present in every chemical reaction, but not all chemical reactions are of interest in thermodynamics. I discuss chemical energy related to common combustion processes in Chapter 16.

- **Nuclear energy:** This type of energy is one of the strongest forms of molecular energy because it's associated with the nucleus of an atom. This energy form isn't discussed in thermodynamics at this level, but at least you know it exists.

Calculating total energy

The total amount of energy that a system contains includes both the macroscopic and microscopic forms of energy associated with it. The following equations for total energy sum up the kinetic, potential, and internal energies. The extensive form of the total energy of a system includes the mass of the system in each energy component. The intensive form is on a per unit mass basis.

$$E = KE + PE + U = \tfrac{1}{2}m \cdot V^2 + m \cdot g \cdot z + m \cdot u \text{ (extensive form)}$$

$$e = ke + pe + u = \tfrac{1}{2}V^2 + g \cdot z + u \text{ (intensive form)}$$

Nearly every thermodynamic analysis uses one of these two equations. You use them to determine the amount of energy that occurs in the form of work and heat for many thermodynamic processes. Because chemical and nuclear energy aren't usually dealt with in introductory thermodynamics, these forms of energy are omitted from the equations calculating the total energy of a system.

Enthalpy

Many thermodynamic processes involve a fluid flowing through a device and undergoing a change in internal energy. Any time a fluid flows into a system, it does work on the system, and any time a fluid flows out of a system, the system does work on the fluid. This occurs when hot water flows though an automobile radiator, for example. The work associated with the fluid flowing into or out of the system is represented by the product of the pressure (P) and the volume (V) of the fluid. This happens in so many situations that a new property, *enthalpy* (H), is used to combine the change in internal energy with this flow work, to make calculations more convenient. Enthalpy is defined by the following equation: $H = U + PV$. The units for enthalpy are the same as those for internal energy. The units for the pressure-volume (PV) product in the SI system are kilopascals-cubic meter ($kPa \cdot m^3$) which is equivalent to kilojoules (kJ). You use enthalpy on thermodynamic systems such as turbines, compressors, nozzles, and heat exchangers. I discuss these nifty devices in Chapter 6.

Specific heat

Whether you watch it or not, a large pot of water on the stove can take a long time to reach a boil. A thermodynamic property called *specific heat* determines how much a material heats up if you add a given amount of energy to it.

Heating up solids and liquids

Some materials can change temperature quickly because they don't need a lot of energy to heat up; others take a lot of energy to heat up. For solid and liquid materials, the following equation shows how you use the mass (m) of the material and its specific heat (c) to determine how much energy is required to change its temperature ($T_2 - T_1$): $U_2 - U_1 = m \cdot c(T_2 - T_1)$. The amount of energy that goes into the material changes the internal energy ($U_2 - U_1$) of the material. I list the specific heat of a few commonly used liquids and solids in Table A-10 of the appendix.

When you heat up (or cool down) a material, the initial temperature and internal energy of the material are represented by T_1 and U_1, respectively. The final temperature and internal energy are T_2 and U_2. In the SI system, the units for specific heat are kilojoules per kilogram-Kelvin (kJ/kg-K).

Specific heat changes with temperature, so this equation loses some accuracy over large differences between T_1 and T_2. Accuracy can be improved by using the specific heat at the average temperature between T_1 and T_2.

You can calculate the amount of energy you need to heat up a cup of water for tea. Take a cup of water with a mass of 200 grams (0.2 kilogram) and heat it up from 22 to 100 degrees Celsius. The specific heat of water is 4.18 kilojoules per kilogram-Kelvin. The change in internal energy of the water is equal to the amount of energy you put into it, so the result is $U_2 - U_1 = (0.2 \text{ kg})(4.18 \text{ kJ/kgK})(100 - 22)°C = 65.2 \text{ kJ}$.

Heating up gases

Gases have two different kinds of specific heat, and the version you use depends on the conditions under which energy is added or removed from the gas. This differs from solids and liquids, which only have one value of specific heat. The equations for the specific heat for gases are very similar to those of a solid or liquid material.

- **Constant-volume specific heat:** If you heat up a gas in a rigid container, the volume remains constant, and you use the constant-volume specific heat (c_v). The constant-volume specific heat relates a temperature change in a process to the change in internal energy in the following equation.

$$U_2 - U_1 = m \cdot c_v(T_2 - T_1)$$

   This equation assumes the specific heat remains constant during a process. The accuracy of the equation improves if you use the specific heat at the average process temperature.

- **Constant-pressure specific heat:** If you heat up a gas in a process that has a constant pressure, you use the constant-pressure specific heat (c_p). For a constant-pressure process, the enthalpy $(H_2 - H_1)$ of the gas changes because the gas must do work in addition to changing internal energy. The constant-pressure specific heat relates a temperature change in a process to the change in enthalpy in the following equation.

$$H_2 - H_1 = m \cdot c_p(T_2 - T_1)$$

   This equation assumes that the specific heat remains constant. The accuracy of the equation improves if you use the specific heat at the average process temperature.

The two different specific heats are related to each other by the gas constant (R). The gas constant for air is 0.287 kilojoules per kilogram-Kelvin. The gas constant is defined by the ideal-gas law and is discussed in detail in Chapter 3. The following equation shows this relationship between c_v and c_p: $c_p = c_v + R$.

Another important ideal-gas property is called the *specific heat ratio* (k). You use this ratio for evaluating thermodynamic processes for compressors, turbines, and reciprocating engines. This property is defined by the following equation:

$$k = \frac{c_p}{c_v}$$

For air, the specific heat ratio is about 1.4. The specific heat ratio of a material is not a constant value property; it usually increases with temperature because specific heat changes with temperature.

Entropy

Entropy is a measure of order or disorder, depending on whether you're a half-full or half-empty kind of person. Entropy is perhaps the most unusual thermodynamic property because it can be applied to so many nonthermodynamic situations. Your desk may be a perfect setting for entropy. Papers may collect there, and as the piles grow taller, the desk becomes more cluttered — an increase in entropy. Reversing the effects of entropy usually takes organized effort. Somebody has to do some work to reduce the entropy on your desk.

When something is highly organized, entropy is low. After cleaning your room, you've put it into a state of low entropy. Conversely, high entropy indicates a high degree of chaos or disorder. Thermodynamically, entropy is usually low at low temperatures and high at high temperatures. Think of molecules: When they're at low temperatures, they don't move around very much and they're highly organized. Entropy is low. If things heat up, the molecules start moving as if they're at a dance party, and entropy increases. Entropy is evaluated using the second law of thermodynamics (see Chapters 8 and 9).

Pressure also affects entropy. High pressure tends to reduce entropy compared to low pressure. So entropy is a bit complicated because it depends on both temperature and pressure. Entropy tells you how effectively energy is used in thermodynamic systems or processes. An ineffective system would generate a lot of entropy, which can be thought of as generating a lot of disorder.

Understanding Thermodynamic Processes

When you pump up a bicycle tire, brew a pot of coffee, or fill a bathtub with hot water, you perform a thermodynamic process. When you drive your car, a series of thermodynamic processes takes place in your engine. When you turn on the lights in your home, a series of thermodynamic processes goes on back at the power plant to give you electricity.

A thermodynamic *process* is simply a means of describing how a system changes from one state to another state. A *system* is the quantity of material that you study or analyze in a problem. A *state* describes the properties of the material at the beginning and at the end of a process. Every system that undergoes a thermodynamic process does so in an environment, also called the *surroundings*. An imaginary line divides the system and its surroundings;

this line is called a *boundary*. A boundary can be a physical surface or an imaginary surface. Depending on the process, the boundary can remain stationary or it can move.

Processes help you correctly analyze thermodynamic problems. If you do a thermodynamic analysis on inflating a bicycle tire, you define the air inside the tire as the system. Everything external to the tire constitutes the surroundings. Before you begin to inflate the tire, the air inside the tire is at a certain temperature and pressure. It has a unique value of internal energy, which is based on the temperature of the air. These conditions describe the initial state of the process. When you have finished inflating the tire, it will have a new temperature, pressure, and internal energy. These conditions describe the final state of the process. The tire itself can be the boundary between the air inside and the surroundings. Inflating the tire causes it to change shape, so the boundary moves in this process.

The upcoming sections tell you more about thermodynamic processes.

Creating a path for a process

If you want to analyze a thermodynamic process, you need to be able to define the path the process is going to take. In thermodynamics, a *path* describes how a process changes a system from one state to another state. Imagine you're going to take a bath. One way to do it is to first heat the water before you fill the tub, and then fill the tub with hot water. Figure 2-2 shows you this process, labeled A. Another way is to fill the tub with cold water, and then heat the water after the tub is full (Process B in Figure 2-2). The starting point and the endpoint in Figure 2-2 are the same, but each process takes a different path from start to finish.

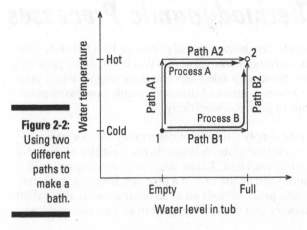

Figure 2-2: Using two different paths to make a bath.

Analysis in thermodynamics is easiest when a path follows a property of the system that remains constant. For example, the graph in Figure 2-2 defines two properties: the fill level of the tub and the water temperature. Process A starts with the tub empty, and the water is heated before going into the tub. Process A begins with Path A1. On Path A1, the water level remains constant; that is, the tub remains empty as the water is heated. After the water is heated, Process A follows Path A2, where the tub is filled with hot water. On this path, the water temperature remains constant as the tub is filled.

You use many of the following types of paths in your thermodynamics analyses of heat engines and refrigeration machines in Chapters 10–13:

- ✔ An *isochoric* or *isometric* path has constant volume.

- ✔ An *isobaric* path has constant pressure.

- ✔ An *isothermal* path has constant temperature.

- ✔ An *isenthalpic* path has constant enthalpy.

- ✔ An *isentropic* path has constant entropy.

- ✔ An *adiabatic* path has no heat transfer.

- ✔ A *reversible* path returns to the initial state unchanged.

In a thermodynamic analysis, you assume each of these different paths occurs under ideal circumstances, meaning that properties remain uniform in a system during a process. You can imagine that when processes happen quickly, the properties aren't necessarily uniform throughout the system. Think of your automobile engine. The engine spins rapidly, and the cylinders burn fuel in quick order. When you analyze the automobile engine, you assume the temperature and pressure inside the cylinder are uniform, but you know that this isn't really true. It's an idealization to make analysis simpler.

Finding the state at each end of a path: The state postulate

Robert Frost is famous for his poetic statement:

> *Two roads diverged in a wood, and I —*
>
> *I took the one less traveled by,*
>
> *And that has made all the difference.*

This poem conjures up an image of starting down a path and arriving at a destination. You may be in one state of mind at the beginning of the road and arrive with a completely different state of mind at the end. This idea shows up in every process in thermodynamics.

A state defines the properties (such as temperature and pressure) of a system or substance at each endpoint of a thermodynamic process. It describes what condition the system is in before the process starts and the condition of the system when the process is finished. Every process has an initial state and a final state. Without a state, you don't know where you're starting from or where you're ending up with a process. The number of properties you need to describe the state of a system is spelled out in the *state postulate*. The state postulate says: Two independent intensive properties are necessary to completely define the state of a simple compressible system.

A simple compressible system doesn't involve kinetic or potential energy or energy from magnetic or electric fields. If these energy forms are involved in a process, then you need to specify properties related to those forms of energy in addition to the two properties required by the state postulate.

Properties are independent if one property can change while the other one is held constant. You see this in Figure 2-2, where each path allows one property to vary while the other one remains constant. Among the many properties in thermodynamics, temperature and specific volume are always independent of each other. Temperature and pressure are independent in most cases, the exception being when a phase change occurs in a constant-pressure process. If the pressure is constant during a phase-change process, then the temperature is constant. Specific volume changes during a constant-pressure phase change, so it can be the other independent property. For a phase change in a rigid container, the specific volume is constant and both the pressure and temperature vary, so either temperature or pressure can be used as an independent property.

One implication of the state postulate is that if you know any two independent properties that describe the state of a material, it's possible to find all the other thermodynamic properties (such as internal energy, enthalpy, and entropy) of the material. For example, in the preceding sections on internal energy and enthalpy, you see equations that allow you to determine these energy values from temperature. In Chapter 3 you can see how temperature, pressure, and volume are related to each other in gases with the ideal-gas law. And in Chapter 8, you can see how entropy is related to both temperature and pressure.

Connecting processes to make a cycle

In any thermodynamic cycle, the fluid goes through several different processes and returns to its initial state at the end of a cycle. Thermodynamic cycles use processes that either involve work or move heat, and Figure 2-3 shows a cycle with four typical processes:

- ✔ Work input
- ✔ Heat input
- ✔ Work output
- ✔ Heat output

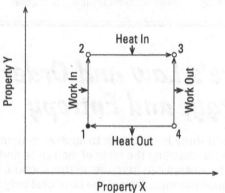

Figure 2-3:
A thermodynamic cycle connects several processes together in a loop.

The processes in a thermodynamic cycle are connected to each other at the initial and final states of each process; that is, the final state of one process is the initial state of the following process. In a typical thermodynamic analysis of a cycle, you need to find the state of the system at the end of each process in the cycle.

In the modern steam engine, for example, water goes through these thermodynamic processes to complete an ideal cycle:

1. **A compression process using a work input increases the water pressure.**

2. **A heat input process generates steam from liquid water.**

3. **An expansion process in a piston or turbine gets work out of the engine.**

4. **A condensation process gets heat out of the system and turns the steam to its liquid state.**

After Step 4, the water is ready to go around the cycle again.

In Chapters 10–13, I cover the thermodynamic analysis of many different cycles for engines and refrigeration.

Incomplete cycles waste energy

Early steam engines were terribly inefficient. A fire boiled water and generated steam. The steam was piped to a piston and then exhausted to the air. In those early engines, the water didn't go through a complete thermodynamic cycle because it wasn't returned back to the boiler so it could go through the engine continuously.

James Watt improved the steam engine's efficiency by adding a condenser after the piston, so the water would feed back into the boiler and complete a thermodynamic cycle within the engine. The water was therefore returned to its initial state in the boiler after completing a series of thermodynamic processes.

Discovering Nature's Law and Order on Temperature, Energy, and Entropy

In thermodynamics, you need three basic things to analyze systems: properties of materials, processes for changing the state of materials, and laws to define how properties and processes can function. In this section, I tell you about the four laws of thermodynamics. Two of the laws deal only with the concept of temperature and aren't really used for performing analyses. They exist to define what constitutes thermal equilibrium and give the basis for the concept of the absolute temperature scale. The laws that you will use most often are known as the first and second laws of thermodynamics. These laws are used to analyze the flow of energy (first law) and the generation of entropy (second law) in a process or a cycle.

The zeroth law on temperature

After a hard day at work, you come home from the office, kick off your shoes, and rest your weary feet on your nice leather ottoman. The leather feels cool to your feet, and it refreshes you. After a while, the leather under your feet warms up and stops feeling quite so refreshing. Your feet and the leather have reached *thermal equilibrium;* you and the leather have the same temperature.

Heat can be transferred from one object to another only if there's a temperature difference between them. The rest of the ottoman is at thermal equilibrium with your living room, so no heat transfer takes place between the ottoman and the room. Take a photo of the ottoman with an infrared camera, and you'll see an image of where your legs and feet were on the ottoman.

When two objects are at the same temperature, they do not transfer heat. This is a statement of the *zeroth law of thermodynamics*. The zeroth (pronounced like zero with a th) law says that if two bodies are independently in thermal equilibrium with a third body, they're in thermal equilibrium with each other. This is stating the obvious. It's almost like saying that if two boys look like a third boy, you're probably looking at triplets. The implication of the zeroth law is that it validates the use of a thermometer for measuring temperature, with the thermometer being the so-called third body.

The first law on energy conservation

When you hear the phrase *energy conservation,* you may think of turning up the thermostat in your house during the summer or riding your bike to the store instead of driving there. In thermodynamics, energy conservation has a completely different meaning. The *first law of thermodynamics* states that energy cannot be created or destroyed; it can only change form. Fundamentally, you study thermodynamics to understand how energy changes form — how heat can be used to do work and how work can be used to move heat from one place to another.

When you use the first law to analyze a system, you need to account for how energy changes in a system. This accounting is known as doing an *energy balance* on a system. Keeping track of what happens to energy in a system is like balancing your checkbook. When you get paid, your checkbook balance goes up; this is like energy coming into a system. You spend money from your account on various things, and your balance goes down. In a thermodynamic system, the energy goes out to do different things, like perform work.

The principle of conservation of energy says that whenever energy enters a system, it must either leave the system in some way or change the energy of the system. Energy can enter and leave a system in any of three ways:

- ✔ **As a form of work:** For example when you compress air with a pump to inflate a bicycle tire, that's work. Your muscles supply the energy to do the work of compression.

- ✔ **As a form of heat:** When you start your car, the fuel burning inside the engine releases heat to the air inside the cylinders, allowing the engine to do work for you.

- ✔ **By adding or removing mass from the system because all mass possesses energy:** If you turn on your hot-water faucet, hot water leaves your water heater and is replaced by cold water. Energy is lost from the water heater by the mass of water with lower energy entering into the system.

You can balance your energy checkbook by using one of these equations:

$$\left[\begin{array}{c}\text{Total energy}\\\text{entering a system}\end{array}\right]-\left[\begin{array}{c}\text{Total energy}\\\text{leaving a system}\end{array}\right]=\left[\begin{array}{c}\text{Change in total}\\\text{energy of a system}\end{array}\right]$$

$$E_{in}-E_{out}=\Delta E_{system}$$

$$E_{in}-E_{out}=\left(Q_{in}-Q_{out}\right)+\left(W_{in}-W_{out}\right)+\left(E_{mass,in}-E_{mass,out}\right)=\Delta E_{system}$$

These three equations express the *energy balance* of a system. The last equation is a generalized form of the energy balance equation. An energy balance equation can apply to both a process and a cycle. When you write an energy balance equation for a process, you usually have either the energy-in or the energy-out term. Therefore, the amount of energy you put into or remove from a process equals the change in energy of the system in the process. Mathematically, you can write it like this:

$$\Delta E_{process}=E_{final}-E_{initial}=E_2-E_1$$

When you write an energy balance for a cycle, the cycle always returns to the initial state at the end of the last process, so the mass terms and the energy change term in the general energy balance equation are equal to zero. You can simplify the general energy balance equation into the following equation, which shows that the net heat transfer in a cycle is equal to the net work you get out of the cycle:

$$\left(Q_{in}-Q_{out}\right)=\left(W_{out}-W_{in}\right)$$

The second law on entropy

If you want to heat up a bowl of noodles for lunch, you don't just let it sit on the table and expect the noodles to cook on their own. Even though the table may have several kilojoules of energy because it has a certain mass and a certain temperature, the quality of the energy isn't sufficient to cook the noodles. The quality of the energy of the table, room, and bowl of soup are all the same, so no energy can be exchanged between them. If you let a bowl of hot noodles sit on the table, it cools off. Energy flows in a certain direction, just as water naturally flows downhill. In order to provide heating, you need a high-quality source of energy relative to the system receiving the energy. The heat source needs to have a higher temperature than the material receiving the heat.

Unlike the other three laws, the second law of thermodynamics isn't summed up in a single, simple statement. Instead, the second law is a collection of concepts about the quality of energy for performing a process. The basic idea is that for a typical thermodynamic heat-addition process, energy flows *from* a high-quality (high-temperature) energy reservoir *into* the process. Conversely, for a typical heat-removal process, energy flows *from* the process *into* a low-quality (low-temperature) energy reservoir. All thermodynamic cycles that provide work or refrigeration operate between high- and low-temperature reservoirs. The high-temperature reservoir is known as the *heat source,* and the low temperature reservoir is known as the *heat sink.*

When you perform a second law analysis on a system, you evaluate the quality of the energy used in the process as well as how much the quality degrades during the process. When you evaluate the quality of energy, you determine the amount of entropy generated in the process. Processes that generate large quantities of entropy in a process are inefficient. A second law analysis is helpful in looking at thermodynamic cycles and identifying where efficiency improvements are most beneficial.

The second law provides a basis for determining the maximum theoretical limit for efficiency of a cycle. No actual system can possibly reach the theoretical efficiency limit because of the effects of heat transfer through a temperature difference and friction. Both of these realistic behaviors generate entropy. In Chapter 7, I discuss the two different statements of the second law of thermodynamics as it applies to heat engines to produce work and to refrigeration cycles to move heat.

The third law on absolute zero

You've probably heard the saying "as slow as molasses in January." This expression means that molasses, which is quite thick at room temperature, gets very thick when it gets cold. The molecules in molasses don't move as much when they're cold as they do when they're warm.

When atoms and molecules in a substance (like molasses) vibrate, it shows they have energy. Even molasses in January has energy. As the temperature of a material decreases, the amount of molecular movement also decreases. If it were possible to bring a material down to absolute zero temperature, all molecular motion would stop, and you'd have a perfectly ordered material. Reaching absolute zero is theoretically impossible because the second law of thermodynamics dictates you need a low temperature reservoir slightly less than absolute zero to bring a process down to absolute zero.

When a pure crystalline solid material is perfectly ordered, you can conclude the entropy of the material must be zero. The energy of a material under this condition is also zero. The concept of zero entropy at zero temperature is the essence of the *third law of thermodynamics,* which serves as an absolute reference point for scaling the entropy of material.

Chapter 3

Working with Phases and Properties of Substances

● ●

In This Chapter

▶ Understanding phase changes in materials

▶ Reading phase diagrams

▶ Interpolating thermodynamic tables to find properties

▶ Finding pressure, volume, and temperature with the ideal-gas equation

● ●

I f you've ever had an iced coffee, you've enjoyed a drink that has gone through several phase changes. A *phase* is a state, such as solid, liquid, or gas, in which a material exists under certain temperature and pressure conditions. The iced coffee you drank started out as liquid water that was boiled, and some of that water turned to a vapor during the boiling process. Then the hot coffee was mixed with ice — solid water. A phase change occurs in a material when it boils, condenses, freezes, or melts.

In this chapter, I introduce you to the different phases of materials and provide diagrams that help you see what conditions a phase exists in. I also show you ways to find properties of materials in different phases, such as by using thermodynamic property tables and the ideal-gas equation. You need to find properties such as pressure, temperature, specific volume, and internal energy to calculate how much heat or work you can get out of a process (more on heat and work in Chapter 4).

It's Just a Phase: Describing Solids, Liquids, and Gases

On a cold winter day, you may see ice or snow outside, which is water in a solid phase. If you're drinking a cup of hot coffee to keep warm, the water is in a liquid phase. You may see vapors — water in a vapor phase — coming from your breath or from the hot coffee. Under different conditions, a material may exist in different phases. Phases depend on two properties, such as temperature and pressure. For example, at sea level, the standard atmospheric pressure is 101.325 kilopascals and water boils at 100 degrees Celsius. But in Denver, Colorado (the mile-high city), the atmospheric pressure is only 84 kilopascals and water boils at around 95 degrees Celsius. This difference in altitude is why some recipes include different directions for high-altitude cooking.

Sometimes pressure is expressed in a unit called *atmospheres,* abbreviated as atm. One atm pressure equals 101.325 kilopascals.

Figure 3-1 shows two diagrams illustrating the pressure, volume, and temperature (*P-v-T*) dependence of the solid-liquid-vapor phases of a material. When you make a 3-D plot of three variables like *P-v-T*, as shown in Figure 3-1, you form a surface in the plot. Think of taking a 2-D plot of two variables, say *X* and *Y*, that form a line or a curve and extending it in a third dimension; the curve becomes a surface. Figure 3-1a is for a material that contracts upon freezing, such as oil, paraffin, aluminum, copper, iron, or gold. Figure 3-1b is for a material, such as water, that expands upon freezing. Water is the most well-known substance that expands when it freezes.

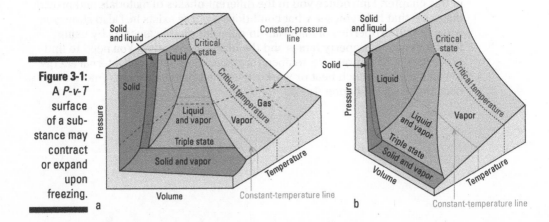

Figure 3-1: A *P-v-T* surface of a substance may contract or expand upon freezing.

At low temperatures, many materials exist as solids. Some materials remain solids even at very high temperatures. As indicated in the *P-v-T* diagrams in Figure 3-1, a solid can change directly into a vapor, although this is usually a slow process. The solid-to-vapor phase-transition process is called *sublimation*. Dry ice and mothballs are two examples of materials that sublimate. Dry ice is often used to ship items that need to remain cold, because the package doesn't get wet with liquid. Regular ice sublimates in your freezer, too. If you keep an ice cube in the freezer for several weeks, you can see that it gets smaller.

Sometimes interpreting a 3-D surface is a bit difficult, especially when it looks as complicated as the ones shown in Figure 3-1. So in the following sections, you look at the *P-v-T* surface in a 2-D perspective, using only two variables at a time: *P-T*, *T-v*, and *P-v*.

The phase diagram

If you take the *P-v-T* surface shown in Figure 3-1 and look at it from the pressure-temperature (*P-T*) point of view, you get what's called a *phase diagram,* as shown in Figure 3-2. For substances that contract when they freeze, the slope of the phase-change line between the solid and liquid in Figure 3-2 is positive. For substances that expand upon freezing, the slope of the phase-change line is negative, as shown by the dashed line in Figure 3-2. You can turn a material that contracts during freezing into a solid if you increase the pressure on the material. Increasing the pressure on a substance that expands during freezing can turn it into a liquid.

Figure 3-2:
A phase diagram tells you whether a material is a solid, liquid, or gas at a given temperature and pressure.

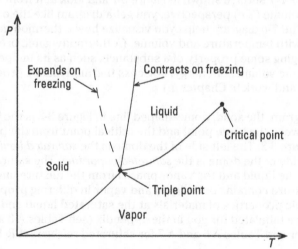

Expands on freezing

Contracts on freezing

Liquid

Critical point

Solid

Triple point

Vapor

The phase-change line between liquid and gas is called the *vaporization line*. The vaporization line starts at the *triple point,* where all three phases coexist. The three-phase mixture can have any proportion of solid, liquid, and gas, as long as the mixture is at the triple-point temperature and pressure. For water, the triple point is at 273.16 Kelvin and 0.6113 kilopascals. The vaporization line ends at the *critical point* where the distinction between a liquid and a vapor disappears. At pressures above the critical pressure, the liquid no longer undergoes a phase transition to a vapor; instead, the liquid is a dense fluid called a *supercritical fluid.*

A vapor is defined as a substance that is in the gaseous phase below the critical temperature.

The T-v diagram

Phase transitions between liquid and vapor happen around you every day. Here are some examples:

- ✔ Water evaporates readily into the air.

- ✔ The nozzles at gasoline pumps have vapor-recovery devices to prevent pollution.

- ✔ Power plants have boilers that generate steam from water.

- ✔ Air conditioners vaporize refrigerant by absorbing heat from your house.

If you take the *P-v-T* surface shown in Figure 3-1 and look at it from the temperature-volume (*T-v*) perspective, you get a diagram like the one shown in Figure 3-3. The *T-v* diagram helps you visualize how a thermodynamic process changes with temperature and volume. (A thermodynamic process is a means of changing some property of a substance, such as its temperature, pressure, specific volume, or energy. I discuss thermodynamic processes involving heat and work in Chapter 4.)

On the *T-v* diagram, the solid, dome-shaped line in Figure 3-3 is the liquid-vapor line between the triple point and the critical point from the phase diagram in Figure 3-2. The left side of the dome is the *saturated liquid line,* and the right side of the dome is the *saturated vapor line.* The saturation lines separate the liquid and the vapor phases from the mixture under the dome. The mixture contains both liquid and vapor in differing proportions. Thermodynamic properties of materials at the saturated liquid and saturated vapor states are tabulated for you in the appendix (see Tables A-3 and A-4 for saturated water and Tables A-6 and A-7 for saturated refrigerant R-134a).

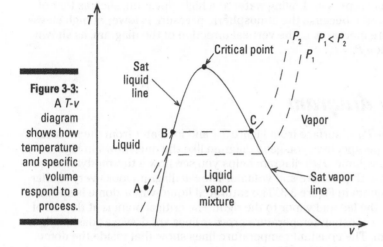

Figure 3-3:
A *T-v*
diagram
shows how
temperature
and specific
volume
respond to a
process.

On the left side of the saturation dome, you have 100-percent liquid, and on the right side of the dome, you have 100-percent vapor. The liquid region is called the *subcooled-liquid* state because the liquid temperature is lower than the temperature of a saturated liquid at the same pressure. This region is also called the *compressed liquid* state. You can think of the liquid as being at a higher pressure than that of a saturated liquid at the same temperature. The terms "subcooled liquid" and "compressed liquid" can be used interchangeably. At the top of the dome is the critical point. At pressures above the critical point, the material is a supercritical fluid.

Two dashed lines are drawn on the *T-v* diagram in Figure 3-3, showing two different constant-pressure processes occurring: one at pressure P_1 and the other at pressure P_2. A pot of water boiling on a stove is an example of a constant-pressure process. The local atmospheric pressure remains constant during the water-heating process.

Suppose the line P_2 represents a process of boiling water at sea-level pressure. The temperature and specific volume of the water start out at Point A labeled in Figure 3-3. As you add heat to the pot, it begins to boil at Point B. Both the temperature and the specific volume increase as you add heat. Notice that the temperature of a pure substance doesn't change during the phase-change process. The boiling water stays at the same temperature until all the liquid has turned to vapor at Point C. The specific volume of the vapor at Point C is significantly greater than the specific volume at Point B.

The line P_1 may represent boiling water at a high elevation, such as that of Denver, Colorado, because the atmospheric pressure is lower at high elevations. Pressure increases in the vertical direction of the diagram, as shown by the equation $P_1 < P_2$.

The P-v diagram

If you take the P-v-T surface from Figure 3-1 and look at it from the pressure-volume (P-v) perspective, you get a diagram like the one shown in Figure 3-4. The pressure-volume (P-v) diagram helps you see how a thermodynamic process changes with pressure and volume. The P-v diagram looks very similar to the T-v diagram in Figure 3-3. The saturated liquid-vapor dome is shown with liquid to the left and vapor to the right. The critical point is at the top of the dome. Two constant-temperature process lines are drawn as dashed lines in the diagram. The constant-temperature lines show that inside the dome, the phase transition between liquid and vapor is a constant-pressure and constant-temperature process.

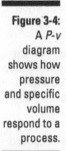

Figure 3-4:
A *P-v* diagram shows how pressure and specific volume respond to a process.

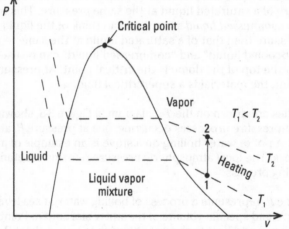

Inside the vapor dome of a P-v or T-v diagram, temperature and pressure are dependent on each other. For a given fluid temperature inside the vapor dome, there's a corresponding saturation pressure. In a similar fashion, for a given fluid pressure inside the vapor dome, there's a corresponding saturation temperature.

Temperature increases in the vertical direction of the diagram in Figure 3-4, as shown by the equation $T_1 < T_2$. Figure 3-4 shows a constant-volume heating process between lines T_1 and T_2 in the superheated vapor region of the diagram.

This process can be illustrated using your car tires as an example. Suppose you check the pressure in your tires one morning when the temperature of the tires is at T_1. You drive the car around a while, and the tires warm up to temperature T_2. If you check the tire pressure again, you'll see that the pressure increases. Although the volume of the tire may increase slightly with temperature, you can assume that volume basically remains constant as the tires warm up.

Knowing How Phase Changes Occur

A phase change in a material is one of the most frequently encountered processes in thermodynamics. Melting and freezing between solid and liquid phases are found in thermal energy storage systems, which are used for energy conservation purposes. In an electric power plant, water is boiled in a steam generator and condensed back into a liquid in the condenser (see Chapter 12). The refrigerator in your kitchen has a refrigerant that boils in the evaporator and condenses from a vapor to a liquid in the condenser (see Chapter 13).

Figure 3-5 shows a phase change from liquid to vapor on a temperature-volume (*T-v*) diagram. As you add heat to a liquid, both the temperature and volume of the liquid increase until the liquid reaches the boiling point. During boiling under constant pressure, the temperature remains constant until all the liquid is turned into vapor. The volume for a given mass of a fluid increases a lot as it becomes a vapor. Often the volume of a vapor is 1,000 times more than the volume of a liquid. Additional heating of the vapor continues to increase both temperature and volume. In this section, I focus on the liquid-to-vapor phase change and introduce you to thermodynamic lingo on phase changes so you can talk like an engineer.

Figure 3-5:
A *T-v* diagram showing a constant-pressure boiling process changing a liquid into a gas.

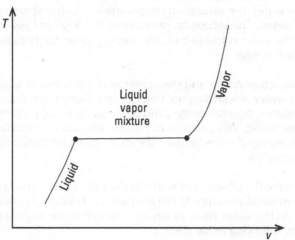

Volume changes due to a temperature change are small in a liquid compared to a vapor.

From compressed liquid to saturated liquid

Have you ever made a boiled egg? To properly prepare a boiled egg, you start with just enough water to cover the egg. The water in the pot is considered a *compressed liquid*. Even though nothing appears to be compressing the water, atmospheric pressure does a great job of compressing it. Without atmospheric pressure, the water in the pot would boil immediately and continue boiling until it all evaporated (without being heated at all).

Whether the water is in a pot on your stove or in a pipe feeding the boiler in a power plant, it's a compressed liquid. As the water in your pot warms up, the volume of the water increases slightly. You can try this at home to see that it's true. If the cold water barely covers the eggs, you'll see that the eggs are completely submerged when the water is really hot because the volume of the water increases.

When the water is boiling, it's a saturated liquid. A *saturated liquid* is a liquid that's able to generate vapor if you add more heat to it. So you see bubbles forming in the liquid. The pressure inside the vapor bubbles is equal to the pressure that's compressing the liquid. This pressure is called the *saturation pressure*. The temperature of the water when vapor bubbles are formed is called the *saturation temperature*.

The difference between saturation pressure and saturation temperature can be a bit tricky. Saturation pressure corresponds to a particular temperature, and saturation temperature corresponds to a particular pressure. You can think of it this way: For water, the saturation temperature at 1 atmosphere pressure is 100 degrees Celsius. The saturation pressure at 100 degrees Celsius is 1 atmosphere. After the water starts to boil, set your egg timer for 10 minutes to get perfect hard-boiled eggs.

If you know the temperature and the pressure of the water at any instant in time while the water is warming up, you can find thermodynamic properties like specific volume, internal energy, and enthalpy by using property tables like those in the appendix. You can also use thermodynamic properties to figure out how much energy it takes to make the water reach the boiling point, as shown in this example.

Suppose you have 0.5 kilogram of water in the pot to make your boiled eggs and the local ambient pressure is 100 kilopascals. How much energy does it take to warm up the water from 20 degrees Celsius to the boiling point? You can find out by following these steps:

1. **Write an equation for the change in energy (ΔE) of the water.**

$$\Delta E = m(\Delta u) = m(u_2 - u_1)$$

You find out about writing energy equations for thermodynamic pro-cesses in Chapter 4. In this equation, m is the mass of water in the pot, u_1 is the internal energy of the water at the beginning of the process at 20 degrees Celsius, and u_2 is the internal energy of the water when it reaches the boiling point.

2. **Find the internal energy (u_1) of the water at 20 degrees Celsius and 0.1 megapascal (100 kilopascals), using Table A-2 in the appendix.**

$$u_1 = 83.9 \text{ kJ/kg}$$

Table A-2 has properties for compressed liquid water as a function of both temperature and pressure.

3. **Find the internal energy (u_2) of the saturated liquid water at 0.1 mega-pascal (100 kilopascals), using Table A-2 in the appendix.**

$$u_2 = 417.3 \text{ kJ/kg}$$

The properties of water at the saturation pressure are shown in the first line of each of the four pressure entries in Table A-2. Alternatively, you can find the saturation properties of water in Table A-4 for many more pressure entries. Notice the saturation temperature (or boiling point) is at 99.6 degrees Celsius. This is because the pressure is slightly less than the standard atmospheric pressure of 101.325 kilopascals.

4. **Calculate the change in energy required to bring the water up to the boiling point, using the equation given in Step 1.**

$$\Delta E = m(u_2 - u_1) = (0.5 \text{ kg})(417.3 - 83.9) \text{ kJ/kg} = 166.7 \text{ kJ}$$

From saturated liquid to saturated vapor

Suppose you do something really nerdy like putting a thermometer in the pan of water while you make your boiled eggs. You notice the water gets warmer until it reaches the boiling point. At the boiling point, the liquid water and the water vapor are at the same temperature and pressure. The thermometer shows the same temperature until all the water is boiled out of the pan.

While the eggs in your pot of boiling water are cooking, you observe steam rising from the water. This steam is a saturated vapor. A *saturated vapor* is at the same temperature and pressure as a saturated liquid in the boiling process. Another way to think about a saturated vapor is to consider that you can condense it back into a liquid without changing the temperature or pressure, even though heat is removed from it. So if you hold the pot lid over the steam, you'll see moisture condensing into water droplets on the lid because the temperature of the lid is lower than the saturation temperature.

The properties of a saturated liquid or vapor depend on temperature or pressure. The pressure of the liquid determines the saturation temperature. In thermodynamic problems involving saturated liquids and vapors, you usually know either the saturation pressure or the saturation temperature of the process. The appendix includes two tables for saturation properties. You use Table A-3 if you know the saturation temperature for water or Table A-4 if you know the saturation pressure for water. Similar tables for refrigerant R-134a are found in Tables A-6 and A-7 of the appendix.

Working with saturated liquids and vapors

Evaporating all the water out of a pan can take a long time, even if the stove is turned on high. The amount of energy required to boil the water completely to a vapor is called the *latent heat of vaporization* or the *enthalpy of vaporization*. Condensing water vapor into a liquid takes just as much energy as evaporating water into a vapor. The enthalpy of vaporization varies with pressure, and it's different for every material.

The enthalpy of vaporization (h_{fg}) is the difference between the enthalpy of saturated liquid (h_f) and the enthalpy of saturated vapor (h_g), as shown in this equation:

$$h_{fg} = h_g - h_f$$

You use the saturation temperature or saturation pressure to find h_g and h_f. For example, you can determine the enthalpy of vaporization for water at 100 kilopascals pressure by following these steps:

1. **Find the enthalpy of the saturated liquid h_f and the saturated vapor h_g for water at 0.1 megapascal (100 kilopascals) pressure.**

   Use the saturated pressure table, A-4, in the appendix to find the following:

   h_f = 417.4 kJ/kg (the enthalpy of the saturated liquid)

   h_g = 2,675 kJ/kg (the enthalpy of the saturated vapor)

2. **Calculate the enthalpy of vaporization (h_{fg}), using this equation:**

   $h_{fg} = h_g - h_f = (2,675 - 417.4)$ kJ/kg = 2,257.6 kJ/kg

Figuring out quality in the vapor dome

If you want to know how much energy you put into the boiling water when only part of the water has turned to vapor, you need to know more than temperature and pressure. Because temperature and pressure are not independent during the boiling process, you need to know the value of a property that changes while the water boils. Mass and volume are two measurable properties that can help you determine how much energy has been used in boiling away part of the water.

A thermodynamic property called *quality* (x) uses the mass fraction (or percentage) of vapor in a liquid-vapor mixture to determine the energy of the mixture between the saturated liquid and the saturated vapor states. Liquid-vapor mixture quality is calculated with the following equation:

$$x = \frac{m_g}{m_f + m_g}$$

In this equation, the quality is x, the mass of vapor in the mixture is m_g, and the mass of liquid is m_f. Quality is a dimensionless property. When the quality is zero, the mixture is saturated liquid; when the quality is one, the mixture is saturated vapor. Quality can have neither a value greater than one nor less than zero. You can find out how to use quality to determine the energy of a liquid-vapor mixture in the following example.

Imagine that you measure the mass of water in the pan after you boil the eggs for ten minutes, and you find that the pan contains 0.44 kilogram of water. What is the quality (x) of a liquid-vapor mixture that has 0.44 kilogram of liquid and 0.06 kilogram of vapor?

You calculate the quality (x) of the liquid-vapor mixture as follows:

$$x = \frac{m_g}{m_f + m_g} = \frac{0.06 \text{ kg}}{0.5 \text{ kg}} = 0.12$$

Getting properties in the vapor dome using quality

You can use the quality to determine many thermodynamic properties, such as the specific volume, internal energy, enthalpy, and entropy of a liquid-vapor mixture. (I discuss the thermodynamic property of entropy in Chapter 8.) The following equation shows you how to calculate the enthalpy of a liquid-vapor mixture, using the quality and enthalpy values from the preceding example in this section:

$$h = x(h_g - h_f) + h_f = 0.12(2,675 - 417.4) \text{ kJ/kg} + 417.4 \text{ kJ/kg} = 688.3 \text{ kJ/kg}$$

This equation can be written in similar ways to find the internal energy (u), specific volume (v), or entropy (s) of a mixture using quality, as shown in these equations:

- **Internal energy:** $u = x(u_g - u_f) + u_f$
- **Specific volume:** $v = x(v_g - v_f) + v_f$
- **Entropy:** $s = x(s_g - s_f) + s_f$

From saturated vapor to superheated vapor

Imagine you want to do something crazy like capture the steam (a saturated vapor) that's coming from your pot of boiling water and heat it up even more. For liability purposes, I'm not giving you any ideas on how to do this. Adding heat to the steam causes its temperature to increase. Steam that's hotter than the saturation temperature is called *superheated steam* or *superheated vapor.* Please be careful with superheated steam, because it will burn you, and you can't see it! As the temperature of a superheated vapor increases, its specific volume also increases as long as the pressure remains constant.

Making superheated vapor isn't really a crazy idea, except when it's done in your kitchen. In fact, you're no doubt breathing a superheated vapor right now. Yes, air is a superheated vapor because it's hotter than liquid air at the same pressure. Superheated vapors don't have to be hot, even though they sound hot. Superheated vapors run around inside your air-conditioning system at home and in your car (see Chapter 13). Power plants use superheated vapors all the time. A steam generator or a boiler creates superheated steam so it can be used to turn a turbine and make electricity. I discuss the thermodynamic analysis of a power plant in Chapter 12.

You can find the energy in a superheated vapor from its temperature and pressure. I show you how to do this in the section "Interpolating with two variables."

Finding Thermodynamic Properties from Tables

You need to know the level of energy that exists in a material to analyze a thermodynamic process or system. You can calculate the internal energy, enthalpy, or entropy (see Chapter 8) of a substance by using complex mathematical equations, but using those equations for every thermodynamics problem would take all the fun out of learning. Fortunately, the complex math has been done for you, and thermodynamic properties have been summarized in tables.

The appendix of this book contains all the property data you need to work the example problems in this book. In this section, I show you how to read and use the thermodynamic property tables in the appendix.

Figuring out linear interpolation

Here's an example that shows you how to interpolate a table to find the value of a property that's in between values listed in a table. You need to know how to do this in many examples throughout this book, unless you have access to a software program that calculates thermodynamic properties for you. Many thermodynamic textbooks include a software package for properties.

You can use temperature and pressure for compressed liquids to determine specific volume, internal energy, and enthalpy. For example, you can find the enthalpy of water at 22 degrees Celsius and 0.1 megapascal pressure using the thermodynamic properties of compressed liquid water as shown in Table A-2 of the appendix. The following steps show you how to find the enthalpy by doing a linear interpolation of the data in the appendix:

1. **Look at Table A-2 in the appendix.**

   It lists four different pressures: 0.01, 0.1, 1.0, and 10.0 MPa.

2. **Choose the section of the table for 0.1 MPa.**

3. **Look for the temperature of 22°C.**

   It isn't listed; you have to interpolate the table between 20° and 30°C.

4. **Find the value of the enthalpy h_{20} at 20°C.**

   $h_{20} = 84.03$ kJ/kg

5. **Find the value of the enthalpy h_{30} at 30°C.**

   $h_{30} = 125.9$ kJ/kg

6. **Use the following relationship between enthalpy and temperature:**

   $$\frac{h_{22} - h_{20}}{h_{30} - h_{20}} = \frac{(22 - 20)°C}{(30 - 20)°C}$$

This equation can be written in a similar way to find internal energy, entropy, or specific volume.

7. **Rearrange the equation to solve for the enthalpy at 22°C, h_{22}, and substitute known enthalpy values.**

$$h_{22} = \frac{(22-20)°C}{(30-20)°C}(h_{30} - h_{20}) + h_{20}$$

$$h_{22} = \frac{(22-20)°C}{(30-20)°C}(125.9 - 84.0)\ kJ/kg + 84.0\ kJ/kg = 92.4\ kJ/kg$$

Interpolating with two variables

Many times you need to interpolate between two different variables, such as temperature and pressure, to find properties such as internal energy, enthalpy, and so forth. Interpolating with two variables is a bit trickier than linear interpolation of one variable. Here's an example that shows you how to do bilinear interpolation with two variables in a data table.

Suppose you take a trip to a power plant and discover that the boiler makes steam at 8 megapascals pressure and 560 degrees Celsius. You can find the enthalpy of the steam from Table A-5 in the appendix. The pressure and temperature aren't listed in the table, so you have to interpolate both variables. You have to do interpolation three times. To find the enthalpy of the steam (h_1) at these conditions, follow these steps:

1. **Find the enthalpy of the steam at pressures and temperatures above and below the desired conditions, using Table A-5 in the appendix.**

   T_{high} = 600°C, P_{high} = 10 MPa, h_a = 3,625 kJ/kg

   T_{high} = 600°C, P_{low} = 1.0 MPa, h_b = 3,698 kJ/kg

   T_{low} = 500°C, P_{high} = 10 MPa, h_c = 3,374 kJ/kg

   T_{low} = 500°C, P_{low} = 1.0 MPa, h_d = 3,478 kJ/kg

   Figure 3-6 illustrates the pressure, temperature, and enthalpy relationship of this example, showing the locations of the table values relative to the state value of the steam.

2. **Find the enthalpy h_{high} of the steam at 600°C and 8 MPa by interpolating the 600°C data.**

   $$h_{high} = \frac{(8-1)\ MPa}{(10-1)\ MPa}(h_a - h_b) + h_b$$

   $$h_{high} = \frac{(8-1)\ MPa}{(10-1)\ MPa}(3,625 - 3,698)\ kJ/kg + 3,698\ kJ/kg = 3,641\ kJ/kg$$

3. **Find the enthalpy h_{low} of the steam at 500°C and 8 MPa by interpolating the 500°C data.**

$$h_{low} = \frac{(8-1)\ \text{MPa}}{(10-1)\ \text{MPa}}(h_c - h_d) + h_d$$

$$h_{low} = \frac{(8-1)\ \text{MPa}}{(10-1)\ \text{MPa}}(3,374 - 3,478)\ \text{kJ/kg} + 3,478\ \text{kJ/kg} = 3,397\ \text{kJ/kg}$$

4. **Find the enthalpy h_1 of the steam at 560°C and 8 MPa by interpolating the 8 MPa data.**

$$h_1 = \frac{(560-500)\ °\text{C}}{(600-500)\ °\text{C}}(h_{high} - h_{low}) + h_{low}$$

$$h_1 = \frac{(560-500)\ °\text{C}}{(600-500)\ °\text{C}}(3,641 - 3,397)\ \text{kJ/kg} + 3,397\ \text{kJ/kg} = 3,543\ \text{kJ/kg}$$

Even though the data tables in the appendix are coarse for the sake of brevity, reasonably accurate values can be obtained by interpolation. The actual value for the enthalpy at 560 degrees Celsius and 8 megapascals is 3,545 kilojoules per kilogram.

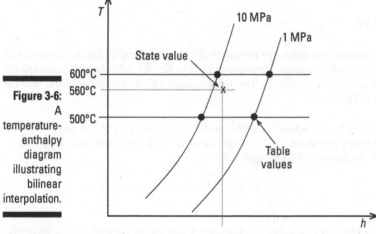

Figure 3-6:
A temperature-enthalpy diagram illustrating bilinear interpolation.

You can find the internal energy, entropy, or specific volume of a mixture at an "in-between" place in a table by using the steps for interpolation given in this example.

Good Gases Have Ideal Behavior

Under certain conditions, you can use a very simple equation to calculate the pressure, temperature, or specific volume of a gas without using property tables. This equation is called the *ideal-gas law* or *ideal-gas equation of state*. An equation of state describes the relationship between the pressure, temperature, and specific volume of a substance. Any gas that follows the ideal-gas law is called an *ideal gas*.

If the pressure, specific volume, or temperature of a gas can't be accurately predicted by the ideal-gas law, then the gas is called a *real gas*. Several other equations of state are available for predicting the *P-v-T* behavior of real gases. Because these equations of state tend to be complicated, many real gases have tables for their thermodynamic properties. For example, the properties of superheated steam and refrigerant R-134a are found in Tables A-5 and A-8, respectively, in the appendix.

The ideal-gas law on a per unit mass basis is written with the following equation:

$$Pv = RT$$

You can multiply this equation by the mass of the system. Then the specific volume (v) becomes the total volume (V), as shown in this equation:

$$PV = mRT$$

In these equations, the units are pressure (P) in kilopascals, specific volume (v) in cubic meters per kilogram, temperature (T) in Kelvin, and the gas constant (R) in kilojoules per kilogram-Kelvin. The total volume (V) is in cubic meters, and the mass (m) is in kilograms.

The gas constant (R) depends on the molecular mass (M) of the gas. You can calculate the gas constant by dividing the universal gas constant $\left(\bar{R}\right)$ by the molar mass, as shown in this equation:

$$R = \frac{\bar{R}}{M}$$

The units for the gas constant are kilojoules per kilogram-Kelvin; for molecular mass, the units are kilograms per kilomole. The molecular masses of several ideal gases are listed in Table 3-1. The universal gas constant is the same for every gas and has the value of $\bar{R} = 8.314$ kJ/kmol·K.

The ideal-gas law can be written in molar form as well, using these two equations: $P\bar{v} = \bar{R}T$ with the specific molar volume ($\bar{v}$), and $PV = n\bar{R}T$ for total system volume. The variable n is the number of kilomoles of gas in the system.

Table 3-1 Molecular Mass and Gas Constant of Selected Ideal Gases

Gas	Molecular Mass M, kg/kmol	Gas Constant R, kJ/kg · K
Air (equivalent)	28.97	0.2870
Argon	39.94	0.2082
Carbon dioxide	44.01	0.1889
Helium	4.00	2.0785
Hydrogen	2.02	4.1158
Nitrogen	28.01	0.2968
Oxygen	32.00	0.2598
Water vapor	18.02	0.4614

How can you tell whether you have an ideal gas? The most basic character-istic of an ideal gas is low density. A gas has low density at low pressure or at high temperature. In practice, many common gases are ideal, such as air, nitrogen, oxygen, helium, hydrogen, carbon dioxide, and water vapor. When water vapor is in the air, it behaves like an ideal gas. When water vapor is under high-pressure conditions, as it is in a steam power plant, it doesn't behave like an ideal gas. I discuss the distinction between ideal and real gases in Chapter 14.

Here's an example that shows you how to use the ideal-gas law. Suppose you pump up a bicycle tire to a pressure of 400 kilopascals (about 60 pounds per square inch) on a day when the ambient temperature is 27 degrees Celsius. You can find the specific volume of the air in the tire by following these steps:

1. **Look up the molecular mass M of air in Table 3-1.**

    M = 28.97 kg/kmol

2. **Calculate the gas constant for air using the universal gas constant.**

    $$R = \frac{8.314 \text{ kJ/kmol} \cdot \text{K}}{28.97 \text{ kg/kmol}} = 0.287 \text{ kJ/kg} \cdot \text{K}$$

    If you check Table 3-1, you'll find this value is already listed for air.

3. **Convert the temperature to absolute units.**

    T = 27°C + 273 = 300 K

Always use absolute temperature units with the ideal-gas equation. Otherwise, you'll get the wrong answer.

4. Rearrange the ideal-gas equation to solve for specific volume. Then substitute in the pressure, temperature, and gas constant values.

$$v = \frac{RT}{P} = \frac{(0.287 \text{ kJ/kg} \cdot \text{K})(300 \text{ K})}{400 \text{ kPa}} \left(\frac{1 \text{ kPa} \cdot \text{m}^3}{1 \text{ kJ}} \right) = 0.215 \text{ m}^3/\text{kg}$$

Note the units of 1 kJ and 1 kPa · m³ are equivalent.

The ideal-gas equation is one of the most commonly used equations in all of thermodynamics. Have fun with it!

Chapter 4

Work and Heat Go Together Like Macaroni and Cheese

*B*efore the industrial revolution, work and heat were separate entities. For example, people built fires to provide heat for warming their dwelling, cooking their food, melting metals, or firing clay pottery. People did their own work while making things, whether it was forming a clay pot, weaving a cloth, or using domesticated animals to pull a plow or turn a mill. As separate entities, work and heat have limited capabilities. When work and heat are put together as a team, however, they can accomplish almost anything — even sending a man to the moon.

I discuss the basic concepts behind work and heat as separate forms of energy transfer in this chapter. Work can be obtained from kinetic energy, potential energy, elastic energy, and internal energy, among other energy forms. For example, the kinetic energy of wind can do work using a wind turbine, the potential energy of a pendulum can do work to operate a clock, the elastic energy stored in a spring can do work to move a wind-up toy, and the internal energy of steam can do work through a steam turbine. I discuss several different forms of energy in Chapter 2.

In thermodynamics, work and heat are often put together to create heat engines or refrigeration machines. A heat engine is a device that uses heat to produce work: automobile engines, jet engines, and electric power plants are examples of heat engines. I discuss various kinds of heat engines and refrigeration systems in Chapters 10–13. Refrigerators and heat engines require special equipment to use heat energy. Removing or adding heat is done by boilers, condensers, evaporators, and combustion chambers. Within these kinds of devices, heat is moved into or out of fluids.

Work Can Do Great Things

For most people, work is a place you go for about 40 hours a week to collect a paycheck. In thermodynamics, work is a form of energy that is organized in a way to do something useful, like move a car, turn an electric generator, run a mechanical clock, or mix a milkshake in a blender. A common thread in each example in this list is motion. Work means energy makes something move. If you're stuck between a rock and a hard place and you push with all your might against the rock until you're exhausted but the rock doesn't move, you're not doing any work at all — even though your body will disagree. If you move the rock by only a millimeter, you're doing work.

Work is defined by applying a force to an object and moving it over a distance. Figure 4-1 shows an object pushed by a force, F, over a distance, s. The work, W, done in the process of moving the object is calculated by the following integral equation:

$$W = \int_1^2 F \, ds$$

To evaluate this integral, you must know how the force varies with distance. In the simplest case, force is constant. The object shown in Figure 4-1 has a constant force applied to it as long as it moves with constant velocity. Integrating a constant force in the work integral results in this equation for work:

$$W = F \cdot s.$$

WARNING!

If the force is applied at an angle different from the direction of motion, use only the force component in the direction of motion in calculating work with this equation.

In this equation, the units of work are kilojoules, the force is in kilonewtons and the distance is in meters. One kilojoule equals 1 kilonewton-meter. The capital letter W indicates the total work; the lowercase letter w indicates work per unit mass with the units of kilojoules per kilogram. Power is the amount of work per unit of time. Sometimes power is designated with the letter P, but this can be confused with pressure, so often W with an over-dot ($\dot{W}$) is used as the symbol for the rate at which work is done (in other words, power). The units for power are kilojoules per second or kilowatts.

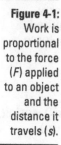

Figure 4-1:
Work is proportional to the force (*F*) applied to an object and the distance it travels (*s*).

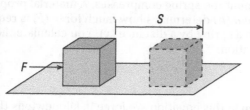

To illustrate how you can use this equation, imagine your car is stuck in a snowdrift. You and a friend dig it out and push it for a distance of 3 meters. If you and your friend together weigh about 150 kilograms and you can put half your weight into pushing the vehicle, you can calculate the work using the following steps:

1. **Calculate the force used for pushing the car**.

   Assume the force comes from half the mass of you and your friend: 75 kilograms. Use the acceleration of gravity, 9.81 m/s^2, to convert mass to a force:

   $$F = (75 \text{ kg})(9.81 \text{ m/s}^2) = 736 \text{ N}$$

2. **Calculate the work done by multiplying the force times the distance over which the force is applied:**

   $$W = (736 \text{ N})(3 \text{ m})(1 \text{ kN}/1{,}000 \text{ N}) = 2.2 \text{ kJ}$$

Now offer your friend a cup of hot cocoa for helping you get unstuck.

In this section, I discuss several forms of mechanical work that define different relationships between a force and distance.

Working with springs

Springs have been used to store energy and do work for a long time — although they're generally used for doing small bits of work like running a clock, a wristwatch, or a wind-up toy. Work is defined by applying a force over a distance. When you push on a linear spring or twist a spiral spring, you apply work to do work on the spring. The spring is able to store energy until it's released for doing work.

You may have noticed that when you push against a spring, the force increases with the amount the spring compresses. A material property called the *spring constant (k)* determines how much force (F) is required to compress or stretch a spring by a distance (x). You calculate the force with the following equation:

$$F = k \cdot x.$$

The units for each term in this equation are force in kilonewtons (kN), spring constant in kilonewtons per meter (kN/m), and distance in meters (m).

You calculate the work of the spring (W_{spring}) using the following equation:

$$W_{spring} = \int F \, dx = \frac{1}{2} k \left(x_2^2 - x_1^2 \right)$$

Figure 4-2 shows that x_1 and x_2 are the initial and final positions of the spring relative to the relaxed length of the spring. The spring is at the relaxed length when the force on it equals zero. If the work starts with the spring at the relaxed position, then x_1 is zero. The units of work are in kilojoules (kJ).

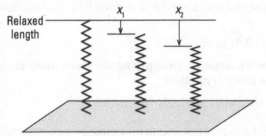

Figure 4-2:
A spring can do work if it is compressed or stretched.

Here's an example of a spring doing work. A 1,500-kilogram car has a spring at each wheel. Each spring is compressed by 0.05 meter due to the weight of the car. Suppose the car hits a speed bump and the springs compress an additional 0.03 meter. You can figure out how much work is done on the springs by following these steps:

1. **Determine the force on the springs due to the weight of the car using the mass of the car and the acceleration due to gravity in this equation:**

   $$F = m \cdot g = (1{,}500 \text{ kg})(9.81 \text{ m/s}^2) = 14{,}700 \text{ N}$$

   Use this unit conversion to get the final answer: $1 \text{ kg} \cdot \text{m/s}^2 = 1 \text{ N}$.

2. **Calculate the spring constant (k), using the distance (x_1) the springs are compressed under the weight of the car, with this equation:**

   $$k = F/x_1 = (14{,}700 \text{ N})/(0.05 \text{ m})(1 \text{ kN}/1{,}000 \text{ N}) = 294 \text{ kN/m}$$

3. **Calculate the work done on the springs using the following equation (with the unit conversion of 1 kN · m = 1 kJ).**

$$W_{spring} = (1/2)(294 \text{ kN/m})(0.08^2 - 0.05^2)\text{m}^2 = 0.57 \text{ kJ}$$

The final displacement of the spring x_2 is 0.05 + 0.03 = 0.08 meter.

Turning a shaft

A shaft is the most common way of putting work into motion. An automobile engine has pistons that are connected to a crankshaft that converts the reciprocating motion of the pistons into rotary motion used by the tires. A jet engine has a shaft that connects a turbine to a compressor so the turbine can turn the compressor. An electric motor has a shaft that rotates to do all kinds of work, like operating an elevator or mixing cake batter.

A shaft does not create work; it only carries work with rotary motion. The work carried by a shaft still uses a force moving over a distance. Figure 4-3 shows a perpendicular force (F) applied to a shaft at the radius (r) from the center axis to create torque (T).

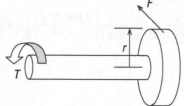

Figure 4-3:
A shaft does
work when
it is turned.

The torque on a shaft is related to the force by this equation:

$$T = F \cdot r$$

The units for torque are newton-meters, force is expressed in newtons, and the radius is expressed in meters. The radius of the applied force is not the distance used in determining the work carried by the shaft. The work distance (s) is determined by multiplying the circumference ($2\pi r$) by the number of rotations (n) turned by the shaft. The work distance is calculated using this equation:

$$s = (2\pi r)n$$

The work transmitted by a shaft (W_{shaft}) is calculated using the torque and the distance with this equation:

$$W_{shaft} = F \cdot s = (T/r)(2\pi r)n = 2\pi \cdot n \cdot T$$

The units of work are kilojoules, so you have to use the following unit conversion in this equation:

$$1 \text{ kJ} = 1,000 \text{ N} \cdot \text{m}$$

If a force is continuously applied to a shaft, it will rotate at a certain speed. The power transmitted by a shaft is determined by the rate at which work is done. You calculate the shaft power ($\dot{W}_{shaft}$), using the number of revolutions ($\dot{n}$) per second, in this equation:

$$\dot{W}_{shaft} = 2\pi \dot{n} T$$

The units for power are kilowatts.

This example shows you how to find the torque on an axle for a car. At 60 miles per hour, a car needs 40 horsepower (or 30 kilowatts). Suppose the wheels spin at 750 revolutions per minute. You can find the torque on the axle with this equation:

$$T = \frac{\dot{W}_{shaft}}{2\pi \dot{n}} = \frac{30 \text{ kW}}{2\pi\left(750 \text{ rev/min}\right)\left(1 \text{ min/60 sec}\right)}\left(\frac{1,000 \text{ N} \cdot \text{m}}{1 \text{ kJ}}\right) = 382 \text{ N} \cdot \text{m}$$

Accelerating a car

A top fuel dragster can accelerate to more than 300 miles per hour in less than a quarter mile. How much power does it take to accelerate a dragster to that speed? Figuring out the answer is quite simple. Accelerating a mass takes work because you must apply a force to make the mass move over a distance. The rate of work tells you how much power is used. You use kinetic energy to analyze the work associated with acceleration. The work required to accelerate a mass (it doesn't have to be a vehicle) is determined with this equation:

$$W_a = \frac{1}{2}m\left(V_2^2 - V_1^2\right)$$

In this equation, V_1 is the initial velocity of the mass when the force is applied to accelerate the mass. If it starts at rest, the initial velocity is zero. V_2 is the final velocity of the mass when the accelerating force is removed. The units used in this equation are kilojoules for the acceleration work, kilograms for the mass, and meters per second for the velocity.

You can calculate the power required to accelerate a 1,000-kilogram dragster to 300 miles per hour in only 4.5 seconds by using the following steps:

1. **Convert the speed to meters per second:**

$$V_2 = \left(\frac{300 \text{ miles}}{\text{hour}}\right)\left(\frac{1,610 \text{ m}}{1 \text{ mile}}\right)\left(\frac{1 \text{ hour}}{3,600 \text{ sec}}\right) = 134 \text{ m/s}$$

2. **Calculate the work due to acceleration.**

Because the dragster starts from rest, V_1 is zero.

$$W_a = \tfrac{1}{2}(1,000 \text{ kg})(134 \text{ m/s})^2 = 9,000 \text{ kJ}$$

The unit conversion of $1 \text{ kg} \cdot (\text{m/s})^2 = 1 \text{ kJ}$ is used to get the answer in kilojoules.

3. **Calculate the average power used to accelerate the dragster.**

Power is the rate at which work is done.

$$\dot{W}_a = \frac{W_a}{\Delta t} = \frac{9,000 \text{ kJ}}{4.5 \text{ sec}} = 2,000 \text{ kW}$$

That's enough power to burn 20,000 light bulbs if each is rated at 100 watts. Or, if you prefer to get a sense of magnitude using horsepower units, 1 horsepower = 0.746 kilowatt. So the dragster uses 2,680 horsepower for acceleration. This calculation neglects friction, rolling resistance of the tires, and aerodynamic drag. The power and acceleration of a real dragster aren't constant, either. An actual top fuel dragster uses up to 7,000 or 8,000 horsepower to overcome all these additional effects.

Moving with pistons

If a gas inside a piston-cylinder device expands against the piston to make it move, it does work on the piston. The pressure of the gas provides a force to displace the piston, which defines work. The gas inside the cylinder is considered a system, and an imaginary boundary separates the system from the surroundings, which consist of the piston and the cylinder (and everything else in the universe). Because the system boundary moves or changes with the motion of the piston, the work of a piston-cylinder device is often called *moving boundary work*. The amount of work done by or on the piston-cylinder depends on the relationship between pressure and volume in the cylinder during a work process.

The moving boundary work (W_b) is defined by the following integral of pressure (P) and volume (V):

$$W_b = \int_1^2 P\, dV$$

The integral is evaluated using the proper relationship between pressure and volume for a process. There are several different processes a piston-cylinder device may follow to do work as shown in Figure 4-4. Path A is a constant-pressure process, Path B is a constant-temperature process, and Path C is a reversible-adiabatic process. A polytropic process can take any path on the diagram as long as it's below Path A. (I discuss these processes in detail in the next bulleted list.)

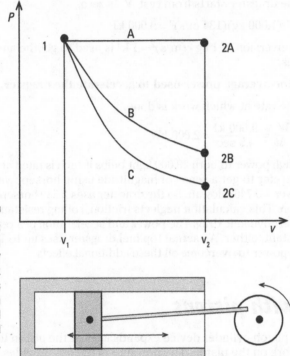

Figure 4-4:
Moving boundary work depends on the path taken and the end states.

The boundary work of a process equals the area under the path between Endpoints 1 and 2. The amount of work extracted from Path A in Figure 4-4 is more than that extracted from Paths B and C because the area under Path A is greater than the area under Path B or C. You can see that the pressure during the constant-pressure process is more than the average pressure during the constant-temperature and reversible-adiabatic processes. Higher pressure provides a greater force and a greater amount of work in a process if the force is applied through the same distance.

Boundary work is positive if work is done by the fluid on the piston; this process is expansion. Boundary work is negative if work is done on the fluid by the piston; this process is compression.

The moving boundary work processes shown in Figure 4-4 are described as follows.

- ✔ **A constant-pressure process:** Imagine the piston shown in Figure 4-4 is frictionless and has no mass. If you add heat to the gas inside the cylinder, the gas expands and makes the piston move. The gas does work on the piston. The gas temperature in the cylinder changes in this process. The process works in reverse by removing heat from the gas. For a constant-pressure process, integrating the equation defining boundary work gives you this result:

$$W_b = P(V_2 - V_1)$$

This equation is valid for both real and ideal gases in a constant-pressure process.

The units of boundary work are kilojoules, pressure is measured in kilopascals, and volume is measured in cubic meters. The following units are equivalent: $1\ kJ = 1\ kPa \cdot m^3$.

- ✔ **A constant-temperature process:** Suppose you push on the piston shown in Figure 4-4 to compress the gas. If you don't remove any heat, the gas temperature will rise. But you can maintain a constant temperature in the gas by removing just the right amount of heat as the gas is compressed. The gas pressure changes in the cylinder in this process. The process works in reverse by heating gas during expansion. For a constant-temperature process with an ideal gas, you can substitute the ideal-gas relationship, $PV = mRT$, into the preceding equation that defines boundary work. I discuss the ideal-gas relationship in Chapter 3. Completing the integration gives the following set of equations that are equal to each other for the boundary work:

$$W_b = P_1 V_1 \ln\frac{V_2}{V_1} = P_1 V_1 \ln\frac{P_1}{P_2} = mRT_1 \ln\frac{P_1}{P_2}$$

- ✔ **A reversible-adiabatic process:** You can make the piston-cylinder device shown in Figure 4-4 adiabatic by putting perfect insulation around the cylinder. *Adiabatic* means no heat is transferred across the system boundary between the gas inside the cylinder and the environment. (For this case, the piston-cylinder device becomes part of the system.) *Reversible* means the process can go in the reverse direction to return to the initial state without creating a change in the system or the surroundings (see Chapter 9).

As you push or pull on the piston, both the gas temperature and pressure in the cylinder change in this process. If the process is both adiabatic and reversible, the pressure-volume relationship is $PV^k = C$, where k is the specific heat ratio ($k = c_p/c_v$) and C is a constant. (Don't worry about the value of the constant for now; it drops out of sight when you use this equation to find the boundary work.)

If you insert the pressure-volume relationship for a reversible-adiabatic process into the boundary work equation and complete the integration, you get the following equation:

$$W_b = \frac{P_2 V_2 - P_1 V_1}{1 - k}$$

If the gas in the cylinder is air, $k = 1.4$, assuming the specific heat doesn't change with temperature during the process.

✔ **A polytropic process:** If you push or pull on the piston shown in Figure 4-4 and allow the temperature and pressure to change and allow heat transfer across the system boundary as in real systems, the pressure-volume relationship can be described by the equation $Pv^n = C$, where n and C are constants. This pressure-volume relationship equation best simulates the behavior of actual piston-cylinder processes.

If you insert the pressure-volume relationship for a polytropic process into the integral of the boundary work equation and complete the integration, you get the following equation:

$$W_b = \frac{P_2 V_2 - P_1 V_1}{1 - n}$$

For the case of $n = 0$, this equation gives the boundary work for a constant-pressure process. For an isothermal process with a gas that obeys the ideal-gas law, $n = 1$ and $C = RT$. But you can't use this equation because it divides by zero; you must use the constant-temperature equation defined in the previous bullet. For a reversible and adiabatic process, the coefficient n equals k, as shown in the previous bullet.

✔ **A constant volume process:** If you hold the piston shown in Figure 4-4 in a locked position and add or remove heat from the system, you increase or decrease the temperature and the pressure of the gas. However, because the piston doesn't move, no work is done, so $W_b = 0$ for this process.

Figuring out boundary work

The following example shows you how to calculate the work for a moving boundary system with expansion of a gas in a cylinder for three different processes: constant-pressure, constant-temperature, and reversible adiabatic. A cylinder filled with air has an initial volume of 0.1 liter at 3,000 kilopascals pressure. It's expanded until the final volume of 1.0 liter is reached. For the first case, the air expands in the cylinder at constant pressure. You calculate the work done by the air against the piston as follows:

$$W_b = (3{,}000 \text{ kPa})[(1.0 - 0.1) \text{ liter}](1 \text{ m}^3/1{,}000 \text{ liters}) = 2.7 \text{ kJ}$$

In the second case, the air expands in the cylinder isothermally. You can find the work done by the air against the piston via the following equation:

$$W_b = (3,000 \text{ kPa})(0.1 \text{ liter})\left(\frac{1 \text{ m}^3}{1,000 \text{ liters}}\right)\ln\left(\frac{1.0 \text{ liter}}{0.1 \text{ liter}}\right)\left(\frac{1 \text{ kJ}}{1 \text{ kPa} \cdot \text{m}^3}\right) = 0.69 \text{ kJ}$$

In the third case, the air expands in the cylinder reversibly and adiabatically. You calculate the work done by the air against the piston as follows:

1. **Write the pressure-volume relationship for a reversible, adiabatic process and let the constant C represent the final state, $P_2 V_2$:**

$$P_1 V_1^k = P_2 V_2^k$$

2. **Rearrange the pressure-volume relationship and calculate the pressure at the final volume, P_2, with $k = 1.4$:**

$$P_2 = P_1\left(\frac{V_1}{V_2}\right)^k = (3,000 \text{ kPa})\left(\frac{0.1 \text{ liter}}{1.0 \text{ liter}}\right)^{1.4} = 119 \text{ kPa}$$

3. **Calculate the boundary work for a reversible, adiabatic process:**

$$W_b = \frac{(119 \text{ kPa})(1.0 \text{ liter}) - (3,000 \text{ kPa})(0.1 \text{ liter})}{1 - 1.4}\left(\frac{1 \text{ m}^3}{1,000 \text{ liters}}\right)\left(\frac{1 \text{ kJ}}{1 \text{ kPa} \cdot \text{m}^3}\right) = 0.45 \text{ kJ}$$

You can see from this example that the amount of work varies depending on the path taken between states for a process.

Heating Things Up, Cooling Things Down

Heating and cooling are used in thermodynamics to add or remove thermal energy from a system. Thermal energy is disorganized energy, and it must be harnessed in certain ways to be turned into work. Otherwise, thermal energy can only heat or cool things. You can place a fan in a room and add heat to the room all you want, but the fan blades will never turn just because you've added thermal energy to the room. If your goal is to operate a fan using electricity, you must organize thermal energy to generate electricity in some kind of device. (Actually, a better solution is to just let the folks at the power plant organize thermal energy and convert it to electricity for you!)

In this section, I discuss various ways to organize thermal energy so it can be used by heat engines and refrigeration systems. Combustion chambers, steam generators, and evaporators typically add thermal energy to a system. Combustion chambers are used by gas turbine engines, automobile engines, and diesel engines. I discuss how to analyze these engines using the Brayton,

Otto, and diesel cycles, respectively, in Chapters 10 and 11. Steam generators (also known as boilers) are used in steam power plants to make electricity in the Rankine cycle. I discuss the analysis of the Rankine cycle in Chapter 12. Evaporators are used to absorb thermal energy in refrigeration systems. I discuss the analysis of refrigeration systems in Chapter 13.

Thermal energy is usually either removed from a system by a condenser or simply exhausted into the air. A steam power plant condenses steam back to liquid water to remove waste heat from the system. A refrigeration system uses a condenser to reject heat that was absorbed from the evaporator. The waste heat for gas turbine, automobile, and diesel engines is removed by the exhaust.

The capital letter Q stands for the total amount of heat transferred in a process, which is represented in kilojoules. The lowercase letter q stands for the amount of heat transferred per unit mass in kilojoules per kilogram. The heat transfer rate in a process uses Q with an over-dot ($\dot{Q}$) and is stated in kilojoules per second or kilowatts.

Pieces of equipment called *heat exchangers* organize thermal energy to do work. As the name implies, these devices absorb heat from a source, such as a combustion process, and transfer it into a fluid. Heat exchangers aren't used just to absorb heat; they're also used to reject heat. Energy from a hot fluid can be transferred to the air in a heat exchanger. Steam generators or boilers, condensers, and evaporators are all different forms of heat exchangers.

Getting hot with boilers

The ideal steam power plant is analyzed with the Rankine cycle in Chapter 12. Heat is added to the Rankine cycle in a boiler. A fossil-fueled boiler burns coal, oil, or natural gas. A nuclear boiler uses heat generated from nuclear fission. Figure 4-5 shows a fossil-fueled boiler.

Water enters the boiler and is called *feedwater* which collects in a large pipe that supplies water to many smaller tubes lining the walls of the firebox.

Heat from the combustion of fuel boils the water to generate steam. Steam is collected in another header at the top of the boiler. The boiler usually contains a separate heat exchanger that heats the steam above the boiling point to create superheated steam. This heat exchanger is located in the hottest part of the boiler. The steam then flows to the next process in the power plant, which is a steam turbine to generate electricity.

The amount of heat that a boiler transfers to the water flowing through it is calculated using this equation:

$$\dot{Q} = \dot{m}(h_{out} - h_{in})$$

The mass flow rate of the water ($\dot{m}$) is in kilograms per second. The enthalpy of the (liquid) water entering the boiler is h_{in}, and the enthalpy of the steam leaving the boiler is h_{out}. The units for enthalpy are kilojoules per kilogram. Water enters the boiler as a compressed liquid (I discuss compressed liquids in Chapter 3). The thermodynamic properties of compressed water are given in Table A-2 in the appendix. Steam usually leaves the boiler as a superheated vapor; you can find the thermodynamic properties in Table A-5 of the appendix. If the steam leaves the boiler as a saturated vapor, use the thermodynamic properties in Tables A-3 and A-4 of the appendix.

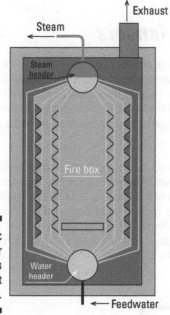

Figure 4-5:
A boiler makes steam out of water.

Consider this example, which shows you how to calculate the heat transfer rate in a boiler. A boiler receives feedwater at 40 degrees Celsius and 10 megapascals pressure. It generates steam at 600 degrees Celsius at the same pressure. The water flow rate through the boiler is 20 kilograms per second. Follow these steps to determine the heat transfer rate of the boiler:

1. **Find the enthalpy h_{in} of the feedwater entering the boiler, using Table A-2 in the appendix:**

   h_{in} = 176 kJ/kg

2. **Find the enthalpy h_{out} of the superheated steam leaving the boiler, using Table A-5 in the appendix:**

   $h_{out} = 3,625$ kJ/kg

3. **Calculate the heat transfer rate as follows:**

   $$\dot{Q} = (20 \text{ kg/sec})[(3,625 - 176) \text{ kJ/kg}](1 \text{ MW}/1,000 \text{ kW}) = 69 \text{ MW}$$

Heat transfer to the fluid is a positive quantity; heat transfer from a fluid is a negative quantity.

Cooling off with condensers

Every heat engine or refrigeration cycle must reject heat — there's no way around it. For power plants and refrigeration cycles, a condenser is a heat exchanger that removes heat from the fluid and sends it to the environment. In a power plant, steam is condensed into liquid water after it leaves the turbine. In a refrigeration cycle, refrigerant vapor is condensed into a liquid after it leaves the compressor.

Many varieties of heat exchangers condense vapors into liquids. Figure 4-6 shows a steam condenser where steam enters from the top and flows over cool tubes that are filled with relatively cold water. The water inside the tubes may only be room temperature, but the tubes are still colder than the steam. The steam condenses on the tubes and drips to a collecting pan at the bottom of the heat exchanger. A pump removes water from the collecting pan and sends it back to the boiler.

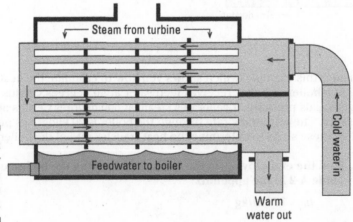

Figure 4-6: A condenser changes a vapor into a liquid.

Steam from turbine

Feedwater to boiler

Cold water in

Warm water out

For a large commercial refrigeration system, the condenser may be very similar to the one just described for a power plant, where refrigerant vapor condenses on water-filled tubes. Small refrigeration systems like residential air conditioners and refrigerators use ambient air to condense the vapor into a liquid. The temperature of the refrigerant vapor entering the condenser is much warmer than the ambient air temperature. If you look on the back of your refrigerator, you can find the condenser. It feels warm to the touch.

The heat transfer rate in a condenser is calculated using the same equation shown for a boiler:

$$\dot{Q} = \dot{m}(h_{out} - h_{in})$$

The only difference is that it's negative because heat is removed from the fluid. The enthalpy of the fluid at the condenser inlet is h_{in}, and the enthalpy at the condenser exit is h_{out}. The following example shows you how to calculate the heat transfer rate from a refrigerant condenser.

A refrigerator condenser has superheated R-134a entering at 1,000 kilopascals pressure and 50 degrees Celsius. Refrigerant leaves the condenser as a saturated liquid. A saturated liquid is in a liquid state but at the boiling point. The refrigerant mass flow rate is 0.005 kilogram per second. You can find the heat transfer rate of the condenser with the following steps:

1. **Find the enthalpy of the superheated refrigerant vapor entering the condenser (h_{in}), using Table A-8 of the appendix:**

   h_{in} = 430.9 kJ/kg

2. **Find the enthalpy of the saturated liquid leaving the condenser (h_{out}), using Table A-7 of the appendix.**

   The pressure is at 1,000 kilopascals.

   h_{out} = 255.5 kJ/kg

3. **Calculate the heat transfer rate as follows:**

   $$\dot{Q} = (0.005 \ \text{kg/sec})\big[(255.5 - 430.9) \ \text{kJ/kg}\big] = -0.88 \ \text{kW}$$

Heat transfer from a fluid is a negative quantity.

Chilling with evaporators

A refrigeration system absorbs heat from a cold environment using a heat exchanger called an *evaporator*. Figure 4-7 shows a liquid-vapor refrigerant mixture that is colder than the local ambient environment entering the evaporator and boiling to become a superheated vapor. Evaporators are only found in refrigeration (and heat pump) systems; they aren't used in heat engines. A fan is used to draw air over the evaporator coils. The refrigerant flows through a tube that's bonded to fins. The fins help improve heat transfer from the ambient air to the refrigerant. The heat transfer rate in an evaporator is found using the same equation used for a condenser:

$$\dot{Q} = \dot{m}(h_{out} - h_{in})$$

The only difference is that the result is positive because heat is absorbed by the fluid. The enthalpy of the fluid at the evaporator inlet is h_{in}, and the enthalpy at the evaporator exit is h_{out}.

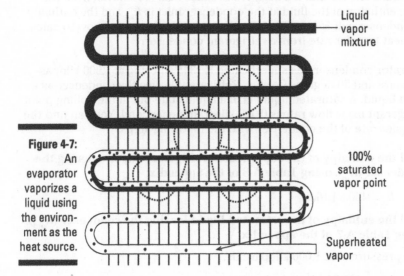

Liquid vapor mixture

100% saturated vapor point

Superheated vapor

Figure 4-7:
An evaporator vaporizes a liquid using the environment as the heat source.

Consider this example, which shows you how to calculate the heat transfer rate from a refrigerant evaporator. A refrigerator evaporator has an R-134a liquid-vapor mixture entering at 250 kilopascals pressure with a 20-percent quality (I discuss quality in Chapter 3). Refrigerant leaves the evaporator as a superheated vapor at 0 degrees Celsius. The refrigerant mass flow rate is 0.005 kilogram per second. You can find the heat transfer rate of the evaporator with the following steps:

1. **Find the liquid enthalpy, h_f, and the vapor enthalpy, h_g, at 250 kPa for the refrigerant liquid-vapor mixture entering the evaporator.**

   By using Table A-7 in the appendix, you find that h_f = 194.3 kJ/kg and h_g = 396.1 kJ/kg.

2. **Calculate the enthalpy of the liquid vapor mixture, h_{in}, at 20-percent quality by using the following equations:**

$$h_{in} = h_f + x(h_g - h_f)$$

$$h_{in} = 194.3 \text{ kJ/kg} + 0.2(396.1 - 194.3) \text{ kJ/kg} = 234.7 \text{ kJ/kg}$$

3. **Find the enthalpy h_{out} of the superheated vapor leaving the evaporator, using Table A-8 in the appendix:**

$$h_{out} = 399.8 \text{ kJ/kg}$$

4. **Calculate the heat transfer rate as follows:**

$$\dot{Q} = (0.005 \text{ kg/sec})\left[(399.8 - 234.7) \text{ kJ/kg}\right] = 0.83 \text{ kW}$$

Heat transfer to a fluid is a positive quantity.

Part II
Employing the Laws of Thermodynamics

The 5th Wave   By Rich Tennant

THE 4th LAW OF THERMODYNAMICS: Don't try to explain the first 3 on a blind date.

In this part . . .

You dig deeper into the first and second laws of thermodynamics. Here, you open up the hood and get your hands dirty using the first law of thermo to figure out how much energy basic engine parts like pistons and cylinders, turbines and pumps, and boilers and condensers use. Then you can use the second law of thermo to see how efficiently things like power plants or refrigerators use energy.

Chapter 5

Using the First Law in Closed Systems

- -

In This Chapter

▶ Working with constant mass systems

▶ Understanding the conservation of energy principle

▶ Applying conservation of energy to ideal-gas processes

▶ Solving energy balance problems for solids and liquids

- -

A thermodynamic system can be as simple as making iced tea or as complex as operating a power plant. When you look at a complex system, breaking it down into individual processes first simplifies your analysis. Then you can eat the elephant one bite at a time, so to speak. This chapter introduces you to thermodynamic analysis of some simple processes involving heat and work interactions. After you understand how to calculate the heat and work quantities for an individual process, you can take the next step and begin solving more complex systems such as automobile engines, gas turbine engines, refrigerators, and power plants, as I discuss in Chapters 10–13.

Thermodynamic analysis of a process uses only a few fundamental rules. These rules include the *conservation of mass* and the *conservation of energy*. In this chapter I discuss how the conservation of mass and energy apply to simple systems. A third rule, called the *ideal-gas law,* relates pressure, temperature, and volume to each other when an ideal gas is used in a process.

Conserving Mass in a Closed System

Every thermodynamic analysis begins by defining a system. A *system* describes the mass or volume you use for analysis. For example, a system can define the amount of gas contained within a piston and cylinder, the amount of air inside a football, or the amount of iced tea in a glass.

There are two basic categories of systems in thermodynamics. In a *closed system,* mass neither enters nor leaves the system during a process. In an *open system,* mass can enter and/or leave the system. This chapter focuses on closed systems. Chapter 6 addresses open systems.

Understanding how to define a system for thermodynamic analysis is important. Say you're defining a system for a glass of iced tea. If you specify only the tea and the ice as the system, a process for that system may involve melting the ice to cool the tea. If you define the system as the ice, the tea, and the glass, a process may include melting the ice to cool the tea and the glass.

Mass, like energy, can be neither created nor destroyed, but it can change form. A solid mass can melt into a liquid, and a liquid can evaporate into a gas. Even in chemical reactions, the mass of the reactants equals the mass of the products (see Chapter 16). In each process, the mass doesn't change. The principle of conservation of mass can be summed up as this: *The net mass transferred into or out of a system equals the change in mass of a system.* Mathematically, this is written as follows:

$$m_{in} - m_{out} = \Delta m_{sys}$$

The units of mass in the SI system are kilograms or grams. The conservation of mass can also be written on a rate basis with this equation:

$$\dot{m}_{in} - \dot{m}_{out} = \frac{dm_{sys}}{dt}$$

When you see a "dot" over a variable like mass ($\dot{m}$), it means the variable is on a rate basis or per unit time. The units for mass flow rate are kilograms per second.

Because no mass flows in or out of a closed system during a thermodynamic process, the conservation of mass equation simplifies to the following equation: m_{sys} = constant.

Balancing Energy in a Closed System

I introduce the first law of thermodynamics in Chapter 2; in this chapter, you get to use it. The first law of thermodynamics is a consequence of the concept of the *conservation of energy,* which states that energy cannot be created or destroyed; it can only change form.

Conservation of energy means that all the energy entering a system (E_{in}) minus the amount of energy leaving it (E_{out}) equals the change in the amount of energy (ΔE_{sys}) within the system, as shown in Figure 5-1. You write the conservation of energy equation for a system as follows:

$$E_{in} - E_{out} = \Delta E_{sys}$$

Figure 5-1:
Conserva-
tion of
energy for
a closed
system.

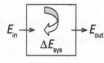

You can see the conservation of energy equation is very similar in form to the conservation of mass equation. The SI units for energy are kilojoules.

The left-hand side of the conservation of energy equation ($E_{in} - E_{out}$) represents the net amount of energy transfer in a system by heat (Q), work (W), and mass (m). For a closed system, no mass can enter or leave the system, so there's no energy transfer by mass in a closed system. Traditionally, the direction of heat transfer (Q) is *to the system* and work (W) is *done by the system*. Work in a system can take many forms, as described in Chapter 4, including boundary work, shaft work, and electrical work. For a closed system, the net energy transfer to the system is written as follows:

$$E_{in} - E_{out} = Q - W$$

The right-hand side of the conservation of energy equation (ΔE_{sys}) represents the change in internal energy (U), kinetic energy (KE), potential energy (PE), and other energy forms (such as magnetic energy, if used) in a system. A system has kinetic energy if it's moving relative to an external frame of reference. A system has potential energy from its position in the earth's gravitational field. I discuss kinetic and potential energy in Chapter 1. For a closed system, the change in the total energy of a system is written like this:

$$\Delta E_{sys} = \Delta U + \Delta KE + \Delta PE$$

If your system is stationary, you don't have to worry about the kinetic or potential energy terms, so the conservation of energy equation is simplified to the following equation for a closed system: $Q - W = \Delta U$.

The conservation of energy can be written on a per unit mass basis with units of kilojoules per kilogram, as follows:

$$q - w = \Delta u$$

You can also write the conservation of energy equation on a per unit time basis, turning it into a rate equation. Energy per unit time is power, so the units of the conservation of energy rate equation are either watts or kilowatts. The conservation of energy rate equation is written as follows, using the over-dot symbol to indicate that it's a rate equation:

$$\dot{Q} - \dot{W} = \frac{dU}{dt}$$

Applying the conservation of energy to a system is sometimes called *writing an energy balance* because you balance the energy in a system.

You can use the first law of thermodynamics to figure out how long it takes to warm up some soup for your lunch. Suppose the stove provides a heat transfer rate ($\dot{Q}$) of 400 watts to the pot of soup. The soup starts at room temperature, and you heat it up by 50 degrees Celsius. The mass (m) of the soup is 0.5 kilogram. You can solve this problem with the following steps:

1. **Write the energy equation on a rate basis for this process because the heat transfer is on a rate basis.**

   This process doesn't involve any work, only heat transfer and the change in internal energy.

   $$\dot{Q} = \frac{dU}{dt}$$

2. **Integrate the energy rate equation with respect to time and replace internal energy with specific heat and temperature, as shown in Chapter 2:**

   $$\dot{Q}\Delta t = \Delta U = m \cdot c \cdot \Delta T$$

3. **Rearrange the energy equation to solve for time (Δt):**

   $$\Delta t = \frac{m \cdot c \cdot \Delta T}{\dot{Q}}$$

4. **Look up the specific heat of water, *c*, in Table A-10 of the appendix:**

   $$c = 4.18 \text{ kJ/kg} \cdot \text{K}$$

5. **Substitute values into the energy equation to find time (Δt):**

   $$\Delta t = \frac{(0.5 \text{ kg})(4.18 \text{ kJ/kg} \cdot \text{K})(50°\text{C})}{400 \text{ W}} \left(\frac{1,000 \text{ W}}{1 \text{ kJ/s}} \right) = 261 \text{ sec}$$

Your soup will be ready to eat in less than five minutes.

Applying the First Law to Ideal-Gas Processes

Every day you breathe in an ideal gas; it may contain a bit of smog, dust, and pollen, but it's still an ideal gas. Air is just one of many gases that relate temperature, pressure, and specific volume to each other with the following equation:

$$PV = mRT$$

This is the *ideal-gas law*. The units for each variable are pressure (P) in kilopascals, total volume (V) in cubic meters, mass (m) in kilograms, the gas constant (R) in kilojoules per kilogram-Kelvin, and temperature (T) in Kelvin. You must use absolute temperatures with the ideal-gas law. I discuss the ideal-gas law in Chapter 3.

The ideal-gas law is very useful in determining the pressure-volume-temperature (P-v-T) relationships for a constant temperature, constant-volume, or constant-pressure process. You can use the ideal-gas law to determine P-v-T properties between the initial State 1 and the final State 2 in a constant temperature, volume, or pressure process, as shown by this equation:

$$\frac{P_1 V_1}{T_1} = \frac{P_2 V_2}{T_2}$$

Note that the mass (m) and the gas constant (R) of the ideal-gas law drop out because they don't change during a closed-system process.

This section examines several thermodynamic processes that involve the ideal-gas law along with the conservation of mass and the conservation of energy. I systematically change Q, W, and ΔU in the energy equation so that one term equals zero for different thermodynamic processes. For a constant-volume process, the work term (W) is zero. In a constant-temperature process, the change in internal energy (ΔU) can be zero as long as a phase change doesn't happen. In an adiabatic process, the heat transfer term (Q) is zero. I use the boundary work problems from Chapter 4 in the examples shown in the following sections.

Working with constant volume

If you heat or cool a closed system in a constant-volume process, you can't get any work out of the system because the boundary doesn't move. You only have changes in the temperature and pressure of the gas in the system. Because work (W) equals zero, the energy equation for this process is simply $Q = \Delta U$ for a stationary system.

The following example shows you how to use the ideal-gas law and the conservation of energy to determine the heat transfer associated with a constant-volume process for a closed system. Super Bowl VI in New Orleans had the coldest kickoff temperature (4 degrees Celsius) of all Super Bowl games. If the football pressure of 190 kilopascals was checked at 25 degrees Celsius (State 1), what was the pressure in the ball at kickoff (State 2)? How much heat transferred from the football by the time the ball was placed on the field? The volume of a standard NFL football is about 4.75×10^{-3} cubic meters. The following steps show you how to find the answers:

1. **Write out the energy equation for the heat loss from the football.**

   This process doesn't involve any work, only heat transfer and a change in internal energy.

   $$Q = \Delta U = m(u_2 - u_1) = m \cdot c_v(T_2 - T_1)$$

   At this point, you don't know the mass (m), so you can't find Q yet. You can use the ideal-gas law to find m.

2. **Use the ideal-gas law to determine the mass of air inside the football.**

   $$m = \frac{PV}{RT} = \frac{(190\ \text{kPa})(4.75 \times 10^{-3}\,\text{m}^3)}{(0.287\ \text{kJ/kg} \cdot \text{K})(298\ \text{K})}\left(\frac{1\ \text{kJ}}{1\ \text{kPa} \cdot \text{m}^3}\right) = 0.0106\ \text{kg}$$

3. **Calculate the heat loss from the football using the energy equation.**

   $$Q = (0.0106\ \text{kg})(0.719\ \text{kJ/kg·K})(277 - 298)\ \text{K} = -0.16\ \text{kJ}$$

   The negative value for Q means that heat is removed from the system.

4. **Use the ideal-gas law to find the pressure at State 2.** For a constant-volume process, the volume cancels out of the ideal-gas law between States 1 and 2. Convert temperatures to absolute temperature in Kelvin.

   $$\frac{P_1 V_1}{T_1} = \frac{P_2 V_2}{T_2} \text{ or } P_2 = P_1\left(\frac{T_2}{T_1}\right) = 190\ \text{kPa}\frac{(4 + 273)\ \text{K}}{(25 + 273)\ \text{K}} = 177\ \text{kPa}$$

Working with constant pressure

If you heat an ideal gas in a cylinder with a weighted piston, as shown in Figure 5-2, you can extract work from the moving piston. The piston maintains a constant pressure inside the cylinder because the weight of the piston applies a constant force on the gas. The constant-pressure process between States 1 and 2 is shown by the thick arrow in the P-v diagram on the right side of Figure 5-2.

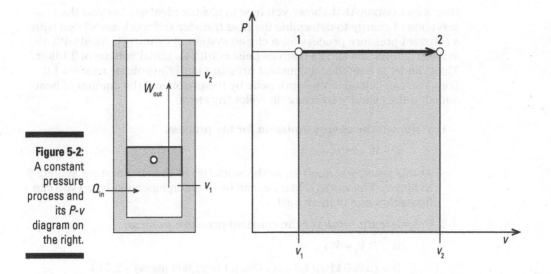

Figure 5-2:
A constant pressure process and its *P-v* diagram on the right.

The energy equation for a constant-pressure process in a stationary closed system can have heat and work interactions and is written as follows:

$$Q - W = \Delta U = m(u_2 - u_1) = m \cdot c_v(T_2 - T_1)$$

For a constant-pressure process, the work (*W*) is calculated by integrating *PdV*, as shown in Chapter 4. This is the area under the process line in the *P-v* diagram in Figure 5-2. The work for a constant-pressure process is determined by the following equation:

$$W = P(V_2 - V_1)$$

If you substitute this expression for work into the energy equation, you have

$$Q - P(V_2 - V_1) = (U_2 - U_1)$$

You can now use the definition of enthalpy, $h = u + Pv$ (see Chapter 3), to express the heat transfer in terms of enthalpy, like this:

$$Q = H_2 - H_1$$

This equation is only valid for a constant-pressure process in a closed system.

The ideal-gas law relates the temperature (*T*) and volume (*V*) of the cylinder between States 1 and 2. Because the pressure is constant, it cancels out of the *P-v-T* relationship, and the ideal-gas law is simplified as follows:

$$\frac{\cancel{P_1}V_1}{T_1} = \frac{\cancel{P_2}V_2}{T_2} \quad \text{or} \quad \frac{V_1}{T_1} = \frac{V_2}{T_2}$$

Here's an example that shows you how to use the ideal-gas law and the conservation of energy to determine the heat transfer and work associated with a constant-pressure process for a closed system. A cylinder is filled with air at 3,000 kilopascals and 25 degrees Celsius with an initial volume of 0.1 liter. The cylinder is heated at a constant pressure until the volume reaches 1.0 liter. You can calculate the work done by the piston and the amount of heat added to the cylinder by using the following steps:

1. **Write out the energy equation for the process.**

   $$Q - W = m(u_2 - u_1)$$

   At this point, you don't know the work (W) or the temperature at State 2 to find u_2. The work (W) and u_2 can be found independently, so you can find either one of them next.

2. **Calculate the work for the constant-pressure process.**

   $$W = P(V_2 - V_1)$$

   $$W = (3,000 \text{ kPa})(1.0 - 0.1 \text{ liter})(1 \text{ m}^3/1,000 \text{ liters}) = 2.7 \text{ kJ}$$

3. **Use the ideal-gas-law equation to determine the temperature of the air at State 2. Convert the initial temperature to absolute temperature:** $T_1 = 25 + 273 = 298$ K.

   $$T_2 = T_1\left(\frac{V_2}{V_1}\right) = 298 \text{ K}\left(\frac{1.0 \text{ liter}}{0.1 \text{ liter}}\right) = 2,980 \text{ K}$$

4. **Look up the values for internal energy of air at T_1 and T_2, using Table A-1 in the appendix.**

   $u_1 = 213 \text{ kJ/kg}$

   $u_2 = 2,664 \text{ kJ/kg}$

5. **Calculate the mass (m) of the air in the cylinder using the ideal-gas-law equation.**

   $$m = \frac{(3,000 \text{ kPa})(0.0001 \text{ m}^3)}{(0.287 \text{ kJ/kg} \cdot \text{K})(298 \text{ K})}\left(\frac{1 \text{ kJ}}{1 \text{ kPa} \cdot \text{m}^3}\right) = 3.51 \times 10^{-3} \text{ kg}$$

6. **Calculate the amount of heat added (Q) to the cylinder during the process, using the energy equation:**

   $$Q = W + m(u_2 - u_1)$$

   $$Q = 2.7 \text{ kJ} + (3.51 \times 10^{-3} \text{ kg})(2,664 - 213) \text{ kJ/kg} = 11.3 \text{ kJ}$$

Alternatively, you can use constant-volume specific heat and temperature in place of internal energy to find the heat transfer of the system.

Working with constant temperature

Heating and cooling an ideal gas at a constant temperature is kind of tricky. If you allow a high-pressure gas to expand in a piston-cylinder to produce work, the gas temperature will decrease. You can maintain a constant temperature in the gas by adding just the right amount of heat to it during the expansion process. The Stirling cycle and the Ericsson cycle (see Chapter 18) have isothermal compression and expansion processes.

In an isothermal process with an ideal gas, the internal energy in the system doesn't change because the initial and final temperatures are the same. Because ΔU equals zero, the energy equation simplifies to $Q = W$.

You calculate the work done by a piston in a closed system for an isothermal process using the following equation, introduced in Chapter 4:

$$W = P_1 V_1 \ln\frac{V_2}{V_1}$$

The ideal-gas law is used to determine the pressure and volume relationship between the initial and final states of a constant-temperature process. Because the temperature is constant, the ideal-gas law relationship reduces to

$$\frac{P_1 V_1}{T_1} = \frac{P_2 V_2}{T_2}$$

or

$$P_1 V_1 = P_2 V_2$$

A cylinder starts with a volume of 0.1 liter at 3,000 kilopascals and 25 degrees Celsius. The gas expands until the volume is 1.0 liter. Heat is added to the cylinder to keep the gas at 25 degrees Celsius. During the process, the pressure inside the cylinder decreases. Figure 5-3 shows this process as the thick arrow from States 1 to 2 on the *P-v* diagram on the right side of the figure. The area under the *P-v* curve is the amount of work in the process.

You can find the work output, the heat transfer, and the final pressure in the cylinder by following these steps:

1. **Write the energy equation for the process.**

   $$Q = W$$

   You can calculate the work, and you know that the heat transfer equals the work.

 2. **Calculate the work for the isothermal process in a closed system.**

$$W = P_1 V_1 \ln \frac{V_2}{V_1}$$

$$W = \left(3{,}000 \text{ kPa}\right)\left(\frac{(0.1 \text{ liter})1 \text{ m}^3}{1{,}000 \text{ liters}}\right)\ln\left(\frac{1.0 \text{ liter}}{0.1 \text{ liter}}\right)\left(\frac{1 \text{ kJ}}{1 \text{ kPa} \cdot \text{m}^3}\right) = 0.69 \text{ kJ}$$

Because the work out equals the heat in, this is also the value for Q.

 3. **Calculate the final pressure in the cylinder using the ideal-gas-law equation.**

$$P_2 = P_1\left(\frac{V_1}{V_2}\right) = 3{,}000 \text{ kPa}\left(\frac{0.1 \text{ liter}}{1.0 \text{ liter}}\right) = 300 \text{ kPa}$$

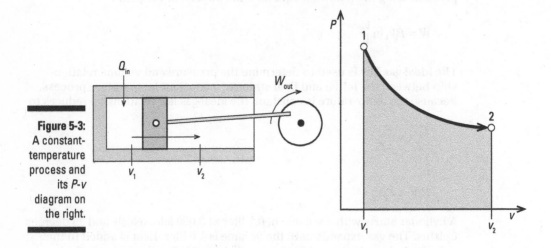

Figure 5-3:
A constant-
temperature
process and
its *P-v*
diagram on
the right.

Working with an adiabatic process

In previous sections, I set W or ΔU to zero in the energy equation for different thermodynamic processes. Now it's time for Q to be zero. When heat doesn't enter or leave a system, the process is adiabatic. The energy equation for a closed system simplifies to this result for an ideal gas:

$$W = -\Delta U = m(u_1 - u_2) = m \cdot c_v(T_1 - T_2)$$

You calculate the work done by a frictionless piston in a closed system for a reversible and adiabatic process using the following equation, introduced in Chapter 4. A process is reversible if both the system and the surroundings return to their initial state when the process proceeds in the reverse direction. The following equation assumes the specific heat is constant during the process. You can have better accuracy in calculating the work if you use the specific heats of the gas at the average process temperature.

$$W = \frac{P_2 V_2 - P_1 V_1}{1 - k}$$

In this equation, k is the ratio of constant-pressure (c_p) and the constant-volume (c_v) specific heat of the gas:

$$k = \frac{c_p}{c_v}$$

For air around room temperature, k equals 1.4. The value of k changes with temperature. You can look up the specific heat of air at different temperatures in Table A-1 of the appendix to calculate the value of k at different temperatures. You find the constant-volume specific heat (c_v) using the constant-pressure specific heat (c_p) and the gas constant (R) in this equation:

$$c_v = c_p - R$$

The ideal-gas law is used to determine the P-v-T relationships between States 1 and 2 for a process. For a reversible, adiabatic process, P-v-T relationships are written in several ways, as shown here:

$$\frac{P_2}{P_1} = \left(\frac{V_1}{V_2}\right)^k \text{ or } \frac{T_2}{T_1} = \left(\frac{V_1}{V_2}\right)^{k-1} = \left(\frac{P_2}{P_1}\right)^{(k-1)/k}$$

The cylinder shown in Figure 5-4 is insulated so that no heat can enter or leave it. The cylinder starts with a volume of 0.1 liter at 3,000 kilopascals and 25 degrees Celsius. The gas expands until the volume inside the cylinder is 1 liter. The right side of Figure 5-4 shows the process on a P-v diagram. The area under the P-v curve, shown by the thick line with an arrow from States 1 to 2, equals the work of the process.

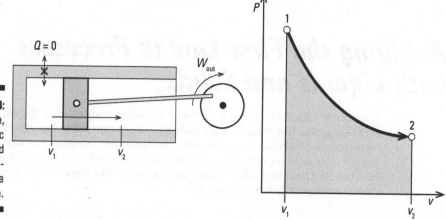

Figure 5-4:
A reversible, adiabatic process and its P-v diagram on the right side.

You can find the temperature and pressure inside the cylinder at State 2 and the work by following these steps:

1. **Write the energy equation for the process.**

$$W = m(u_1 - u_2) = m \cdot c_v(T_1 - T_2)$$

You can calculate the work by finding the temperature T_2 at State 2.

2. **Find the temperature T_2 at State 2 using the ideal-gas equation for a reversible adiabatic process.**

$$T_2 = T_1\left(\frac{V_1}{V_2}\right)^{k-1} = 298 \text{ K}\left(\frac{0.1 \text{ liter}}{1.0 \text{ liter}}\right)^{1.4-1} = 119 \text{ K}$$

3. **Calculate the work by using the change in temperature from States 1 to 2 and the constant-volume specific heat.**

$$W = m \cdot c_v(T_1 - T_2)$$

$$W = (3.51 \times 10^{-3} \text{ kg})(0.719 \text{ kJ/kg·K})(298 - 119) \text{ K} = 0.45 \text{ kJ}$$

4. **Find the pressure P_2 at State 2, using the ideal-gas equation for a reversible adiabatic process.**

$$P_2 = P_1\left(\frac{V_1}{V_2}\right)^k = (3,000 \text{ kPa})\left(\frac{0.1 \text{ liter}}{1.0 \text{ liter}}\right)^{1.4} = 119 \text{ kPa}$$

5. **You can also calculate the work done by pushing the piston, by using the boundary work equation for a reversible adiabatic process, as shown in Chapter 4.**

$$W = \frac{P_2 V_2 - P_1 V_1}{1-k}$$

$$W = \frac{(119 \text{ kPa})(1.0 \text{ liter}) - (3,000 \text{ kPa})(0.1 \text{ liter})}{1-1.4}\left(\frac{1 \text{ m}^3}{1,000 \text{ liters}}\right)\left(\frac{1 \text{ kJ}}{1 \text{ kPa} \cdot \text{m}^3}\right) = 0.45 \text{ kJ}$$

Applying the First Law to Processes with Liquids and Solids

The conservation of energy equation for liquids and solids is the same as that for ideal gases. But liquids and solids aren't compressible like gases, so no relationship exists between pressure, temperature, and volume like the ideal-gas law. Because liquids and solids are incompressible, they aren't able to do work by changing volume the way a gas can. But they can do work using kinetic energy or potential energy.

The internal energy of a solid or a liquid changes when its temperature changes. There's no distinction between a constant-pressure and a constant-volume specific heat for solids or liquids. So the specific heat is just called the specific heat (c). The internal energy of a system is related to the mass (m), the specific heat (c), and the temperatures of a process, as follows:

$$U_2 - U_1 = m \cdot c \cdot (T_2 - T_1)$$

Here's an example of a closed system that shows how mass and energy are conserved for liquids and solids. A restaurant owner advertises that he has the best iced tea in town. A server fills a glass with 350 grams of ice at 0 degrees Celsius and 150 grams of freshly brewed tea at 75 degrees Celsius. Some of the ice melts to cool the tea to 0 degrees Celsius. It takes 333 joules of energy per gram of ice to melt at 0 degrees Celsius. How much ice melts to cool the tea down to 0 degrees Celsius? Find out by following these steps:

1. **Define the closed system to include only the mass of the liquid and the ice, as shown in Figure 5-5.**

   Neglect the mass of the glass.

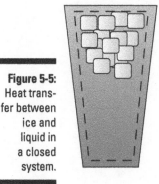

Figure 5-5: Heat transfer between ice and liquid in a closed system.

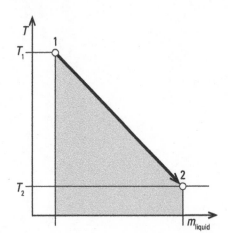

2. **Write an energy balance equation for the melting process.**

   The amount of energy required to cool the liquid equals the amount of energy given up by the ice and determines how much ice melts. The right side of Figure 5-5 shows the change in temperature of the liquid as the mass of liquid increases due to the ice melting.

   $$(Q_{ice})_{in} = (Q_{liquid})_{out}$$

3. Write expressions for the heat transfer from the liquid and to the ice.

$$(Q_{ice})_{in} = m_{melt} \cdot h_{sf}$$

$$(Q_{liquid})_{out} = m_{liquid} \cdot c \cdot (T_1 - T_2)$$

In this equation, m_{melt} is the mass of ice that melts; h_{sf} is the latent heat of fusion of the ice when it melts from a solid to a liquid at 0 degrees Celsius. The amount of liquid poured into the glass is m_{liquid}.

4. Solve for the mass of ice that melts from the energy equation:

$$m_{melt} = \frac{m_{liquid} c (T_1 - T_2)}{h_{sf}} = \frac{(150 \text{ g})(4.18 \text{ J/g} \cdot \text{K})(75-0)°\text{C}}{(333 \text{ J/g})} = 141 \text{ g}$$

Almost half of the ice in the glass melts to cool the hot tea.

Chapter 6

Using the First Law in Open Systems

. . . .

In This Chapter

▶ Working with open systems (also called control volume systems)

▶ Developing the conservation of mass and energy principles for steady flow processes

▶ Applying conservation of energy to four common steady flow processes

▶ Using conservation of mass and energy on non-steady flow processes

. . . .

*I*f you've ever pulled into a gas station and filled up a tire that was looking a little deflated, you've used an open system. See, you've used thermodynamics without even realizing it!

This chapter introduces you to thermodynamic analysis of some simple processes involving heat and work for open systems. The concepts using the first law of thermodynamics I discuss here are similar to the ones presented in Chapter 5 for closed systems. The difference between an open system and a closed system is whether or not a fluid is allowed to flow into or out of a system. You may be able to guess by their names that a fluid can't flow in a closed system, but it can flow in an open system.

Some examples of simple open systems include heat exchangers, pumps, compressors, turbines, and nozzles. These devices are used in more complicated systems such as power plants, air-conditioning systems, and jet engines, which I discuss in Chapters 10–13.

Conserving Mass in an Open System

The best way to begin every thermodynamic analysis is by defining a system. A *system* describes a region enclosed by an imaginary boundary (which may be fixed or flexible) that contains a mass or volume to use for analysis. A system that doesn't allow mass to enter or leave is called a *closed system*. The mass inside a closed system is often called the *control mass*. A system that allows mass to enter and leave is called an *open system*. The volume of an open system is often called the *control volume*.

This chapter focuses on thermodynamic analysis using the conservation of mass and conservation of energy for open systems. *Conservation of mass* means that the mass flow rate of material entering a system minus the mass flow rate leaving equals the mass that may accumulate within the system, as described by this equation:

$$\dot{m}_{\text{in}} - \dot{m}_{\text{out}} = \frac{dm_{\text{sys}}}{dt}$$

When you see a "dot" over a variable like mass ($\dot{m}$), the dot means that the variable is on a rate basis or per unit time. The units for mass flow rate are kilograms per second.

Defining mass and volumetric flow rates

The size of the inlets and outlets of some open systems, such as nozzles and diffusers in jet engines, is important because they're sized to take advantage of changes in kinetic energy. The mass flow rate entering or leaving an open system is related to the area of the opening (A), the average fluid velocity normal to the inlet ($\mathbf{V}$), and the fluid density (ρ), as shown in this equation:

$$\dot{m} = \rho \mathbf{V} A$$

I use bold font for velocity ($\mathbf{V}$) and italicized font for total volume (V) to distinguish between these two variables throughout this book.

In this equation, the units of area are in square meters, velocity is in meters per second, and density is in kilograms per cubic meter.

The volumetric flow rate is related to the mass flow rate and is calculated either by using the average fluid velocity ($\mathbf{V}$) and the area (A) of the opening, or by dividing the mass flow rate by the fluid density, as shown here:

$$\dot{V} = \mathbf{V}A = \frac{\dot{m}}{\rho}$$

The units for volumetric flow rate are cubic meters per second.

Applying conservation of mass to a system

Here's an example that shows you how to use the conservation of mass principle for an open system. Figure 6-1 shows a jet engine mounted on an aircraft. The system is defined by the dashed line around the engine. The system has two inlets, one for air and the other for fuel. The system has one outlet for exhaust.

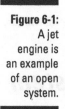

Figure 6-1:
A jet
engine is
an example
of an open
system.

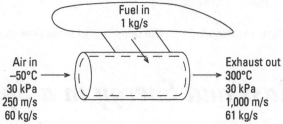

Fuel in
1 kg/s

Air in
−50°C
30 kPa
250 m/s
60 kg/s

Exhaust out
300°C
30 kPa
1,000 m/s
61 kg/s

The aircraft is flying at 250 meters per second. The air temperature is −50 degrees Celsius, and the pressure is 30 kilopascals. The air mass flow rate into the engine is 60 kilograms per second, and the fuel mass flow rate is 1 kilogram per second. The exhaust is 300 degrees Celsius and has a velocity of 1,000 meters per second. You can analyze this system to determine the volumetric flow rates into and out of the engine by following these steps:

1. **Write the conservation of mass equation for the system.**

   No mass accumulates within the system, so the mass flow in equals the mass flow out.

   $$\dot{m}_{air} + \dot{m}_{fuel} = \dot{m}_{exhaust} = (60 + 1) \text{ kg/s} = 61 \text{ kg/s}$$

   To determine the volumetric flow rates, you need to find the gas density at the inlet and the exhaust.

2. **Find the density ρ_{in} of the air at the inlet, using the ideal-gas-law equation.**

   $$\rho_{in} = \frac{P}{RT_{in}} = \frac{30 \text{ kPa}}{(0.287 \text{ kJ/kg} \cdot \text{K})(223 \text{ K})}\left(\frac{1 \text{ kJ}}{1 \text{ kPa} \cdot \text{m}^3}\right) = 0.469 \text{ kg/m}^3$$

3. **Find the density ρ_{out} of the exhaust at the exit, using the ideal-gas-law equation. Assume the exhaust has the properties of air.**

   $$\rho_{out} = \frac{P}{RT_{out}} = \frac{30 \text{ kPa}}{(0.287 \text{ kJ/kg} \cdot \text{K})(573 \text{ K})}\left(\frac{1 \text{ kJ}}{1 \text{ kPa} \cdot \text{m}^3}\right) = 0.182 \text{ kg/m}^3$$

  4. **Calculate the volumetric flow rate of air at the inlet, using the mass flow rate of the incoming air: 60 kg/s.**

$$\dot{V}_{in} = \frac{\dot{m}_{air}}{\rho_{in}} = \frac{60 \text{ kg/s}}{0.469 \text{ kg/m}^3} = 128 \text{ m}^3/\text{sec}$$

  5. **Calculate the volumetric flow rate of the exhaust at the exit, using the total mass flow rate of fuel plus air.**

$$\dot{V}_{out} = \frac{\dot{m}_{exhaust}}{\rho_{out}} = \frac{61 \text{ kg/s}}{0.182 \text{ kg/m}^3} = 335 \text{ m}^3/\text{sec}$$

The volumetric flow rate depends on the density and mass flow rate of the air.

Balancing Mass and Energy in a System

Thermodynamic analysis of open systems requires using the principles of conservation of mass and conservation of energy. When you apply these two principles to a system, you're balancing the mass and energy entering and leaving a system with any change in mass and energy in the system.

Although some systems may have multiple inlets and outlets, as shown in the jet engine example in the preceding section, many systems have only one inlet and one outlet, which simplifies things. For a single-stream system with one inlet and one exit and no change in mass within the system, the conservation of mass rate equation is written as follows:

$$\dot{m}_{in} - \dot{m}_{out} = 0 \text{ or } (\rho VA)_{in} - (\rho VA)_{out} = 0$$

The conservation of energy principle helps you keep track of energy as it flows through a system. Conservation of energy means the energy coming into the system (E_{in}) minus the energy leaving (E_{out}) equals the change in energy within the system (ΔE_{sys}). The conservation of energy equation for any system is written as $E_{in} - E_{out} = \Delta E_{sys}$. The units for energy are kilojoules.

The left-hand side of the conservation of energy equation ($E_{in} - E_{out}$), represents the net amount of energy transfer in a system by heat (Q), work (W), and mass (m). Work in a system can take many forms, as described in Chapter 4, including boundary work, shaft work, and electrical work. Energy transfer by mass includes the internal (U), kinetic (KE), and potential (PE) energy of the flow.

It takes work to move mass into or out of a control volume. This work is called *flow work,* and it's the product of the fluid pressure (P) and the fluid specific volume (v). Flow work is written as $w_{flow} = Pv$. Flow work and internal energy are combined into a property called *enthalpy* (see Chapter 2) as defined by this equation: $h = u + Pv$. For an open system, the net energy transfer to the system is written as follows:

$$E_{in} - E_{out} = \left[Q + W + \dot{m}(h + ke + pe) \right]_{in} + \left[Q + W + \dot{m}(h + ke + pe) \right]_{out}$$

The right-hand side of the conservation of energy equation (ΔE_{sys}), represents the change in internal energy (U), kinetic energy (KE), potential energy (PE), and other energy forms (such as magnetic energy, if used) in the system. A system has kinetic energy if it's moving relative to an external frame of reference. A system has potential energy from its position in the earth's gravitational field. I discuss kinetic and potential energy in Chapter 1. For an open system, the change in the total energy of a system is written as follows:

$$\Delta E_{sys} = \Delta U + \Delta KE + \Delta PE$$

If your system is stationary, you don't have to worry about the kinetic or potential energy terms of the whole system, but you may still have kinetic energy or potential energy in the fluid.

The conservation of energy equation can also be written on a per unit mass basis: $e_{in} - e_{out} = \Delta e_{sys}$. Or, you can write it on a per unit time or rate basis: $\dot{E}_{in} - \dot{E}_{out} = \dfrac{dE_{system}}{dt}$. I show you how to write the conservation of energy equation on a rate basis for an open system in the following section.

When Time Stands Still: The Steady State Process

Many thermodynamic systems operate continuously, such as air-conditioning systems, power plants, jet engines, and automobile engines. Each of these sophisticated systems uses a number of components, such as pumps, compressors, turbines, nozzles, and diffusers. Because these components operate in steady conditions, energy isn't accumulated within the devices.

In an open system with steady flow, the mass flow rate of the fluid moving through the system remains constant over time. This constancy means that $\dfrac{dE_{system}}{dt} = 0$ in terms of the energy balance equation, and this process is known as a *steady state* process. In a steady state process, mass and energy flow rates don't change within the system. The conservation of energy equation for a steady flow system with only one inlet and one outlet on a rate basis is $\dot{E}_{in} - \dot{E}_{out} = 0$.

The energy equation for steady flow systems includes terms for heat transfer and work interactions associated with changes in enthalpy, kinetic, and/or potential energy. You can write the energy equation on a rate basis to include these energy forms, as shown here:

$$\left(\dot{Q}_{in} - \dot{Q}_{out}\right) + \left(\dot{W}_{in} - \dot{W}_{out}\right) = \dot{m}\left[h_{out} - h_{in} + \frac{1}{2}\left(V_{out}^2 - V_{in}^2\right) + g\left(z_{out} - z_{in}\right)\right]$$

If you have more than one inlet and/or outlet, you must add additional mass flow rate and energy terms to the energy balance.

The following example shows you how to use the conservation of energy on a steady flow open system. An aircraft is flying at 250 meters per second. The air temperature is –50 degrees Celsius, and the pressure is 30 kilopascals. The air mass flow rate into the engine is 60 kilograms per second, and the fuel mass flow rate is 1 kilogram per second. The exhaust is 300 degrees Celsius and has a velocity of 1,000 meters per second. Figure 6-2 shows the jet engine as an open system.

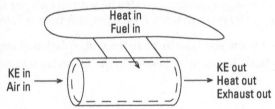

Figure 6-2: An energy balance on a jet engine as a steady-flow open system.

You can write an energy balance on the jet engine to calculate the amount of heat transfer into the engine and the change in kinetic energy by following this example.

1. **Write the energy balance equation for the steady-flow open system:**

$$\dot{Q}_{in} = \dot{m}_{out}h_{out} - \dot{m}_{in}h_{in} + \frac{1}{2}\left(\dot{m}_{out}V_{out}^2 - \dot{m}_{in}V_{in}^2\right)$$

Inside the jet engine, work is done by the turbine to drive the compressor, but because the work doesn't leave (cross) the system boundary, the work term is zero. (See Chapter 10 for additional description and analysis of jet engines.) Potential energy is neglected because the elevation doesn't differ between the inlet and outlet of the engine.

2. **Look up the enthalpy of air h_{in} at the inlet temperature $T_{in} = -50°C$ using Table A-1 in the appendix.**

You need to interpolate the table; I show you how in Chapter 3.

$h_{in} = 223.4$ kJ/kg

3. **Look up the enthalpy of air h_{out} at the outlet temperature T_{out} = 300°C using Table A-1 in the appendix.**

   You need to interpolate the table.

   h_{out} = 579.2 kJ/kg

4. **Calculate the change in kinetic energy (ΔKE), using the kinetic energy terms in the energy balance equation.**

   The mass flow rate at the exit, $\dot{m}_{out}$, equals the mass flow rate of air plus the mass flow rate of fuel. Assume the kinetic energy of the fuel entering the engine is zero, because it has very little mass and velocity compared to the air.

   $$\Delta KE = \frac{1}{2}\left(\dot{m}_{out}V_{out}^2 - \dot{m}_{in}V_{in}^2\right)$$

   $$\Delta KE = \frac{1}{2}\left[\left(61 \text{ kg/s}\right)\left(1,000 \text{ m/s}\right)^2 - \left(60 \text{ kg/s}\right)\left(250 \text{ m/s}\right)^2\right]$$

   $$_{...}\left(\frac{1 \text{ kJ}}{1,000 \text{ m}^2/\text{s}^2}\right) = 28,600 \text{ kJ/s}$$

5. **Calculate the heat transfer to the engine using the energy balance equation.**

   $$\dot{Q}_{in} = \left(61 \text{ kg/s}\right)\left(579.2 \text{ kJ/kg}\right) - \left(60 \text{ kg/s}\right)\left(223.4 \text{ kJ/kg}\right) + 28,600 \text{ kJ/s}$$

   $$\dot{Q}_{in} = 50,500 \text{ kJ/s}$$

Of the 50.5 megawatts of heat entering the engine, only 28.6 megawatts are converted to kinetic energy to propel the aircraft using thrust. The rest of the heat input to the engine leaves as waste heat because the temperature of the exhaust is well above the ambient air temperature.

Using the First Law on Four Common Open-System Processes

Four common open-system processes are analyzed using the energy equation in this section:

- Nozzles and diffusers
- Pumps, compressors, and turbines
- Heat exchangers
- Throttling valves

The energy equation contains terms for heat transfer, work, enthalpy, and kinetic energy as needed for each type of device.

Flowing through nozzles and diffusers

If you take a garden hose and cover most of the opening with your thumb, the water squirts out quite a distance. When you do this, you've made a *nozzle.* (Of course you can buy good nozzles for your garden hose at a hardware store and give your thumb a rest.) A nozzle, as shown in Figure 6-3, is a mechanical device that accelerates a fluid as it flows through a pipe, hose, or duct. Changing the velocity of a fluid changes its kinetic energy. You can use the first law of thermodynamics to determine the effects of changing the kinetic energy of a fluid.

Another device that's similar to a nozzle is called a *diffuser.* A diffuser, as shown in Figure 6-3, is simply a nozzle with the fluid flowing in the reverse direction. The fluid decelerates as it flows through a diffuser. Heating, ventilating, and air-conditioning systems use diffusers to slow down the air coming from the fan so that it gently flows into a room. Jet engines use diffusers to slow down the air entering the engine and use nozzles to accelerate the exhaust to provide greater thrust.

Figure 6-3:
A nozzle or a diffuser is used to change the velocity (and kinetic energy) of a fluid.

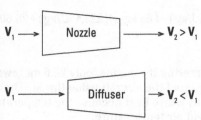

Making assumptions for nozzles and diffusers

When you apply the first law of thermodynamics to a nozzle or a diffuser, you usually make the following assumptions:

✔ No heat is transferred between the nozzle or diffuser and the environment. If the temperature of the fluid is significantly different than the temperature of the environment, insulation is used to minimize heat transfer.

✔ No work is done on or by a nozzle or diffuser because they don't have any moving parts that can be used for work.

✔ No change in potential energy occurs between the inlet and the outlet.

✔ The only energy forms that apply to nozzles and diffusers are changes in enthalpy and kinetic energy.

✔ Kinetic energy is neglected at the inlet of a nozzle and at the outlet of a diffuser if the flow velocity is low. When you square a large number, it becomes huge; when you square a small number, it becomes tiny.

Writing the energy balance for nozzles and diffusers

Using these assumptions, the energy balance for nozzles or diffusers is written as follows:

$$\dot{m}\left(h_{in} + \frac{1}{2}V_{in}^2\right) = \dot{m}\left(h_{out} + \frac{1}{2}V_{out}^2\right)$$

The total energy entering a nozzle or diffuser equals the total energy leaving. As a fluid accelerates or decelerates, the kinetic energy changes the enthalpy of the fluid. Because the mass flow rate entering the device equals the mass leaving, mass can be dropped from the energy balance and the equation can be rearranged to determine the exit enthalpy (h_{out}), as follows:

$$h_{out} = h_{in} - \frac{V_{out}^2 - V_{in}^2}{2}$$

A change in enthalpy results in a change in temperature and pressure. In a diffuser, a high-velocity fluid decelerates and the fluid temperature rises. In a nozzle, the fluid temperature decreases.

Analyzing a diffuser

You can use the energy equation to determine the increase in the temperature of the air flowing through the diffuser of a jet engine. The air temperature entering the diffuser is –50 degrees Celsius. The enthalpy of inlet air is 223.4 kilojoules per kilogram. The air velocity at the diffuser inlet is 250 meters per second, and exit velocity is negligibly small, so you can assume it's nearly zero. The energy equation for a diffuser with zero velocity at the exit is written as follows:

$$h_{out} = h_{in} + \frac{V_{in}^2}{2}$$

You calculate the enthalpy h_{out} of the air at the diffuser exit using the inlet enthalpy h_{in} and velocity V_{in}:

$$h_{out} = 223.4 \text{ kJ/kg} + \frac{(250 \text{ m/s})^2}{2}\left(\frac{1 \text{ kJ/kg}}{1{,}000 \text{ m}^2/\text{s}^2}\right) = 254.7 \text{ kJ/kg}$$

Look up the corresponding air temperature in Table A-1 of the appendix. You need to interpolate the table to find that T_{out} = –18.8°C. The rise in air temperature is due to kinetic energy changing into internal energy in the air.

Working with pumps, compressors, and turbines

Pumps are used in many applications, ranging from circulating water through filters in a swimming pool to providing drinking water for a city. Pumps use a work input to increase the pressure in a liquid and make it circulate in a network of pipes. In thermodynamics, the most common pump application is the use of a Rankine cycle in a power plant (see Chapter 12) to pressurize and circulate water through a boiler.

A compressor is similar to a pump except it pressurizes and circulates a gas instead of a liquid. Figure 6-4 shows a diagram of a compressor that's found in gas turbine engines, jet engines, and industrial facilities (see Chapter 10). The shape of the diagram indicates that the specific volume of the gas decreases as the pressure of the gas increases in the compressor. A compressor has a large number of blades, like a fan, mounted on a shaft. It may have many rows of blades, called *stages*, that increase the pressure step by step from one stage to the next. The blades for each stage get progressively smaller because the specific volume decreases as the gas is compressed. It takes much more work per unit mass to compress a gas than a liquid.

Figure 6-4: Compressors and turbines change the pressure of a fluid.

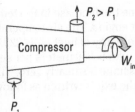
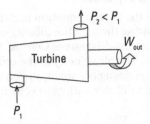

A turbine extracts work from a gas, such as steam in a Rankine cycle power plant (see Chapter 12) or air in a Brayton cycle engine (see Chapter 10). Figure 6-4 shows a diagram of a turbine.

Making assumptions for pumps, compressors, and turbines

When you apply the first law of thermodynamics to a pump, compressor, or turbine, you usually make the following assumptions:

✔ Turbines are usually insulated because they have hot gas flowing through them. Minimizing heat loss provides more work output. Compressors often have cooling to reduce the work input required. Pumps usually aren't insulated because a heat loss or gain doesn't really change the work input.

✔ No change in potential energy occurs between the inlet and the outlet of the machine.

✔ No change in kinetic energy occurs between the inlet and the outlet. For a pump, a liquid is incompressible, so the inlet and outlet velocities are the same. In a turbine, the change in kinetic energy can be sizable, but the change in enthalpy is usually much greater, so any change in kinetic energy is ignored for simplicity.

Writing the energy balance for a compressor, turbine, or pump

Using these assumptions, you can write the energy balance for a compressor, turbine, or pump as follows:

$$\left(\dot{Q}_{in} - \dot{Q}_{out}\right) + \left(\dot{W}_{in} - \dot{W}_{out}\right) = \dot{m}\left(h_{out} - h_{in}\right)$$

A pump and a compressor use the work-in term. A turbine uses the work-out term. Usually, there isn't any heat transfer into a pump, compressor, or turbine, so the heat-in term is zero. A compressor uses the heat-out term if it's cooled. A turbine doesn't have a heat-out term if it's well insulated.

Analyzing a compressor

Here's an example that shows you how to use the conservation of energy equation to determine the work required to operate the compressor of the jet engine. Suppose the compressor inlet enthalpy is $h_{in} = 254.7$ kJ/kg. The compressor exit temperature is 500 Kelvin, and the heat loss (q) from the compressor to the ambient air is 50 kilojoules per kilogram. The air mass flow rate through the compressor is 60 kilograms per second. You can find the work of the compressor as follows:

1. **Write out the energy equation to solve for the rate of compressor work.**

$$\dot{W}_{in} = \dot{m}\left[q_{out} + \left(h_{out} - h_{in}\right)\right]$$

2. **Look up the enthalpy of air for the compressor exit h_{out} at 500 Kelvin in Table A-1 of the appendix.**

$h_2 = 503.5$ kJ/kg

3. **Calculate the rate of compressor work.**

Use the mass flow rate, the heat loss, and the change in enthalpy of the air in the energy equation.

$$\dot{W}_{in} = \left(60 \text{ kg/s}\right)\left[50 + \left(503.5 - 254.7\right) \text{ kJ/kg}\right]\left(\frac{1 \text{ MW}}{1,000 \text{ kJ/s}}\right) = 17.9 \text{ MW}$$

The compressor work is very large. If a pump pressurized and moved the same size mass of water as the mass of air moved by the compressor, the work would be only about 25 kilowatts and the temperature rise of the water would be less than 1 degree Celsius. This difference is due to the high specific volume of the air compared to water.

Moving energy with heat exchangers

Heat exchangers are devices that exchange heat between two different fluids. They're used in a variety of applications. The radiator in a car circulates hot fluid (usually a mix of ethylene or propylene glycol and water) through tubes and air passing over the tubes removes heat from the fluid. A furnace exchanges heat between the burners and the air circulating in a house. An air-conditioning system uses two heat exchangers: one to remove heat from inside the house and another to dump the heat outside. Power plants use heat exchangers to condense steam into liquid water so it can be pressurized and circulated through the boiler with a pump.

Figure 6-5 shows a diagram of one type of heat exchanger. Hot fluid A runs through the center of the heat exchanger giving up heat to cold fluid B, which flows over the outside of fluid A. In this process, the temperature of fluid A decreases while the temperature of fluid B increases. Because heat transfer requires a temperature difference between fluids, the outlet temperature of fluid A is warmer than the outlet temperature of fluid B. In a heat exchanger, the fluids are kept separate from each other.

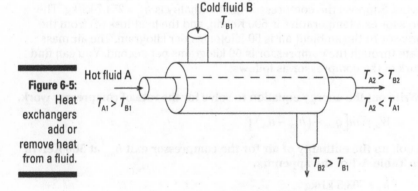

Figure 6-5:
Heat exchangers add or remove heat from a fluid.

Making assumptions for heat exchangers

When you apply the first law of thermodynamics to a heat exchanger system, you usually make the following assumptions:

✔ No work is performed because the device is designed only to transfer heat, not to perform work.

✔ Changes in kinetic and potential energy are small, so they can be ignored.

✔ The energy released by one fluid stream is absorbed by the other fluid stream.

✔ A heat exchanger that doesn't transfer heat to the environment is insulated to minimize heat exchange with the environment. This maximizes heat transfer between the two fluids inside the heat exchanger.

Writing the energy balance for heat exchangers

Following these assumptions, you can write the energy equation for a heat exchanger as follows:

$$\dot{m}_A \left(h_{in} - h_{out} \right) = \dot{m}_B \left(h_{out} - h_{in} \right)$$

The mass flow rates of the different fluid streams can be different, especially if one fluid stream is a gas and the other is a liquid. The difference in specific heat or difference in temperature change of the two fluids in a heat exchanger often accounts for different mass flow rates.

Analyzing a heat exchanger

Here's an example using the energy equation for a heat exchanger. A steam condenser operates at a pressure of 10 kilopascals. The quality of the steam entering the condenser is 0.90. *Quality* is a thermodynamic property that describes the fraction of vapor in a liquid-vapor mixture. I introduce you to quality in Chapter 3. At the exit of the condenser, the water is saturated liquid. The mass flow rate of the steam is 6 kilograms per second. Air at 25 degrees Celsius is used to condense the steam. The air temperature leaving the heat exchanger is at 40 degrees Celsius. Follow these steps to find the mass flow rate of air required to condense the steam:

1. **Write the energy balance equation for the steam condenser heat exchanger.**

   For the air side of the equation, you can use either the enthalpy of the air, which you can look up in Table A-1 of the appendix and interpolate, or the constant-pressure specific heat (c_p) and the air temperatures (T_{in} and T_{out}). The constant-pressure specific heat of air (c_p) is 1.005 kJ/kg · K. Both methods are shown in the following equation.

   $$\dot{m}_{Air} = \dot{m}_{Water} \frac{\left(h_{in} - h_{out} \right)_{Water}}{\left(h_{out} - h_{in} \right)_{Air}} = \dot{m}_{Water} \frac{\left(h_{in} - h_{out} \right)_{Water}}{c_P \left(T_{out} - T_{in} \right)_{Air}}$$

2. **Find the enthalpy of the saturated liquid water (h_f), using the saturated pressure table for water at 10 kPa in Table A-4 of the appendix.**

   h_f is also the enthalpy of the water leaving the condenser (h_{out}). The value is $h_f = h_{out} = 192$ kJ/kg.

3. **Find the enthalpy of the saturated water vapor (h_g), using the saturated pressure table for water in Table A-4 of the appendix.**

   The value of h_g is $h_g = 2{,}585$ kJ/kg.

4. **Calculate the enthalpy of the steam at the condenser inlet (h_{in}), using the steam quality and the enthalpy of the saturated liquid and vapor.**

   $$h_{in} = x(h_g - h_f) + h_f = 0.9(2{,}585 - 192)\ \text{kJ/kg} + 192\ \text{kJ/kg} = 2{,}346\ \text{kJ/kg}$$

5. **Calculate the air mass flow rate using the energy equation from Step 1.**

   $$\dot{m}_{Air} = \frac{(6\ \text{kg/s})(2{,}346 - 192)\ \text{kJ/kg}}{(1.005\ \text{kJ/kg} \cdot \text{K})(40 - 25)°\text{C}} = 857\ \text{kg/s}$$

The air mass flow rate is nearly 15 times more than the water mass flow rate because a tremendous amount of energy is required to condense steam into liquid water. Air-cooled condensers at power plants are very large because so much air is required to condense the steam.

Reducing pressure with throttling valves

When you need to decrease the pressure of a fluid without extracting work in the process, a very simple device is used for the job. This device is called a throttling valve. Throttling valves are used mostly in air-conditioning systems (see Chapter 13), where a hot, high-pressure liquid is "throttled" to a low pressure. This process causes the temperature of the liquid to decrease, just as the temperature of a gas decreases as it flows through a turbine. At low pressures, the liquid may become a liquid-vapor mixture. Under the right conditions, the temperature of the liquid-vapor mixture can be cold enough to absorb heat from, say, your house or the inside of your refrigerator and freezer.

A *throttling valve* restricts the flow of a fluid in a pipe or tube, causing the pressure to decrease. Figure 6-6 shows three different ways fluid flow can be restricted with a throttling valve. The flow can be restricted via the use of

✔ A small orifice

✔ A porous plug

✔ A capillary (small-diameter) tube

Each method of flow restriction makes the fluid squeeze through a small opening, which causes the pressure to decrease.

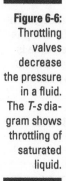

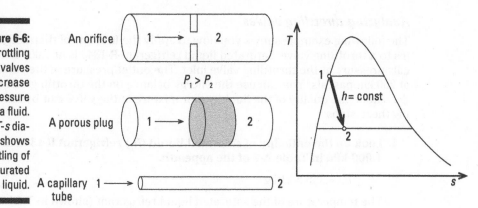

Figure 6-6: Throttling valves decrease the pressure in a fluid. The *T-s* diagram shows throttling of saturated liquid.

Making assumptions for throttling valves

You can make the following assumptions about the energy balance for throttling valves:

- ✔ No heat is transferred between the valve and the environment because the device is usually very small. The surface area of a throttling valve is insufficient for allowing the fluid to lose much heat to the environment. One exception to this assumption is the capillary tube, where the surface area is relatively large compared to the fluid volume so it can transfer heat.

- ✔ No work is done on or by a throttling valve because there are no moving parts that can be used for work.

- ✔ No change in kinetic or potential energy occurs between the inlet and the outlet of the valve.

Applying these assumptions to a throttling valve leaves you with only the enthalpy terms of the energy equation: $h_{in} = h_{out}$.

A throttling valve is a constant enthalpy or isenthalpic device. Figure 6-6 shows a constant enthalpy line drawn on a *T-s* diagram for a refrigerant flowing through a throttling valve. You see that the saturated liquid at State 1 becomes a liquid-vapor mixture at State 2. In the ideal refrigeration cycle (see Chapter 13), the liquid entering a throttling valve is usually a saturated liquid. However, a subcooled (or compressed) liquid can be throttled just as well.

A throttling valve usually isn't used with ideal gases because enthalpy is a function only of temperature for ideal gases. So the temperature of an ideal gas doesn't change as it flows through a throttling valve.

Analyzing throttling valves

The following example shows you how to apply the first law of thermodynamics to a throttling valve. Saturated liquid refrigerant R-134a is at 1,000 kilopascals pressure at the throttling valve inlet. The outlet pressure of the valve is at 250 kilopascals. You can use the energy balance on the throttling valve to determine the quality of the liquid-vapor mixture at the valve exit by following these steps:

1. **Look up the enthalpy of saturated liquid (h_f) refrigerant R-134a at 1,000 kPa in Table A-7 of the appendix.**

   $h_{in} = 255.5$ kJ/kg

   The temperature of the saturated liquid refrigerant (shown in Table A-7) at this pressure is 39.4 degrees Celsius, which is warm enough to reject heat to the outdoor environment in an air-conditioning system.

2. **Look up the enthalpy of saturated liquid (h_f) and saturated vapor (h_g) refrigerant R-134a at 250 kPa in Table A-7 of the appendix.**

   $h_f = 194.3$ kJ/kg

   $h_g = 396.1$ kJ/kg

3. **Calculate the quality (x_{out}) of the refrigerant liquid-vapor mixture.**

   $$x_{out} = \frac{h_{out} - h_f}{h_g - h_f} = \frac{(255.5 - 194.3) \text{ kJ/kg}}{(396.1 - 194.3) \text{ kJ/kg}} = 0.303$$

The temperature of the low-pressure mixture is –4.3 degrees Celsius, as found in Table A-7 of the appendix, which is cold enough to remove heat from inside a refrigerator. Freezer temperatures are usually colder than this.

When Time Is of the Essence: The Transient Process

Suppose you need to determine how much work it takes to inflate a tire, fill an air-conditioning system with refrigerant, or expel exhaust gases from an automobile engine. These kinds of processes are known as *transient processes* or non-steady processes. In a transient process, mass and energy are stored within or released from the system. In this section I show you what happens when ΔE_{sys} is included in the energy equation:

$$E_{in} - E_{out} = \Delta E_{sys}$$

A transient process takes place over a time interval, so mass flow rates and energy flow rates aren't important like they are in a steady state process. Filling an air conditioner with refrigerant or inflating a tire with air is an open system with unsteady flow, because the mass within the system changes over time. Steady flow open systems have a fixed mass within the system for all time.

Figure 6-7 shows an open system with one inlet and one outlet. The system has a piston that can do work. A control volume is drawn as a dashed line to define the boundary for the system.

Figure 6-7: An open system with a transient process has a change in mass and energy within the control volume.

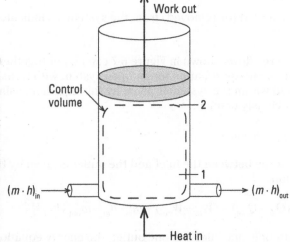

Making assumptions for the energy balance

You can make the following assumptions for the energy balance equation for this system:

- There are no significant changes in kinetic or potential energy between the inlet and outlet of the control volume.

- The amount of mass and energy flowing into the control volume is $(m \cdot h)_{in}$ and the amount leaving the control volume is $(m \cdot h)_{out}$. The enthalpy of the fluid entering and leaving the control volume remain constant during the process.

✔ At the start of the filling process, the piston is located at Position 1 in Figure 6-7, and the mass times energy product within the control volume is $(m_1 \cdot u_1)$.

✔ At the end of the filling process, the piston is at Position 2, and the mass times energy product within the control volume is $m_2 \cdot u_2$.

✔ The size of the control volume in Figure 6-7 changes during the filling process. If the control volume changes size, boundary work can be extracted from the system using a piston. (I discuss boundary work in Chapter 4.) In some systems, the size of the control volume may remain fixed. If the control volume doesn't change in size, it can't do any boundary work.

✔ Heat may be added (or removed) from the system, as indicated in Figure 6-7.

The mass and energy flows shown in Figure 6-7 can be put together to form the conservation of energy equation for an open system with transient flow. The energy stored within the system is given by this equation, using the assumptions previously defined:

$$\Delta E_{sys} = m_2 \cdot u_2 - m_1 \cdot u_1$$

The change in energy between the inlet and the outlet is given by the following equation:

$$E_{in} - E_{out} = (Q_{in} - Q_{out}) + (W_{in} - W_{out}) + m_{in} \cdot h_{in} - m_{out} \cdot h_{out}$$

For a system with only one inlet and one outlet, the energy equation is put together from these two equations and rearranged as follows:

$$(Q_{in} - Q_{out}) + (W_{in} - W_{out}) = m_{out}h_{out} - m_{in}h_{in} + (m_2 u_2 - m_1 u_1)_{system}$$

If you have more than one inlet and outlet, you just add additional mass flow and energy terms to the energy balance.

Analyzing a transient process

You can use the energy equation for a transient process in the following example. An engine exhausts air at 2,500 Kelvin from a cylinder with an initial volume of 1.0 liter and a final volume of 0.1 liter. The exhaust temperature remains constant during the process. The pressure inside the cylinder during the exhaust process remains constant at 1,000 kilopascals. The process is adiabatic. You can find the work done by the piston to push the exhaust out of the cylinder by following these steps:

1. **Write the energy balance equation for the system.** The heat transfer is zero, and no mass flows into the system.

$$W_{out} = (m \cdot h)_{out} - (m_1 u_1 - m_2 u_2)_{sys}$$

2. **Look up the internal energy and the enthalpy of air at 2,500 K in Table A-1 of the appendix.**

Because the temperature of the air remains constant during the exhaust process, $u_1 = u_2$.

$u = 2,166$ kJ/kg

$h = 2,883$ kJ/kg

3. **Calculate the mass of air in the cylinder m_1 at the start of the exhaust process, using the ideal-gas law.**

$$m_1 = \frac{P_1 V_1}{RT_1} = \frac{(1{,}000 \text{ kPa})(1 \text{ liter})}{(0.287 \text{ kJ/kg} \cdot \text{K})(2{,}500 \text{ K})} \left(\frac{1 \text{ m}^3}{1{,}000 \text{ liters}} \right) \left(\frac{1 \text{ kJ}}{1 \text{ kPa} \cdot \text{m}^3} \right)$$

$$= 1.39 \times 10^{-3} \text{ kg}$$

4. **Calculate the mass of air in the cylinder m_2 at the end of the exhaust process, using the ideal-gas law.**

$$m_2 = m_1 \left(\frac{V_2}{V_1} \right) = 1.39 \times 10^{-3} \text{ kg} \left(\frac{0.1 \text{ liter}}{1.0 \text{ liter}} \right) = 1.39 \times 10^{-4} \text{ kg}$$

5. **Find the mass of air exhausted from the cylinder.**

$$m_{out} = m_1 - m_2 = (1.39 \times 10^{-3} - 1.39 \times 10^{-4}) \text{ kg} = 1.25 \times 10^{-3} \text{ kg}$$

6. **Calculate the work done by the piston in moving the exhaust out of the cylinder, using the equation from Step 1 and noting that $u_1 = u_2$:**

$$W_{out} = (1.25 \times 10^{-3} \text{ kg})(2{,}883 \text{ kJ/kg}) - (1.39 \times 10^{-3} - 1.39 \times 10^{-4}) \text{kg}$$
$$\times (2{,}166 \text{ kJ/kg})$$

$$W_{out} = 0.9 \text{ kJ}$$

You can verify the work done by the piston by using the analysis of a constant-pressure process, as shown in this equation (see Chapter 5):

$$W_{out} = P(V_1 - V_2)$$

$$W_{out} = (1{,}000 \text{ kPa})(1.0 - 0.1 \text{ liter}) \left(\frac{1 \text{ m}^3}{1{,}000 \text{ liters}} \right) \left(\frac{1 \text{ kJ}}{1 \text{ kPa} \cdot \text{m}^3} \right) = 0.9 \text{ kJ}$$

You can see the work is the same using either method of analysis.

Chapter 7

Governing Heat Engines and Refrigerators with the Second Law

. . . .

In This Chapter

▶ Figuring out the second law of thermodynamics

▶ Recognizing thermal energy reservoirs

▶ Harnessing heat engines and estimating their efficiency

▶ Regulating refrigerators and calculating coefficients of performance

. . . .

Ralph Waldo Emerson said, "Build a better mousetrap and the world will beat a path to your door." This sentiment has been the mantra of many inventors, even before Emerson came along and expressed it so eloquently. Since the dawn of human civilization, people have thought of countless ways to do more work with less effort. Some have even tried to get work done without any effort at all. People have put a lot of thought and effort into trying to invent perpetual-motion machines, but to no avail.

The second law of thermodynamics is useful for determining what is and isn't possible when it comes to building a better "mousetrap" that uses energy to produce work or work to move heat. Many perpetual motion concepts can obey the first law of thermodynamics, which is the conservation of energy principle, but they still aren't possible. If you fire a bullet into a block of wood, the kinetic energy of the bullet is converted to heat because energy is conserved. But have you ever heard of a bullet lodged in a block of wood cooling off and firing itself out of the wood block? The energy is still conserved, but this process isn't possible. I introduce the first law of thermodynamics in Chapter 2 and apply it to several systems in Chapters 5 and 6.

In this chapter, I introduce you to the second law of thermodynamics. The second law shows there are limits to the way heat is converted to work and the way work moves heat from a cold place to a warm one. The second law of thermodynamics can't easily be defined in a simple, single sentence like the first law. Instead, it describes a set of connected concepts about energy. It's almost like a Bill of Rights for energy rather than just a law. Just as the Bill of Rights describes the limits of the U.S. federal government's power, the second law describes the limits of energy as it's used in thermodynamic processes.

Looking at the Impact of the Second Law

A thermodynamic process is a way of changing energy from one form to another, such as by heating, cooling, compressing, or expanding a liquid or a gas. The second law describes the limits on the use of energy for thermodynamic processes in the following ways:

- ✔ It identifies the natural or spontaneous direction a process will go. For example, heat flows from hot to cold; pressure flows from high to low.

- ✔ It says the quality of energy characterizes the capacity for converting heat into work. Heat from a high-quality energy source can do more work than heat from a low-quality energy source even if they have the same amount of total energy.

- ✔ It requires the heat input processes for heat engines and refrigerators to receive energy from a thermal energy reservoir that has a higher quality than the heat engine or refrigerator.

- ✔ It requires the heat rejection processes for heat engines and refrigerators to reject energy to a thermal energy reservoir that has a lower quality than the heat engine or refrigerator.

- ✔ It describes the degradation in energy quality during a process, which is known as the *decrease in energy availability principle.* I discuss this principle in detail in Chapter 9.

- ✔ It determines theoretical limits on the efficiency of heat engines and refrigerators.

- ✔ It states that the amount of disorder in a system and its surroundings always increases during a thermodynamic process. You can increase the amount of order on a system by doing work on it, but the disorder of the system's surroundings grows even more.

I give you more details about these concepts of the second law in the following sections.

Defining Thermal Energy Reservoirs

The Hoover Dam created a reservoir called Lake Mead out of the Colorado River. Electric generators within the dam provide about 2,080 megawatts of electricity. Lake Mead has an average elevation of 1,200 feet above sea level. Below the dam, the river elevation is about 600 feet above sea level. If you use sea level as the reference point, the Hoover Dam is able to extract only about half of the potential energy available from the Colorado River before the river meets its end as an essentially dry stream bed near the Gulf of California. Formerly, the river made it all the way to the Gulf.

Thermodynamic processes that produce work use energy from a reservoir just like the Hoover Dam generators use the Lake Mead Reservoir. At the end of a process, energy is rejected to another reservoir (like the Colorado River). A thermal energy reservoir that supplies energy to a thermodynamic process is called a *heat source reservoir*. A thermal energy reservoir that receives energy from a thermodynamic process is called a *heat sink reservoir*. Figure 7-1 shows the flow of energy from source to sink reservoirs.

Figure 7-1: A heat source reservoir supplies energy to a process, whereas a heat sink reservoir absorbs it from a process.

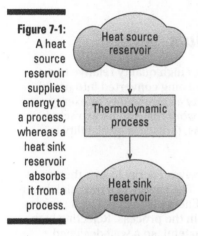

Parameters of a thermal reservoir

A *thermal reservoir* is a body that has a thermal capacity larger than the thermodynamic process it's part of. The temperature of the reservoir doesn't change if you remove or add energy to interact with a thermodynamic process. The thermal capacity of a reservoir is a product of its mass times its specific heat.

Examples of natural thermal reservoirs include the atmosphere, lakes, oceans, and rivers. These thermal reservoirs are usually energy sinks for thermodynamic systems like power plants and automobile engines. A power plant rejects heat into any of these reservoirs. The choice of energy sink depends on economic and environmental considerations. Heat rejection from a power plant can dump a lot of energy, which may disrupt the ecosystem of a small river or lake. So rejecting the heat to the atmosphere may be the best choice in that situation. An automobile rejects heat only into the atmosphere because having it connected to a river or lake isn't practical. I discuss the thermodynamic analysis of automobile engines in Chapter 11 and power plants in Chapter 12.

A thermal reservoir doesn't have to be a large or natural body. The heat source for a power plant or an automobile engine is the combustion of fuel. The combustion process defines the heat source reservoir for the power plant or the engine.

A thermal reservoir can be something as small as the inside of your refrigerator or your kitchen. Your refrigerator absorbs heat from the air inside it, making the inside of the fridge the heat source reservoir. The refrigerator then rejects the heat into the kitchen, which is the heat sink reservoir.

Heat reservoirs come in many forms and don't necessarily have to be large. The only requirement is that they be sufficiently large to maintain a constant temperature even though they give or receive energy with a process.

Considering highs and lows

The energy within a heat source reservoir has high quality relative to a process — meaning that it has a capacity for being converted into useful work. Some textbooks describe energy quality as availability or exergy. *Availability* or *exergy* is the maximum useful work that can be extracted from a reservoir by a thermodynamic process. (I discuss availability in detail in Chapter 9.)

A separate thermal energy reservoir that receives energy from a thermodynamic process is called a *heat sink reservoir.* The energy of a heat sink reservoir has low quality relative to the process. Energy is rejected by a process when it can't perform any more useful work in the process. Rejecting high-quality energy to a heat sink reservoir is wasteful, so a well-designed process should reject energy that's as low-quality as possible.

For example, in a power plant that uses a gas turbine engine to generate electricity, the exhaust of the engine is at a relatively high temperature compared to the ambient air. In some cases, the exhaust may be hot enough to generate steam, which can be used in another turbine to generate additional electricity. So instead of being thrown away, relatively high-quality energy from the gas turbine exhaust can be used as the heat source for a steam plant. This process is called the combined cycle, and I give a brief description of it in Chapter 18.

Working with the Kelvin-Planck Statement on Heat Engines

In an agrarian economy, one way to convert energy into work is to feed a mule, hitch it up to a plow, and crack a whip. The farmer gets his field plowed and fertilized at the same time. In an industrial economy, a heat engine converts energy into work. The industrial farmer fills the tractor with diesel fuel, plows the ground, and cracks open a soda can. The mule now looks leisurely over the field, watching dust clouds form behind the plow.

A *heat engine* is a device that converts heat into work. Examples of heat engines include

- ✔ Gas turbine engines that are used in power plants to make electricity and on aircraft as jet engines. (See Chapter 10.)

- ✔ Internal combustion engines used in automobiles, trucks, boats, trains, lawn mowers, and so on. (See Chapter 11.)

- ✔ Steam power plants for generating electricity. (See Chapter 12.)

Characterizing heat engines

All scientific observations of a heat engine reveal that it must reject a portion of its energy to a low-temperature reservoir to complete its cycle. No heat engine is able to convert all the heat input to work. This observation is expressed as the *Kelvin-Planck statement* of the second law of thermodynamics, which says that it's impossible for any heat engine that operates in a cycle to receive heat only from a single reservoir and convert all the heat into work.

A heat engine must operate in a cycle between high- and low-temperature reservoirs to produce a net work output. Heat engines such as those found in power plants, automobiles, and jet engines absorb heat from a combustion process and reject the heat into the environment. Despite the many varieties of heat engines, Figure 7-2 shows they all have the following characteristics:

- ✔ A heat engine uses a fluid to absorb and reject heat. The fluid can also accept work input and provide work output from the heat engine. The fluid can be either a gas or a liquid. When the fluid is a liquid, the heat absorption process changes it to a vapor, and the heat rejection process changes it back to a liquid. The working fluid is usually air or water, but other fluids can be used in specialized systems.

- ✔ Heat is absorbed from a high-temperature (T_H) energy reservoir, such as a combustion process, nuclear energy, or solar power. Q_{in} is the amount of heat absorbed by the heat engine. The temperature of the reservoir is higher than the temperature of the fluid within the heat engine. The heat input is greater than the heat output, as shown by the thickness of the Q_{in} arrow compared to the Q_{out} arrow in Figure 7-2.

- ✔ A work output is produced, often by rotating a shaft to mechanically operate another device or machine. W_{out} is the amount of work produced by the heat engine. The work output exceeds the work input, as implied by the thickness of the arrow shown in Figure 7-2.

✔ A work input is required to compress or pressurize the working fluid in the engine. W_{in} is the amount of work input to the heat engine. Note that in Figure 7-2, the arrow for the W_{in} is thinner than the one for the W_{out}, because the work in must be less than the work out.

✔ Heat is rejected to a low-temperature (T_L) reservoir, such as the atmosphere or a body of water, to allow the engine to complete a cycle. Q_{out} is the amount of heat rejected by the heat engine. The temperature of the reservoir is lower than the temperature of the fluid within the heat engine. Heat naturally flows from a higher temperature to a lower temperature. Note that in Figure 7-2, the arrow for Q_{out} is thinner than the one for Q_{in}, implying the heat rejected is less than the heat input.

In a cycle, the working fluid is returned to its starting condition after the heat rejection process is complete. The basic heat transfer and work processes of a heat engine are repeated continuously to produce work from a heat source.

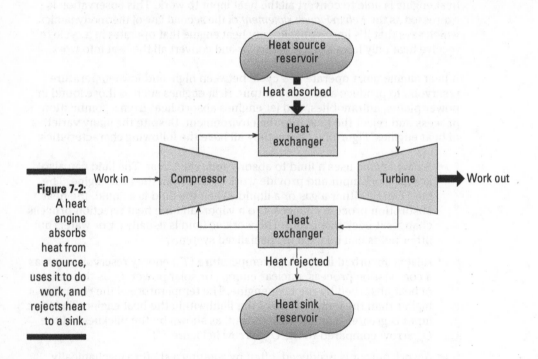

Figure 7-2:
A heat engine absorbs heat from a source, uses it to do work, and rejects heat to a sink.

Determining thermal efficiency

The net work out ($W_{net,out}$) of a heat engine is equal to the difference between the work output (W_{out}) and the work input (W_{in}). In other words, $W_{net,out} = W_{out} - W_{in}$.

The net work out of a heat engine can also be written on a rate basis. The rate at which work is done is power. The dot over the variable indicates it's on a rate basis. The net power output from a heat engine is

$$\dot{W}_{net,out} = \dot{W}_{out} - \dot{W}_{in}$$

The conservation of energy principle shows that the net work out of the heat engine equals the difference between the heat input (Q_{in}) and the heat rejected (Q_{out}) by the engine. That is, $W_{net,out} = Q_{in} - Q_{out}$. This is true on a rate basis as well.

Every heat engine requires some work input and some heat rejection. As the Kelvin-Planck statement says, not all the energy coming into the engine can be converted into useful work. When you compare the net work output to the heat input, you find the *thermal efficiency* of the engine, as shown in this equation:

$$Thermal \ Efficiency = \frac{Net \ work \ output}{Total \ heat \ input}$$

In general, any measure of performance is (what you want)/(what you provide). For a heat engine, what you want is to produce work. What you provide is heat. Calculating thermal efficiency is one way to evaluate the performance of a heat engine. Thermal efficiency reflects how well an engine uses energy to produce work. You calculate the thermal efficiency (η_{th}) of a heat engine using either of the following equations:

$$\eta_{th} = \frac{W_{net,out}}{Q_{in}} = 1 - \frac{Q_{out}}{Q_{in}}$$

Suppose you need to evaluate the efficiency of a power plant. If the amount of heat absorbed (Q_{in}) by the steam from the combustion process is 1,000 kilowatts and the amount of heat rejected (Q_{out}) to the air is 600 kilowatts, you calculate the thermal efficiency as follows:

$$\eta_{th} = 1 - \frac{600 \ kW}{1,000 \ kW} = 40\%$$

Doing good things with waste heat

Naturally, you may wonder whether anything can be done with the waste heat. After all, it seems like such a waste to reject it all to a heat sink. In fact, many good things can be done with waste heat. The possibilities are limited only by your imagination and range from the silly to the serious. On the silly side is the idea of cooking your food with a "carbeque." If you take a long road trip, you may be able to cook a roast and potatoes on your engine block, but I'm not endorsing this idea.

A better use of waste heat is *cogeneration*, a process whereby waste heat from a power plant is used for industrial purposes or for district heating. Large manufacturing facilities that have their own power plants, like paper mills, usually need a lot of heat for various processes. Many cities and college campuses provide heating for downtown and campus buildings from power plants.

Another example is the *combined cycle*, which takes the heat from a gas turbine engine exhaust, which can be quite hot, and generates steam. The steam can then be used in a power plant to generate additional electricity. The heat rejected by one heat engine can be the heat source for another heat engine operating at lower temperatures. (One man's trash is another man's treasure.) I describe cogeneration and combined cycles in a bit more detail in Chapter 18.

Chilling with the Clausius Statement on Refrigeration

Heat naturally flows from hot sources to cold sinks. Until the invention of the first refrigeration machine, you couldn't do much about the natural flow of heat. You just had to sweat it out during the summer until winter came. All that has changed, but moving heat from a cold reservoir up to a warmer reservoir comes with a price, as you may know if you've paid a huge electric bill in the summer for air conditioning.

Thermodynamically, a refrigerator is very much like a heat engine operating in reverse. It uses work to move heat from a low-temperature reservoir, such as the inside of your refrigerator, to a high-temperature reservoir, like your kitchen. Examples of refrigeration machines include air conditioners and heat pumps. A heat pump is basically an air conditioner that can pump heat from the outdoors to the inside of a house during the winter. It works best in areas with mild winters (I explain why in Chapter 13).

The following sections discuss how the second law of thermodynamics applies to refrigeration cycles and provides a brief introduction on how the refrigeration cycle works. I go into more depth and discuss how to analyze refrigeration cycles in Chapter 13.

Characterizing refrigerators

Because heat naturally flows from hot to cold temperatures, making it flow in the opposite direction takes some effort. This observation is expressed as the *Clausius statement* of the second law of thermodynamics: It's impossible for a refrigerator to move heat from a colder reservoir to a warmer reservoir without a work input. Figure 7-3 illustrates the following characteristics of refrigerators:

✔ A refrigerator uses a fluid to absorb and reject heat. The fluid can also accept work input and, in some cases, provide work output in the refrigerator. (The work output of certain kinds of refrigeration cycles is used to offset the work input.) The fluid can be either a gas or a liquid. When the fluid is a liquid, the heat absorption process changes it to a vapor, and the heat rejection process changes it back to a liquid. The working fluid for most refrigeration systems is a liquid refrigerant, but air is used in air-conditioning systems on aircraft.

✔ Heat is absorbed by a type of heat exchanger called an *evaporator* from a low-temperature (T_L) energy reservoir, such as the inside of your refrigerator. Q_{in} is the amount of heat absorbed by the refrigerator. The temperature of the reservoir is higher than the temperature of the fluid within the evaporator of a refrigerator.

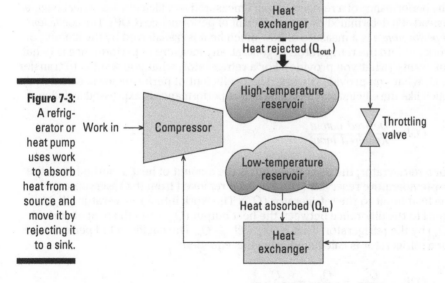

Figure 7-3:
A refrigerator or heat pump uses work to absorb heat from a source and move it by rejecting it to a sink.

> ✔ A work input is required to compress the working fluid in the refrigerator. W_{in} is the amount of work used by the refrigerator.
>
> ✔ Heat is rejected by a type of heat exchanger called a *condenser* to a high-temperature (T_H) reservoir, such as your kitchen in the case of a refrigerator. Q_{out} is the amount of heat rejected by the refrigerator. The temperature of the reservoir is lower than the temperature of the fluid within the refrigerator condenser. Heat naturally flows from a higher temperature to a lower temperature. Note that in Figure 7-3, the arrow for Q_{out} is thicker than the arrow for Q_{in}, implying that the heat rejected exceeds the heat absorbed because the condenser must reject heat added by the work input of the compressor.
>
> ✔ A valve reduces the liquid pressure between the condenser and the evaporator to match that of the compressor. No work is extracted in the liquid refrigerant-based refrigeration cycle.

Finding the coefficient of performance

In a cycle, the working fluid returns to its starting condition after the heat rejection process is complete. The heat transfer and work processes in a refrigeration cycle are repeated continuously to move heat from the low-temperature reservoir to the high-temperature one.

The performance of a refrigerator isn't measured by efficiency like a heat engine. Instead, it's determined by the coefficient of performance (COP). The *coefficient of performance* is a measure of how much heat is transferred by the amount of work put into the refrigerator. In general, any measure of performance is (what you want)/(what you provide). For a refrigerator, what you want is to transfer heat. What you provide is work. The coefficient of performance is calculated much like the efficiency. The coefficient of performance is expressed as follows:

$$COP = \frac{Desired\ output}{Required\ input}$$

For a refrigerator, the desired output is the amount of heat absorbed from the low-temperature reservoir (Q_L). The heat removed from the reservoir equals the heat input to the refrigerator (Q_{in}). The work into a refrigerator (W_{in}) is equal to the difference between the heat output (Q_{out}) and the heat absorbed (Q_{in}) by the refrigerator. That is, $W_{in} = Q_{out} - Q_{in}$. The coefficient of performance for a refrigerator is calculated using this equation:

$$COP_R = \frac{Q_{in}}{W_{net,in}} = \frac{Q_L}{W_{net,in}} = \frac{Q_L}{Q_H - Q_L}$$

For a heat pump, the desired output is the amount of heat rejected to the warm energy reservoir (Q_H), the interior of a house. The heat added to the reservoir equals the heat output of the refrigerator (Q_{out}). You calculate the coefficient of performance for a heat pump by using this equation:

$$COP_{HP} = \frac{Q_{out}}{W_{net,in}} = \frac{Q_H}{W_{net,in}} = \frac{Q_H}{Q_H - Q_L}$$

You can find the coefficient of performance for a refrigerator or a heat pump by working out the following example. Suppose you have a refrigerator that absorbs 1 kilowatt of heat from the cold reservoir and rejects 1.3 kilowatts of heat to the warm reservoir. You can find the actual coefficient of performance for the refrigerator with the following equation:

$$COP_R = \frac{1 \text{ kW}}{(1.3 - 1) \text{ kW}} = 3.3$$

Pondering perpetual-motion machines

In 1918, the U.S. Patent Office declared that it would no longer consider any more patents for a perpetual-motion machine. But that hasn't stopped people from trying to invent one. The first law of thermodynamics states that energy cannot be created or destroyed; it can only change form. But many perpetual-motion machine ideas exist, their purpose being to generate energy for free.

One popular concept, dating from the twelfth century, is the idea of shifting mass in an overbalanced wheel, as shown in the following figure. In one version, the wheel has hollow, curved spokes partially filled with mercury (shown as the dark, shaded portions of the spokes in the figure below). As the wheel turns, the mercury moves around inside the spokes so that one side is always heavier than the other, causing the wheel to continually rotate. This wheel could then theoretically be used to do work. But if this 900-year-old idea had any merit, these wheels would be very popular by now.

(continued)

(continued)

Unfortunately, the amount of energy needed to move the mercury away from the center always equals the amount of energy it takes to move it back towards the center. Any perpetual-motion machine that attempts to violate the first law of thermodynamics is called a *perpetual-motion machine of the first kind.*

Other perpetual-motion machines attempt to adhere to the first law of thermodynamics while violating the second law. You may try to invent a heat engine like the one shown in the following figure. You can see that there's a heat input to take care of the first law, a work output, and a work input. The heat rejection part of the engine is missing. Why bother rejecting any heat when it clearly reduces the efficiency of the engine? For the heat engine shown, the temperature of the whole system would continually rise until it failed because the heat wasn't rejected. A perpetual-motion machine with only one thermal energy reservoir violates the second law and is called a *perpetual-motion machine of the second kind.*

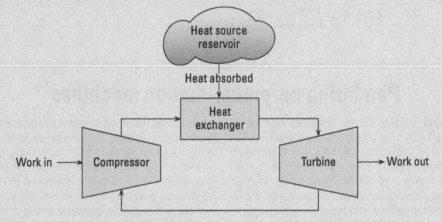

The Clausius statement and the Kelvin-Planck statement are equivalent in their consequences. Only a perpetual-motion machine would be able to prove either of these two statements to be wrong, and no one has been able to produce such a machine to date.

Chapter 8

Entropy Is the Demise of the Universe

- -

In This Chapter

▶ Getting familiar with entropy

▶ Understanding why entropy always increases

▶ Figuring out temperature and entropy relationships

▶ Analyzing the entropy change for thermodynamic processes

▶ Working with isentropic processes

- -

*W*hen the universe was brand spanking new, the quality of the energy in the universe was at its peak. The potential to do work was at its maximum. Every time energy is used to do work or provide heat, whether the activity is natural or man-made, the quality of the energy diminishes. (I discuss the quality of energy in Chapter 7.) As the quality of energy decreases, so does its capacity to do work.

In this chapter I introduce you to a thermodynamic property called *entropy*. The way entropy works is opposite that of the quality of energy, which means that entropy grows as the quality of energy decreases. At the beginning of the universe, entropy was at its minimum. All thermodynamic processes increase the total entropy of the universe. The quality of energy will be at its minimum at the end of the universe, whereas entropy will be at its peak. This is sometimes called the entropy death of the universe.

After introducing you to the concept of entropy, I show you how to calculate the change in entropy for a thermodynamic process, and I discuss some ideal processes that can do work without a change in entropy.

What Is Entropy?

Remember when you first got a new desk? You arranged all your papers and knick-knacks on it. It looked nice and neat. But if you're like most people, things started to pile up on your desk and before you knew it, books, candy wrappers, sticky notes, and empty coffee cups took over. Yes, one aspect of entropy is at work here. Things that start out neat and tidy naturally become disordered. You can picture the universe this way. In the beginning, it was much smaller than it is today; its energy was concentrated into a very small space. But as the universe ages and expands, it becomes more disordered. Making something ordered again takes effort; you have to do some work.

Entropy has many different interpretations. Its definition depends on who you're talking to. In principle, entropy is used by physicists, theologians, engineers, philosophers, information specialists, and economists, among other professionals. Entropy is often thought of as a measure of disorder of a system. But how can you quantify order or disorder? Entropy is a thermodynamic property of a substance that needs to be quantified in order to be useful.

In thermodynamics, you find microscopic and macroscopic perspectives on entropy.

Taking a microscopic view of entropy

On a microscopic level, entropy starts with the third law of thermodynamics, which I discuss in Chapter 2. At absolute zero temperature, the molecules in a substance have no energy to move, vibrate, or rotate. The entropy of the material is zero. As energy is added to a material, the entropy of the molecules increases because they become more energetic and more disorganized — the way your desk gets more cluttered the more you use it.

The entropy of a material increases as its temperature increases. Solid materials have less entropy than liquids, and liquids have less entropy than gases. As molecules in a material increase in temperature, they like to spread out and take up more room; that is, they become more disordered.

Pressure has the opposite effect on entropy of a material. As the pressure of a gas, liquid, or solid increases, the entropy decreases. However, liquids and solids are considered nearly incompressible, so the entropy decrease is minimal. Pressure forces molecules closer together; they become more ordered, so entropy decreases.

Scan through the thermodynamic property tables in the appendix to see how entropy increases with temperature and decreases with pressure.

Looking at entropy on a macroscopic level

On a macroscopic level, entropy is associated with the amount of energy in a system that is transferred by heat and isn't used for doing work. For example, when a system (which could simply be a hot cup of tea sitting on your kitchen table) has an *internally reversible* change in energy by a small amount of heat transfer (δQ), the change in entropy (dS), is equal to the heat transfer in the system divided by the absolute temperature (T) of the system boundary. Mathematically, this definition of entropy is written like this:

$$dS \equiv \left(\frac{\delta Q}{T} \right)_{\text{Int. Rev.}}$$

The units for entropy on a per unit mass basis are kilojoules per kilogram-Kelvin (kJ/kg · K).

The system boundary separates the system from its environment or surroundings. If you define the tea in the cup as the system, the cup itself can be considered the boundary, and the kitchen can be considered the surroundings.

An internally reversible change in energy means the system can go back to its original state if the process goes in the reverse direction. This concept is an idealization of real heat-transfer processes and assumes the temperature of the system and the surroundings are the same during heat transfer. In the real world, heat transfer requires a temperature difference between two objects. As the temperature difference between two objects approaches zero, the heat transfer rate between them decreases. It may require a very long time or a very large area to transfer heat between two objects to the point where it becomes impractical.

In real systems, there are always irreversibilities, such as friction, sudden expansion of a gas, or heat transfer with a temperature difference. The entropy change in a system having a real process is always greater than that of an internally reversible process, which means the entropy change for a system with a real irreversible process is expressed by the following inequality:

$$dS_{\text{Irrev.}} > \left(\frac{\delta Q}{T} \right)_{\text{Int. Rev.}}$$

The terms δQ (pronounced "delta Q") and dS are differential quantities, which means you can integrate them using calculus along a process path to find the entropy change of a system ΔS_{sys}. You can get rid of the inequality in the preceding equation by adding an *entropy generation* term, S_{gen}, that accounts for irreversibility, as shown in the following equation:

$$\Delta S_{\text{sys}} = S_2 - S_1 = \int_1^2 \frac{\delta Q}{T} + S_{\text{gen}}$$

Entropy generation is always a positive quantity, and its value depends on the process used in the system. In an adiabatic process, no heat transfer occurs across the system boundary, so the entropy change in the system is due only to irreversibility.

If energy in the form of heat is removed from a system, the entropy of the system decreases. If energy in the form of heat is added to a system, the system's entropy increases. For the cup of tea in your kitchen, the entropy of the tea naturally decreases as it cools off. Meanwhile, the kitchen receives energy from the tea, so its entropy increases.

Entropy is transferred in the same direction and at the same time as heat is transferred in a system.

If energy is removed from a system by doing work instead of losing heat, it's possible for the entropy of the system to remain unchanged. This is called an *isentropic process*. I discuss isentropic processes in detail in the section "Analyzing Isentropic Processes."

If the temperature of a fluid decreases in a constant-pressure process, the entropy decreases, but if the pressure decreases in a constant-temperature process, the entropy increases. So in an isentropic process, the entropy decrease due to temperature decrease is offset by the entropy increase caused by the pressure decrease. An isentropic process can also occur when both temperature and pressure increase.

Coping with the Increase in Entropy Principle

Every real thermodynamic process generates entropy; it adds a small amount of disorder to the universe. In a thermodynamic process, the total amount of entropy generated includes the entropy change of the system and the entropy change of the surroundings. The total change in entropy (ΔS_{total}) for a system (ΔS_{system}) and its surroundings ($\Delta S_{surroundings}$) is always a positive number. This is expressed mathematically with the following equation:

$$\Delta S_{total} = \Delta S_{system} + \Delta S_{surroundings} \geq 0$$

You can see that the total entropy of a system and its surroundings always increases in your own kitchen. Figure 8-1 shows a refrigerator operating in a kitchen. The refrigerator has a heat exchanger in the back that dissipates heat removed from inside the refrigerator. The heat is dumped into the kitchen. You can define the heat exchanger on the refrigerator as the system and the kitchen as the surroundings.

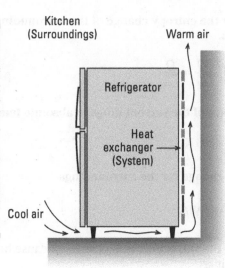

Figure 8-1:
If the entropy of the heat exchanger on a refrigerator decreases, the entropy of its surroundings (the kitchen) increases.

Kitchen (Surroundings)

Warm air

Refrigerator

Heat exchanger (System)

Cool air

Here's an example using a refrigerator in a kitchen that shows you how to find the total entropy of a system and its surroundings. A heat exchanger, called a *condenser,* removes 200 watts of heat from the refrigerator. The condenser is 40 degrees Celsius and is the system. The kitchen is 22 degrees Celsius and is the surroundings. If you assume the heat transfer process is reversible, there is no entropy generation due to irreversibility. You find the total entropy change (ΔS_{total}) of the system and the surroundings by following these steps:

1. **Write the equation for the entropy change of the system with no entropy generation.**

$$\Delta S_{sys} = S_2 - S_1 = \int_1^2 \frac{\delta Q}{T_{sys}} = \frac{Q}{T_{sys}}$$

In this equation, T_{sys} is the boundary temperature of the system.

2. **Convert the system boundary temperature to absolute temperature.**

$$T_{sys} = 40°C + 273 = 313 \text{ K}$$

3. **Calculate the entropy change for the system.**

$$\Delta S_{sys} = \frac{-200 \text{ W}}{313 \text{ K}} = -0.64 \text{ W/K}$$

The entropy change of the system is negative because heat is removed from the system.

4. **Write the equation for the entropy change of the surroundings, with no entropy generation.**

$$\Delta S_{sur} = S_2 - S_1 = \int_1^2 \frac{\delta Q}{T_{sur}} = \frac{Q}{T_{sur}}$$

5. **Convert the temperature of the surroundings to absolute temperature.**

$$T_{sur} = 22°C + 273 = 295 \text{ K}$$

6. **Calculate the entropy change for the surroundings.**

$$\Delta S_{sur} = \frac{200 \text{ W}}{295 \text{ K}} = 0.68 \text{ W/K}$$

The entropy change of the surroundings is positive because heat is added to the surroundings.

7. **Add the entropy change for the systems and the surroundings to find the total entropy change.**

$$\Delta S_{total} = \Delta S_{sys} + \Delta S_{sur} = (-0.64 + 0.68) \text{ W/K} = 0.04 \text{ W/K}$$

This example demonstrates that when you add the entropy change of both the system and the surroundings together, the total entropy increases. This instance demonstrates that even internally reversible thermodynamic processes add entropy to the universe. Don't worry about the entropy death of the universe — it's a long way off.

Working with T-s Diagrams

The temperature-entropy, or *T-s*, diagram helps you visualize a process with respect to the second law of thermodynamics; much like the *P-v* and *T-v* property diagrams I discuss in Chapter 3 help you grasp the first law of thermodynamics. Figure 8-2 shows the *T-s* diagram of water. The diagram shows the saturated liquid and saturated vapor lines in the shape of a dome. Lines of constant pressure (for $P = 1$ and 10 megapascals) and constant volume (for $v = 0.1$ and 0.5 cubic meter per kilogram) on the diagram give you a sense of how constant pressure or constant volume processes may follow along these paths. You usually don't find pressure-entropy, or *P-s*, diagrams in thermodynamics books, because they're not as useful as the *T-s* diagram.

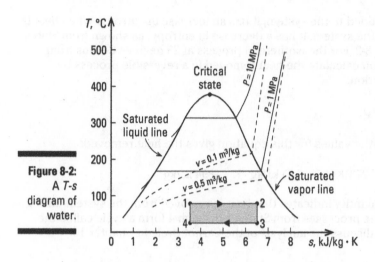

Figure 8-2:
A *T-s*
diagram of
water.

You can find the heat transfer for an internally reversible process by rearranging the equation used to define entropy (see the section "Looking at
entropy on a macroscopic level") and integrating it like this:

$$Q_{\text{Int. Rev.}} = \int_1^2 T\,dS$$

From this equation, you see that heat transfer (Q) is related only to temperature and entropy and equals the area under an internally reversible process
curve drawn on a *T-s* diagram. To integrate this equation, you need the relationship between temperature and entropy for an internally reversible process.
For example, a constant temperature process from States 1 to 2 at 75 degrees
Celsius is shown in Figure 8-2. The entropy at States 1 and 2 is 3.5 and 6.5 kilojoules per kilogram-Kelvin, respectively. Integrating a constant temperature
process gives the following equation for calculating heat transfer:

$$Q = T \cdot \Delta S \text{ or } q = T(s_2 - s_1)$$

As I discuss in Chapter 2, equations written with lowercase variables are on a
unit mass basis (intensive form), whereas equations written with uppercase
variables (extensive form) are used when the mass of the system is known.

You can calculate the heat transfer of an ideal reversible process, using
absolute temperature. The result of this calculation is as follows:

$$q = (75°C + 273)K(6.5 - 3.5) \text{ kJ/kg} \cdot K = 1{,}044 \text{ kJ/kg}$$

When heat is added to the system, it has an increase in entropy. When heat is removed from the system, it has a decrease in entropy, as shown from States 3 to 4 in Figure 8-2. For the isothermal process at 25 degrees Celsius from States 3 to 4, you calculate the heat removed in a reversible process by using this equation:

$$q = T(s_4 - s_3)$$

Substituting in the values for this equation gives the heat removed:

$$q = (25°C + 273)K(3.5 - 6.5) \text{ kJ/kg} \cdot K = -894 \text{ kJ/kg}$$

The negative quantity indicates that heat is removed from the system during the process. The processes from States 1 through 4 form a cycle called the Carnot cycle. I discuss the analysis of the Carnot cycle in Chapter 10.

Using T-ds Relationships

Finding the change in entropy associated with a process requires integration of an appropriate equation for dS. (I know you were hoping to avoid calculus, but here it is.) To perform the integration, you must know the kind of path the process takes. Common paths are isothermal, isobaric, isochoric, isenthalpic, isentropic, and adiabatic. I discuss these process paths in Chapter 2.

The path taken by a real process differs somewhat from these ideal-process, constant-property paths. In the section "Working with *T-s* Diagrams," I show that the integration for an isothermal path is quite easy. When the temperature varies in a process, you need a mathematical relationship between the heat transfer (δQ) and the temperature (T) to perform the integration of dS. Two important relationships, known as the Gibbs equations, are used to define relationships between entropy, heat transfer, and temperature.

The energy equation developed in Chapter 5, written in differential form, gives a mathematical relationship between heat transfer (δQ), internal energy (dU), and work (δW) for an internally reversible process as follows:

$$\delta Q_{rev} = dU + \delta W_{rev}$$

You get the first Gibbs equation by replacing δQ_{rev} with $dU + \delta W_{rev}$ in the equation that defines entropy:

$$dS = \left(\frac{\delta Q}{T}\right)_{Int. Rev.}$$

The first Gibbs equation relates the change in entropy of a system (dS) to internal energy (dU) and boundary work where $\delta W_{rev} = PdV$, as follows:

$$dS = \frac{dU}{T} + \frac{PdV}{T}$$

Internal energy is the energy in a material related to its molecular activity. I discuss internal energy in Chapter 2. *Boundary work* occurs when a boundary in a system moves, such as a piston moving in a cylinder (see Chapter 5).

The second Gibbs equation relates the entropy change of a system to enthalpy and flow work. *Enthalpy* (H) is a property that combines internal energy (U) plus the product of pressure and specific volume (PV) as I discuss in Chapter 2. *Flow work* is associated with the work done by flowing fluids in a process, such as a turbine or a compressor in a gas turbine engine (see Chapter 5). The second Gibbs equation is written as follows:

$$dS = \frac{dH}{T} - \frac{VdP}{T}$$

I discuss how these relationships are used to calculate changes in entropy for several different thermodynamic systems in the following sections.

Calculating Entropy Change

In this section, I discuss how to use and modify the two Gibbs equations from the preceding section to determine the entropy change for processes involving solids, liquids, gases, saturated liquid-vapor mixtures, and ideal gases. You can find the enthalpy change of a process by integrating either of the Gibbs equations between the initial and final states of a process.

To perform the integration of the Gibbs equations, you must know the relationship between internal energy or enthalpy and temperature for a substance. You must also know the relationship between the pressure, volume, and temperature of a substance to complete the integration. For an ideal gas, the ideal-gas law can be used. For other substances, you need to use tabulated data, which can be found in the appendix.

For pure substances

Entropy is a property, and you can find the value of entropy for a substance from thermodynamic tables just like any other property, such as internal energy or enthalpy. You need any two independent intensive properties — for example, temperature and pressure or internal energy and specific volume — to determine the value of entropy for a substance. Figure 8-3 shows you what I mean. You can find the entropy of several different substances from the thermodynamic property tables in the appendix. I discuss intensive properties in Chapter 2.

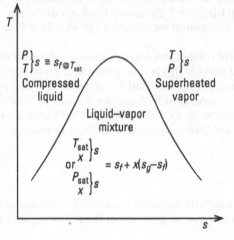

Figure 8-3: The entropy of a substance is found using any two independent intensive properties.

$$\left.\begin{array}{c} P \\ T \end{array}\right\} s \cong s_{f@T_{sat}} \qquad \left.\begin{array}{c} T \\ P \end{array}\right\} s$$

Compressed liquid

Superheated vapor

Liquid–vapor mixture

$$\left.\begin{array}{c} T_{sat} \\ x \end{array}\right\} s$$ or $$\left.\begin{array}{c} P_{sat} \\ x \end{array}\right\} s = s_f + x(s_g - s_f)$$

Although the third law of thermodynamics states the entropy of a substance is zero at absolute zero temperature, thermodynamic tables usually define entropy as being equal to zero at a more convenient reference temperature. This definition means that at temperatures below the reference temperature, entropy can have negative values. The reference temperature for water is 0.01 degree Celsius at the saturated liquid state. For refrigerant R-134a as a saturated liquid, the reference temperature is –40 degrees Celsius. The reference temperature for ideal gases is absolute zero temperature. Negative values for entropy aren't significant, because only the change in entropy for a thermodynamic process is important. Check out the thermodynamic property tables in the appendix to see these reference values for yourself.

Figure 8-3 shows that the entropy of a compressed liquid, $s(T,P)$, at temperature T and pressure P, can be approximated by the entropy of the saturated liquid, $s_f(T)$, at the given temperature T: $s(T,P) \approx s_f(T)$.

For example, you can look up the entropy of saturated liquid water at 10 megapascals pressure and 260 degrees Celsius in Table A-2 of the appendix to find that it's 2.870 kilojoules per kilogram-Kelvin. This calculation requires a bit of interpolation of the table. I show you how to interpolate tables in Chapter 3. Or, you can look up the entropy of saturated water at 260 degrees Celsius in Table A-3 of the appendix and find that it has a value of 2.884 kilojoules per kilogram-Kelvin. You can see that the entropy for the saturated liquid is not that different from the entropy of the compressed liquid at the same temperature, so it's a reasonable approximation.

In the liquid-vapor mixture region shown in Figure 8-3, you can find the entropy using the quality (x) of the mixture and the entropy at the saturated liquid state (s_f) and the saturated vapor state (s_g) with the following equation: $s = s_f + x(s_g - s_f)$.

You can use any two independent intensive properties to find entropy in the liquid-vapor mixture region, such as quality (x) and saturation temperature (T_{sat}) or quality (x) and saturation pressure (P_{sat}). You find values for s_f and s_g using either the saturation temperature or the saturation pressure property tables. Thermodynamic property tables for saturated water and R-134a appear in the appendix.

In the superheated vapor region of Figure 8-3, you can find the entropy using any two independent intensive properties, such as temperature (T) and pressure (P). Thermodynamic property tables for superheated water and R-134a are given in the appendix. Observe that entropy increases with temperature increase and decreases with pressure increase in those tables.

The following equation shows that the total entropy change (ΔS) for a process is calculated using the mass of the material (m) and the entropy at the initial state (s_1) and the final state (s_2). The entropy change for a process is ΔS:

$$\Delta S = m\Delta s = m(s_2 - s_1)$$

This example shows you how to calculate the entropy change for a cooling process. Refrigerant R-134a at State 1 is cooled from 45 degrees Celsius at a constant pressure of 1 megapascal until the quality is 0.10 at State 2, as shown in Figure 8-4.

1. **Find the entropy at State 1 at 45°C and 1 MPa, which is a superheated vapor.**

   Use interpolation to find the value in Table A-8 in the appendix: s_1 is 1.731 kilojoules per kilogram-Kelvin.

2. **Find the entropy at the saturated liquid and saturated vapor conditions for 1 megapascal pressure.**

   Use Table A-7 in the appendix: s_f is 1.188 and s_g is 1.711 kilojoules per kilogram-Kelvin.

3. **Calculate the entropy at State 2 using the following equation:**

   $$s_2 = 1.188 \text{ kJ/kg} \cdot \text{K} + 0.10(1.711 - 1.188) \text{ kJ/kg} \cdot \text{K} = 1.240 \text{ kJ/kg} \cdot \text{K}$$

4. **Calculate the change in entropy for the cooling process using this equation:**

   $$\Delta s = s_2 - s_1 = (1.240 - 1.731) \text{ kJ/kg} \cdot \text{K} = -0.491 \text{ kJ/kg} \cdot \text{K}$$

You see that for cooling processes, the entropy decreases. But the heat absorbed by the surroundings increases the entropy of the surroundings. As a whole, the entropy of the universe increases.

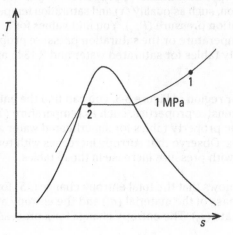

Figure 8-4:
A *T-s*
diagram of
a refrigerant
cooled in a
constant-
pressure
process
from States
1 to 2.

1 MPa

For liquids and solids

The change in entropy for a solid or a liquid in a thermodynamic process is strongly dependent upon temperature. Entropy doesn't change very much with pressure. Because solids and liquids are considered incompressible, increasing the pressure doesn't make the molecules within the material move much closer together; therefore, the molecules don't become significantly more ordered. Remember, when molecules become more ordered in a material, the entropy decreases.

You can use the first Gibbs equation to calculate the entropy change for a liquid or solid in a thermodynamic process. If you assume liquids and solids are incompressible, $dv = 0$. Then the first Gibbs equation simplifies to this equation:

$$ds = \frac{du}{T}$$

I show you in Chapter 2 that the change in internal energy (du) of a solid or liquid material is related to its specific heat (c) and the change in temperature (dT), as long as a phase change doesn't occur: $du = cdT$.

Solids and liquids have only one specific heat property, unlike gases, which have a constant-pressure specific heat and a constant-volume specific heat.

You replace the internal energy term in the first Gibbs equation with specific heat to find the change in entropy for a solid or liquid material, as shown in the following equation:

$$ds = \frac{cdT}{T}$$

By integrating both sides of this equation, you find the change in entropy for a solid or liquid material in a thermodynamic process from States 1 to 2. Because specific heat changes with temperature, you get more accurate results for the entropy change if you use the specific heat (c_{avg}) at the average temperature of the process. You can calculate the change in entropy of a solid or liquid material with the following equation:

$$s_2 - s_1 \cong c_{avg} \ln \frac{T_2}{T_1}$$

You must use absolute temperatures in this equation because the temperatures form a ratio.

The following example shows you how to calculate the entropy change for a liquid in a thermodynamic process. Water at 80 degrees Celsius is heated to 160 degrees Celsius. You can find the change in entropy for this process by following these steps:

1. **Look up the specific heat of water.**

   In Table A-10 of the appendix, you find by interpolation that the value of c_{avg} is about:

   $$c_{avg} = 4.21 \text{ kJ/kg} \cdot \text{K}$$

2. **Convert the temperatures for T_1 and T_2 to Kelvin.**

   This temperature conversion formula appears in Chapter 2.

   $$T_1 = 80°C + 273 = 353 \text{ K}$$

   $$T_2 = 160°C + 273 = 433 \text{ K}$$

3. **Calculate the change in entropy.**

   $$s_2 - s_1 = (4.21 \text{ kJ/kg} \cdot \text{K}) \ln \left(\frac{433 \text{ K}}{353 \text{ K}} \right) = 0.860 \text{ kJ/kg} \cdot \text{K}$$

For ideal gases

In a thermodynamic process with an ideal gas, the change in entropy depends on the changes in both temperature and pressure. Both of the Gibbs equations can be used to calculate the entropy change for an ideal gas in a process. For the first Gibbs equation, you use the constant-volume specific heat to substitute $du = c_v dT$ and the ideal-gas law to substitute $P = RT/v$ (see Chapter 3). Then the first Gibbs equation looks like this:

$$ds = c_v \frac{dT}{T} + R \frac{dv}{v}$$

In the second Gibbs equation, you use the constant-pressure specific heat to replace $dh = c_p dT$ and the ideal-gas law to substitute $v = RT/P$. Then the second Gibbs equation looks like this:

$$ds = c_P \frac{dT}{T} - R \frac{dP}{P}$$

You integrate both of these forms of the Gibbs equations to find an equation for the change in entropy of an ideal gas in a thermodynamic process, as shown here:

$$s_2 - s_1 = \int_1^2 c_v(T) \frac{dT}{T} + R \ln \frac{v_2}{v_1}$$

$$s_2 - s_1 = \int_1^2 c_P(T) \frac{dT}{T} - R \ln \frac{P_2}{P_1}$$

The specific heat for most ideal gases changes with temperature (monatomic gases such as helium, argon, and neon are exceptions), so you can complete the integration in these equations if you know the temperature dependence of specific heat for a gas. In many thermodynamic processes, such as simple compression or expansion of a gas, the temperature change isn't too great, and you can assume the specific heat is constant during the process. For constant specific heat, the integration results in the following equations for calculating the entropy change of an ideal gas:

$$s_2 - s_1 = c_{v,avg} \ln \frac{T_2}{T_1} + R \ln \frac{v_2}{v_1}$$

$$s_2 - s_1 = c_{p,\,avg} \ln \frac{T_2}{T_1} - R \ln \frac{P_2}{P_1}$$

Which equation you use depends on the kind of process you're analyzing. If you know the volume ratio of a process, that is, (v_2/v_1), use the first equation. The volume ratio is used in the analysis of Otto-cycle and diesel-cycle heat engines, as I discuss in Chapter 11. If you know the pressure ratio of a process, that is, (P_2/P_1), use the second equation. You use the pressure ratio in the analysis of compressors and turbines in gas turbine engines, as I discuss in Chapter 10. The temperature ratio (T_2/T_1) in these equations use absolute temperature.

You get more accurate results if you use the specific heat of the gas at the average temperature of the process.

If the temperature change of an ideal gas during a process is large, as in a combustion process, you may get inaccurate results by assuming constant specific heat; to overcome this, you use variable specific heat. The integration of variable specific heat for ideal gases is completed by defining a function $s°$, in terms of temperature, as follows:

$$s° = \int_0^T c_P(T)\frac{dT}{T}$$

The integration begins at absolute zero temperature and ends at the temperature of interest. The results of this integration function are tabulated for air in Table A-1 in the appendix. This function is used exclusively with the entropy change determined by the second Gibbs equation. A separate function would need to be defined and tabulated to use the first Gibbs equation. Because a separate function isn't necessary to determine the entropy change of an ideal gas, you only find $s°$ in thermodynamic property tables. You use the following equation to determine the entropy change for an ideal gas, assuming variable specific heat:

$$s_2 - s_1 = s_2° - s_1° - R\ln\frac{P_2}{P_1}$$

I show you how to calculate the change in entropy for an ideal gas by using constant specific heat and variable specific heat in the following sections.

The constant specific heat assumption is easy to use and provides fairly accurate estimates of the change in entropy when the temperature change of the process is relatively small. Whether the temperature change in a process is large or small, the variable specific heat assumption is more accurate than the constant specific heat method, but may require interpolation of a property table.

Using constant specific heat

Here's an example that shows you how to calculate the entropy change for a process involving an ideal gas. Air is compressed and heated, as the *T-s* diagram in Figure 8-5 shows, from 27 degrees Celsius and 100 kilopascals to 1,127 degrees Celsius and 1 megapascal. Follow these steps to calculate the entropy change for this process:

1. **Convert the temperatures from Celsius to Kelvin to get absolute temperatures.**

$$T_1 = 27°C + 273 = 300 \text{ K}$$

$$T_2 = 1,127°C + 273 = 1,400 \text{ K}$$

2. **Look up the constant-pressure specific heat for air at T_1 = 300 K and at T_2 = 1,400 K.**

In Table A-1 of the appendix, you find the following:

$$c_{p1} = 1.005 \text{ kJ/kg} \cdot \text{K and } c_{p2} = 1.200 \text{ kJ/kg} \cdot \text{K}$$

3. **Find the average specific heat of the air during the expansion process, using this equation:**

$$c_{p,avg} = \frac{1}{2}(1.005 + 1.200) \text{ kJ/kg} \cdot \text{K} = 1.103 \text{ kJ/kg} \cdot \text{K}$$

4. **Calculate the entropy change using the following equation.**

The gas constant (R) for air equals 0.287 kilojoule per kilogram-Kelvin.

$$s_2 - s_1 = (1.103 \text{ kJ/kg} \cdot \text{K}) \ln \frac{1,400 \text{ K}}{300 \text{ K}} - (0.287 \text{ kJ/kg} \cdot \text{K}) \ln \frac{1 \text{ MPa}}{0.1 \text{ MPa}} = 1.04 \text{ kJ/kg} \cdot \text{K}$$

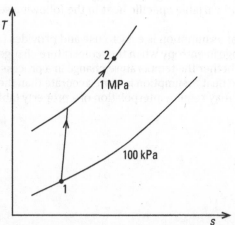

Figure 8-5:
A *T-s* diagram of air being compressed and heated from States 1 to 2.

The constant-volume specific heat isn't listed in the appendix tables. When you need to use the constant-volume specific heat of an ideal gas to determine the change in entropy, you can calculate it using the following equation, which I introduce in Chapter 2: $c_v = c_p - R$.

Using variable specific heat

You can find the change in entropy, assuming variable specific heat from the same example used in the constant specific heat section, by following these steps:

1. **Look up the reference entropy ($s°$) for air at $T_1 = 300$ K and at $T_2 = 1,400$ K.**

   Use Table A-1 in the appendix.

   You need to interpolate for State 1. I show you how to interpolate a table in Chapter 3.

   $$s°_1 = 6.870 \text{ kJ/kg} \cdot \text{K and } s°_2 = 8.529 \text{ kJ/kg} \cdot \text{K}$$

2. **Calculate the entropy change using the following equation.**

   The gas constant (R) for air equals 0.287 kilojoule per kilogram-Kelvin.

   $$s_2 - s_1 = \left(8.529 - 6.870\right) \text{ kJ/kg} \cdot \text{K} - \left(0.287 \text{ kJ/kg} \cdot \text{K}\right) \ln \frac{1 \text{ MPa}}{0.1 \text{ MPa}} = 1.0 \text{ kJ/kg} \cdot \text{K}$$

You can see that the difference in the change in entropy between the constant specific heat method and the variable specific heat method for this example isn't very large (only 4 percent). The result of the variable specific heat method is the more accurate one.

Analyzing Isentropic Processes

You may think that every ideal process creates a change in entropy. Believe it or not, entropy doesn't change in some ideal processes. An ideal compression or expansion process involving an ideal gas is reversible and adiabatic, has no change in entropy, and is called an *isentropic process*.

You can use either the constant specific heat assumption or the variable specific heat assumption when analyzing isentropic processes, as shown in the previous section "Calculating Entropy Change." The constant specific heat method gives satisfactory results when the temperature change isn't large, as in simple compression/expansion processes. The variable specific heat method gives the most accurate results, especially for large temperature changes in a process. I discuss analysis of isentropic processes using both methods in the following sections.

Using constant specific heat

For an isentropic process, the change in entropy equations shown in the previous section are set equal to zero. This gives three mathematical equations to relate temperature, pressure, and specific volume to each other. One equation relates temperature to pressure, the second relates temperature to specific volume, and the third relates pressure to specific volume. These equations are as follows:

$$\left(\frac{T_2}{T_1}\right)=\left(\frac{v_1}{v_2}\right)^{k-1}$$

$$\left(\frac{T_2}{T_1}\right)=\left(\frac{P_2}{P_1}\right)^{(k-1)/k}$$

$$\left(\frac{P_2}{P_1}\right)=\left(\frac{v_1}{v_2}\right)^{k}$$

Because temperature appears as a ratio in these equations, you must use absolute temperatures. The variable k in these equations is called the ratio of specific heats. You can calculate k for an ideal gas using the following equation:

$$k=\frac{c_P}{c_v}$$

For air, k equals 1.4 for processes that are within a few hundred degrees Celsius of room temperature. Because specific heat varies with temperature, you should calculate k using the specific heats at the average process temperature. No units are associated with k because it's a ratio.

Here's an example that shows you how to analyze an isentropic process: the expansion of air in an ideal turbine. The air enters the turbine at 1,127 degrees Celsius and 1 megapascal pressure. It leaves the turbine at 100 kilopascals. Figure 8-6 shows this process on a T-s diagram. You can find the final temperature (T_2) of the air using these steps:

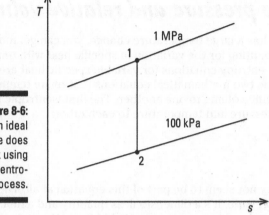

Figure 8-6:
An ideal
turbine does
work using
an isentro-
pic process.

1. **Convert the inlet temperature to absolute temperature.**

$$T_1 = 1{,}127°C + 273 = 1{,}400 \text{ K}$$

2. **Rearrange the isentropic equation relating temperature and pressure to solve for the outlet temperature T_2.**

$$T_2 = T_1 \left(\frac{P_2}{P_1} \right)^{(k-1)/k}$$

3. **Substitute known values for T_1, the pressure ratio, and k for air:**

$$T_2 = 1{,}400 \text{ K} \left(\frac{0.1 \text{ MPa}}{1.0 \text{ MPa}} \right)^{(1.4-1)/1.4} = 725 \text{ K}$$

Alternately, T_2 equals 452 degrees Celsius.

Using relative pressure and relative volume

If an isentropic process has a large temperature change, you can get more accurate results by accounting for the variation in specific heat with temperature. The change in entropy equations for variable specific heat are set equal to zero and provide two mathematical equations that relate temperature, pressure, and specific volume to one another. The first isentropic process equation relates pressure and temperature to each other:

$$\frac{P_2}{P_1} = \frac{P_{r2}}{P_{r1}}$$

At first, temperature may not seem to be part of this equation at all. The term P_r is called the *relative pressure*. It's a dimensionless quantity and a function of temperature only. Thus, temperature is secretly hidden in this equation. You calculate the relative pressure by using this equation:

$$P_r = \exp\left(\frac{s° - s°_{ref}}{R}\right)$$

The relative pressure for air is tabulated as a function of temperature in Table A-1 in the appendix. The $s°_{ref}$ in this equation is a reference state entropy that keeps the exponent from blowing up. Normally, you use relative pressure in the analysis of a compressor or turbine in a gas turbine engine, because gas turbine engines typically use a pressure ratio to define engine operating conditions. However, you can also use relative pressure for any isentropic process for an ideal gas where you either know or can find the initial and final pressures.

The second isentropic process equation relates specific volume and temperature to each other. You use this equation when analyzing an automobile or truck engine with the Otto or diesel cycle. These cycles typically use a compression ratio, which is a volumetric ratio to define engine operating conditions. You can use the equation for any process where you know or can find the initial and final specific volumes. The second isentropic process equation is defined as follows:

$$\frac{v_2}{v_1} = \frac{v_{r2}}{v_{r1}}$$

The term v_r is called the *relative volume*. It's also a dimensionless quantity that's only a function of temperature. The relative volume is calculated using the following equation where the constant C is simply a scaling term:

$$v_r = \frac{C \cdot T}{P_r}$$

You don't need to use this equation to calculate the relative volume; the calculations have already been done for you. The relative volume for air is tabulated as a function of temperature in Table A-1 of the appendix.

You can use the relative pressure relationship to find the temperature at State 2 in the example in the section "Using constant specific heat." Follow these steps:

1. **Look up the relative pressure (P_{r1}) at State 1 at 1,400 K.**

   Using Table A-1 in the appendix, you find that $P_{r1} = 361.8$.

2. **Rearrange the isentropic process equation for relative pressure to solve for P_{r2}.**

$$P_{r2} = P_{r1}\left(\frac{P_2}{P_1}\right)$$

3. **Substitute known values for P_{r1} and the pressure ratio into this equation.**

$$P_{r2} = 361.8\left(\frac{0.1\text{ MPa}}{1.0\text{ MPa}}\right) = 36.2$$

4. **Look up the value of P_{r2} to find T_2.**

   Use Table A-1 in the appendix. You need to interpolate in the table to find $T_2 = 785$ K or 512 degrees Celsius.

Accounting for the variation in specific heat creates quite a bit of difference in the final temperature. The constant specific heat assumption estimated $T_2 = 452$ degrees Celsius. This example illustrates how important it is to use the variable specific heat assumption in your analysis of an isentropic process for an ideal gas when there's a large temperature change in the process.

Balancing Entropy in a System

In a thermodynamic system, energy can enter, leave, or be stored within the system by heat transfer, work, and mass flow. You use an energy balance equation to keep track of the energy flow in a system. Entropy can only enter or leave a system by mass flow and heat transfer. Entropy can be generated within the system by irreversibilities. You can write an entropy balance on a system to keep track of entropy flow, as follows:

$$\Delta S_{system} = S_2 - S_1 = S_{in} - S_{out} + S_{gen}$$

The entropy balance means the change in entropy of a system (ΔS_{system}) during a process equals the difference between the final (S_2) and initial (S_1) entropy of the system.

Entropy generation (S_{gen}) includes only the entropy generated within the system; it doesn't include entropy generated in the surroundings. If the process within the system is internally reversible, the entropy generation is zero.

Heat is a disorganized form of energy, so entropy flows with it. Entropy enters the system (S_{in}) as heat is transferred to the system. Entropy is removed from the system (S_{out}) as heat is transferred from the system. You can calculate the entropy transfer by heat (S_{heat}) in a system by dividing the heat transfer through a system boundary (Q_k) by the absolute temperature (T) of the boundary for each heat transfer process, as shown in the following equation:

$$S_{heat} \cong \sum \frac{Q_k}{T_k} \text{ where } k \text{ is the number of boundaries}$$

A system may have more than one heat transfer process; in fact, many systems have a heat addition and a heat rejection process.

Work is an organized form of energy, so entropy can't be transferred into or out of a system by work. Entropy can still be generated within a system as work is dissipated by irreversibilities, such as friction, which changes work into heat. The entropy transfer by work to a system is zero, or $S_{work} = 0$.

Entropy is an intensive property, which means that mass has entropy associated with it. If mass flows into or out of an open system, entropy flows with it. You can calculate the total entropy (S_{mass}) crossing a system boundary due to mass flow with this equation:

$$S_{mass} = m \cdot s$$

Here, s is the specific entropy of the material in the process at the system boundary, and m is the amount of mass that flowed into the system. Because mass often flows through a system, the entropy transfer by mass can be written on a rate basis:

$$\dot{S}_{mass} = \dot{m} \cdot s$$

A closed system doesn't have any mass flow that crosses a system boundary, so the entropy transfer by mass flow is zero.

Chapter 9

Analyzing Systems Using the Second Law of Thermodynamics

....

In This Chapter

▶ Understanding the concept of energy availability

▶ Estimating changes in energy availability

▶ Discovering reversible work and irreversibility

▶ Determining second-law efficiency

....

The world's energy resources are limited, and whenever an energy shortage is perceived, companies that produce energy are encouraged to squeeze out as much energy as possible. Analyzing a system using the second law of thermodynamics can tell you where you can make the most improvement to a system to more effectively use its energy. I explain the second law in detail in Chapter 7.

The bottom line of the second law analysis is that you must match your resources to your needs as best you can. Suppose you want to boil a pot of water and you have two heat sources available. The temperature of one source is 1,000 degrees Celsius, and the other is 101 degrees Celsius. Which heat source will generate the most entropy to boil the water? If you guessed the high-temperature heat source, you're right. The 1,000-degree heat source generates the most entropy because the quality of its energy is unnecessarily great for the intended purpose. Much of the energy quality would be wasted.

In this chapter, I discuss the concept of energy availability, which is a measure of the quality of energy used for a thermodynamic process or system. A system with high energy availability has the capacity for doing more work than a system with low energy availability. A well-designed thermodynamic system optimizes the availability as a resource for meeting its needs.

I also address the concepts of reversible work and irreversibility. Reversible work is an ideal work process that provides the maximum possible work output or requires the minimum possible work input. Irreversibility comes from things like friction or heat loss to the environment and wastes energy because it decreases the quality of energy as it's used in a process.

Lastly, I discuss another measure of efficiency based on the second law of thermodynamics, which compares the actual performance of a system to that of the theoretically best possible system. This measurement lets you know how well your system compares to the ideal system.

Measuring Work Potential with Energy Availability

Energy availability (or "availability" for short) is a thermodynamic property that depends on the state, such as temperature and pressure, of both the environment and the system. Availability is a measure of the quality of energy. Energy with a high quality relative to the environment can be used for producing work. Low-quality energy isn't as useful for doing work as high-quality energy. *Availability* represents the maximum useful work that can be obtained from a system by a reversible heat engine that operates between an energy source and the environment.

The environment is called the *dead state* because no work can be extracted from the environment. The availability of energy at the temperature and pressure of the environment is zero. In the dead state, a system doesn't have any kinetic or potential energy, either.

Usually, the atmosphere is used as the dead state environment because most thermodynamic processes operate within it. Even though the atmosphere contains a tremendous amount of energy, that energy can't be used for doing work, except in windy areas where the kinetic energy of the atmosphere can turn a wind-powered generator. In the dead state, the atmosphere has no wind.

The thermodynamic properties of a system at the dead state are identified with the subscript zero, such as P_0, T_0, u_0, and so forth. The dead state is defined as 25 degrees Celsius and 100 kilopascals unless otherwise specified.

Some textbooks use the term *exergy* instead of availability as a measure of energy quality. Exergy was coined in the 1950s to give this thermodynamic property a short name that sounds similar to energy and entropy. Although "exergy" is a widely accepted term, it obscures the meaning, so I prefer to use the term "availability" — which was first used to describe energy quality in the 1940s — in this book. Newer isn't always better.

Determining the Change in Availability

The quality of energy decreases in every thermodynamic process as it is expended to do work. In this section, I discuss the analysis of the change in availability (ΔA_{system}) for closed systems and open systems. In a closed system, the mass of the system remains fixed, whereas in an open system, mass is allowed to flow through. The system refers to the fluid (either a liquid or a gas) that's used in a thermodynamic process to move heat or produce work.

The decrease in availability of a system between two states represents the *maximum* amount of useful work output that can be done by the system. If the availability of a system increases between the initial and final states, then it represents the *minimum* amount of work input required by the system. The availability between two states is independent of the type of system used, the type of process in the system, and the type of heat and work interactions with the surroundings.

Calculating availability in closed systems

No mass enters or leaves a closed system during a thermodynamic process. However, energy may enter or leave a closed system by heat transfer or work. A piston in a cylinder is an example of a closed system. Figure 9-1 shows that you can add heat to a piston-cylinder system and extract work from the piston as it moves. As the piston works, the availability of the energy decreases in the process. Not all the heat added to the cylinder is converted to useful work. You can determine how much heat can be converted to work using the second law of thermodynamics.

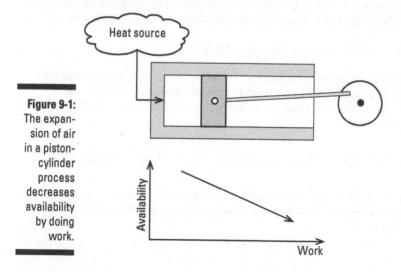

Figure 9-1:
The expansion of air in a piston-cylinder process decreases availability by doing work.

Measuring quality of energy by determining availability

The quality of the energy in a closed system is measured by determining the availability. To get the maximum amount of energy out of a closed system, you have to bring the system reversibly to the dead state, which is at the temperature (T_0) and pressure (P_0) of the atmosphere. You can write an energy balance that includes internal, kinetic, and potential energy on a closed system going to the dead state with the following equation (see Chapter 5):

$$q - w = (u - u_0) + \frac{1}{2}(\mathbf{V}^2 - \mathbf{V}_0^2) + g(z - z_0)$$

In the dead state, both the velocity ($\mathbf{V}_0$) and the elevation (z_0) are zero. The energy equation is used to determine the availability of a system with the following adjustments:

✔ **Boundary work against the atmosphere isn't useful work because it pushes the atmosphere out of the way as the piston or system boundary moves.** The work in moving the boundary is described by the change in volume of a system, such that $\delta w = P dv$. Assuming the atmospheric pressure remains constant during a process as it proceeds to the dead state, the work against the atmosphere is given by integrating this equation so that $w = P_0(v - v_0)$.

✔ **Heat rejected by the system to the dead state isn't used for doing any work.** The heat is rejected reversibly at the dead-state temperature. The heat transfer is described by the change in entropy of the system such that $\delta q = T ds$ (see Chapter 8). This equation is integrated assuming the heat transfer occurs at the atmospheric temperature, which remains a constant temperature during the process, so that $q = T_0(s - s_0)$.

Substituting the adjustments for boundary work (w) against the atmosphere and heat rejection (q) into the preceding energy equation gives you the specific availability (a) for a closed system, as shown in the following equation:

$$a = (u - u_0) + P_0(v - v_0) - T_0(s - s_0) + 1/2\,\mathbf{V}^2 + gz$$

This equation can be used to determine the availability at both the initial State 1 and the final State 2 of a thermodynamic process. You can find the change in availability (Δa) for a closed system process using the following equation:

$$\Delta a = (u_2 - u_1) + P_0(v_2 - v_1) - T_0(s_2 - s_1) + \frac{\mathbf{V}_2^2 - \mathbf{V}_1^2}{2} + g(z_2 - z_1)$$

Note that the dead-state values aren't in this equation (aside from P_0 and T_0) because they cancel each other out when you find the difference between states in a process.

Looking at an example of calculating availability in a closed system

Here's an example that shows you how to calculate the availability of a closed-system process. A piston-cylinder system is initially filled with air at the dead state when the piston is at the bottom of the cylinder at volume V_0. The piston compresses the air to volume V_1, and a combustion process heats the air. The heated air expands in the cylinder to volume V_2 (where $V_2 = V_0$), producing work in an isentropic process. You can determine the availability at the beginning and end of an isentropic work output process in a piston-cylinder system, using the following information.

The properties at the beginning of the isentropic work process, State 1, are

> $T_1 = 2{,}116$ K
>
> $u_1 = 1{,}791$ kJ/kg
>
> $V_1 = 0.0714$ liter

The properties at the end of the isentropic work process, State 2 (based on using variable specific heat), are

> $T_2 = 1{,}110$ K
>
> $u_2 = 854$ kJ/kg
>
> $V_2 = 0.571$ liter

The dead state is at 20 degrees Celsius and 100 kilopascals.

Because the kinetic and potential energy don't change, the availability equation for State 1 simplifies to this equation:

$$a_1 = (u_1 - u_0) + P_0(v_1 - v_0) - T_0(s_1 - s_0)$$

You can determine the availability (a_1) at State 1 by following these steps:

1. **Find the internal energy (u_0) and entropy (s_0) of air at the dead state, using Table A-1 in the appendix.**

   T_0 is 20 degrees Celsius or 293 Kelvin. P_0 is 100 kilopascals. You need to do interpolation, which I explain in Chapter 3.

   > $u_0 = 209$ kJ/kg and $s_0 = 6.846$ kJ/kg · K

2. **Find the specific volume (v_0) of air at the dead state using the ideal-gas law.**

   $$v_o = \frac{RT_0}{P_0} = \frac{(0.287 \text{ kJ/kg} \cdot \text{K})(293 \text{ K})}{100 \text{ kPa}} = 0.841 \text{ m}^3/\text{kg}$$

   The cylinder was initially filled with air at the dead state prior to being compressed and heated in a combustion process.

3. **Calculate the specific volume (v_1) of the air at State 1.**

   The specific volume (v) is related to the total volume (V) by the mass (m) in the system: $V = mv$. You can find the specific volume (v_1) by using the ratio of the total volumes (V_1/V_2) between States 1 and 2 and the specific volume (v_0) at the dead state, because the mass in a closed system is constant, as shown here:

$$v_1 = v_0\left(\frac{V_1}{V_2}\right) = \left(0.841\,\text{m}^3/\text{kg}\right)\left(\frac{0.0714\,\text{liter}}{0.571\,\text{liter}}\right) = 0.105\,\text{m}^3/\text{kg}$$

4. **Find the reference entropy s° at State 1 using the temperature T_1 by interpolation of Table A-1 in the appendix.**

$$s_1^\circ = 9.031\,\text{kJ/kg}\cdot\text{K}$$

5. **Use the following equation to find the change in entropy ($s_1 - s_0$).**

$$s_1 - s_0 = s_1^\circ - s_0^\circ - R\ln\left(\frac{P_1}{P_0}\right)$$

   This equation is discussed in Chapter 8.

6. **Find the pressure at State 1 using the ideal-gas law.**

$$P_1 = \frac{RT_1}{v_1} = \frac{\left(0.287\,\text{kJ/kg}\cdot\text{K}\right)\left(2,116\,\text{K}\right)}{0.105\,\text{m}^3/\text{kg}} = 5,784\,\text{kPa}$$

7. **Calculate the change in entropy ($s_1 - s_0$) using the reference entropy values and the pressure values for State 1 and the dead state.**

$$s_1 - s_0 = \left(9.031 - 6.846\right)\text{kJ/kg}\cdot\text{K} - \left(0.287\,\text{kJ/kg}\cdot\text{K}\right)\ln\left(\frac{5,784\,\text{kPa}}{100\,\text{kPa}}\right) = 1.020\,\text{kJ/kg}\cdot\text{K}$$

8. **Calculate the availability a_1 at the start of the process at State 1, using values for the internal energy, specific volume, and entropy found in Steps 1–7.**

$$a_1 = \left(1,791 - 209\right)\text{kJ/kg} + 100\,\text{kPa}\left(0.105 - 0.841\right)\text{m}^3/\text{kg} - 293\,\text{K}\left(1.020\,\text{kJ/kg}\cdot\text{K}\right)$$
$$a_1 = 1,210\,\text{kJ/kg}$$

REMEMBER

This value describes the quality of the energy available for performing work.

These units are equivalent to each other: $1\,\text{kPa}\cdot\text{m}^3 = 1\,\text{kJ}$ (see Chapter 2) and are used in the pressure-volume term in the preceding equation.

9. **Calculate the availability a_2 at the end of the process at State 2.**

   At the end of the isentropic work process, the piston is back to the bottom of the cylinder, so the specific volume at v_2 is the same as the dead state v_0. Because $v_2 = v_0$, the pressure-volume term in the availability equation is zero. The expansion process is isentropic, which means that $s_1 = s_2$ (see Chapter 8), so $s_1 - s_0 = s_2 - s_0$.

$$a_2 = \left(854 - 209\right)\text{kJ/kg} - 293\,\text{K}\left(1.020\,\text{kJ/kg}\cdot\text{K}\right) = 346\,\text{kJ/kg}$$

10. **Find the change in availability during the expansion process in the cylinder from State 1 to State 2.**

$$\Delta a = a_2 - a_1 = (346 - 1{,}210) \text{ kJ/kg} = -864 \text{ kJ/kg}$$

The negative value means the availability decreased during the expansion process.

This example is related to the Otto cycle, which is discussed further in Chapter 11.

Calculating availability in open systems with steady flow

An open system allows mass and energy to flow through it during a thermodynamic process. Energy may enter or leave an open system by heat transfer or work. A heat exchanger is an example of an open system. As the heat moves from the hot fluid to the cold fluid, the availability of the energy decreases in the process.

Measuring quality of energy by determining availability

The quality of the energy in an open system is measured by determining the availability. To get the maximum amount of energy out of an open system, you have to bring the system reversibly to the dead state, which is at the temperature (T_0) and pressure (P_0) of the atmosphere. You can write an energy balance, including enthalpy, kinetic, and potential energy on an open system going to the dead state with the following equation (see Chapter 6):

$$q - w = (h - h_0) + \frac{1}{2}\left(\mathbf{V}^2 - \mathbf{V}_0^2\right) + g(z - z_0)$$

The energy equation is used to determine the availability of an open system with the following adjustments:

✔ **No boundary work against the atmosphere exists in a steady-flow open system.** The flow work ($w = Pv$; see Chapter 6) of the fluid is accounted for by the enthalpy in the energy equation. Enthalpy (h) is a property that combines internal energy (u) with flow work (Pv).

✔ **Heat rejected by the system to the dead state isn't used for doing any work.** The heat is rejected reversibly at the dead state temperature so that $q = T_0(s - s_0)$. (This adjustment applies to closed systems as well.)

Substituting the adjustment for heat rejection into the energy equation gives you the flow availability (a_f) for an open system in the following equation. (The subscript "f" is used to indicate that the availability is associated with fluid flow in an open system.)

$$a_f = (h - h_0) - T_0(s - s_0) + 1/2\ \mathbf{V}^2 + gz$$

This equation can be used to determine the availability at both the initial State 1 and the final State 2 of a thermodynamic process. You can find the change in flow availability (Δa_f) for an open system process using the following equation:

$$\Delta a_f = \left(h_2 - h_1 \right) - T_0 \left(s_2 - s_1 \right) + \frac{V_2^2 - V_1^2}{2} + g \left(z_2 - z_1 \right)$$

Note that the dead-state values aren't in this equation (aside from T_0) because they cancel each other out when you find the difference between states in a process.

Looking at an example of calculating availability in a steady flow system

Here's an example that shows you how to calculate the change in availability of an open-system process with steady flow. Figure 9-2 shows a feedwater heater, which is an energy recovery device in a power plant. I discuss the use and analysis of a feedwater heater in Chapter 11. Steam enters the feedwater heater at 1 megapascal pressure and 300 degrees Celsius and mixes with water entering at 80 degrees Celsius and 1 megapascal pressure. The steam flow rate is 5 kilograms per second, and the water flow rate is 40 kilograms per second. The feedwater heater is insulated, so no heat is lost to the environment at 25 degrees Celsius. You can neglect changes in kinetic and potential energy of the fluid in the feedwater heater.

Figure 9-2:
Mixing two fluid streams in a feed-water heater decreases availability by heat transfer.

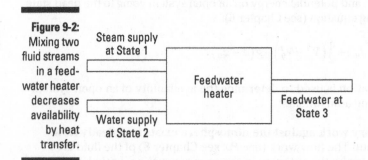

You can calculate the availability entering and leaving the feedwater heater to determine the change in availability for the process by following these steps:

1. **Write an energy balance on the feedwater heater.**

   The energy balance means that the energy coming into the feedwater heater equals the energy leaving the feedwater heater.

The rate of energy entering the feedwater heater is $\dot{E}_{\text{in}} = \dot{m}_1 h_1 + \dot{m}_2 h_2$, where $\dot{m}_1$ is the mass flow rate of the steam and $\dot{m}_2$ is the mass flow rate of the water. The variable h_1 is the enthalpy of the steam, and h_2 is the enthalpy of the water.

The rate of energy leaving the feedwater heater is $\dot{E}_{\text{out}} = (\dot{m}_1 + \dot{m}_2) h_3$, where $(\dot{m}_1 + \dot{m}_2)$ is the total water mass flow rate, and h_3 is the enthalpy of the water leaving.

2. **Write an equation for the enthalpy at State 3, h_3, by setting $\dot{E}_{\text{in}} = \dot{E}_{\text{out}}$ and solving the resulting equation for h_3:**

$$h_3 = \frac{m_1 h_2 + m_2 h_2}{m_1 + m_2}$$

3. **Find the enthalpy h_1 and entropy s_1 of the steam entering the feedwater heater using Table A-5 in the appendix.**

State 1 is superheated steam at 1 megapascal pressure and 300 degrees Celsius:

h_1 = 3,051 kJ/kg, and s_1 = 7.123 kJ/kg · K

4. **Find the enthalpy h_2 and entropy s_2 of the water entering the feedwater heater using Table A-2 in the appendix.**

State 2 is a compressed liquid at 1 megapascal pressure and 80 degrees Celsius:

h_2 = 336 kJ/kg, and s_2 = 1.075 kJ/kg · K

5. **Calculate h_3 using the equation derived in Step 2:**

$$h_3 = \frac{(5\ \text{kg/s})(3{,}051\ \text{kJ/kg}) + (40\ \text{kg/s})(336\ \text{kJ/kg})}{45\ \text{kg/s}} = 638\ \text{kJ/kg}$$

6. **Find the enthalpy h_0 and entropy s_0 of the water at the dead state of 25 degrees Celsius and 100 kilopascals by interpolating Table A-2 of the appendix.**

In the dead state, water is a compressed liquid.

h_0 = 105 kJ/kg, and s_0 = 0.367 kJ/kg · K

7. **Calculate the availability of the steam at State 1, as shown in the following equations.**

Recall that kinetic and potential energy don't change during the process, so those terms are zero in the flow availability equation. (Remember that $\dot{A}$ is the extensive form of a.)

$$\dot{A}_{f1} = \dot{m}_1 [(h_1 - h_0) - T_0(s_1 - s_0)]$$

$\dot{A}_{f1}$ = 5 kg/s[(3,051 – 105) kJ/kg – 298 K(7.123 – 0.367) kJ/kg · K]

$\dot{A}_{f1}$ = 4,663 kJ/s

8. **Calculate the availability of the water at State 2, as shown in the following equations.**

   Assume no changes in kinetic energy or potential energy take place during the process, so those terms are zero in the flow availability equation:

   $$\dot{A}_{f2} = \dot{m}_2\left[(h_2 - h_0) - T_0(s_2 - s_0)\right]$$

   $$\dot{A}_{f2} = 40 \text{ kg/s}[(336 - 105) \text{ kJ/kg} - 298 \text{ K}(1.075 - 0.367) \text{ kJ/kg} \cdot \text{K}]$$

   $$\dot{A}_{f2} = 801 \text{ kJ/s}$$

9. **Calculate the availability of the water at State 3, as shown in the following equations:**

   Find s_3 using h_3 by interpolation in Table A-2 in the appendix.

   $$\dot{A}_{f3} = \dot{m}_3\left[(h_3 - h_0) - T_0(s_3 - s_0)\right]$$

   $$\dot{A}_{f3} = 45 \text{ kg/s}[(638 - 105) \text{ kJ/kg} - 298 \text{ K}(1.854 - 0.367) \text{ kJ/kg} \cdot \text{K}]$$

   $$\dot{A}_{f3} = 4{,}044 \text{ kJ/s}$$

10. **Calculate the change in availability $\dot{A}$ for the feedwater heater using these equations:**

    $$\Delta\dot{A} = \dot{A}_{exit} - \dot{A}_{inlets} = \dot{A}_{f3} - \left(\dot{A}_{f1} + \dot{A}_{f2}\right)$$

    $$\Delta\dot{A} = 4{,}044 \text{ kJ/s} - (4{,}663 + 800) \text{ kJ/s} = -1{,}419 \text{ kJ/s}$$

    The availability decreases in the mixing process because the high-quality energy entering as steam is mixed with low-quality energy in the water.

Calculating availability in open systems with transient flow

Many open systems don't have a steady fluid flow like a pump, a turbine, or a heat exchanger. Filling a scuba tank with air is an example of an open-system process with transient flow. Sometimes a transient flow process is called an unsteady flow process because the flow rate changes with time. Although you don't normally need a thermodynamic analysis of a scuba tank, some industrial processes store energy by filling a tank or pressure vessel. The stored energy can then be used to do work. Many factories use pneumatic tools, which run on a supply of compressed air stored in a tank. You can find the availability of a stored energy supply for an air tank. This calculation tells you the maximum amount of work available for a process that uses energy from a storage tank.

Here's an example that shows you how to determine the change in availability of an open system with transient flow. Figure 9-3 shows a compressed air tank that holds 5 kilograms of air at 1.5 megapascals pressure and at 87 degrees Celsius.

Figure 9-3:
Compressed
air in a
storage tank
has avail-
ability for
doing work.

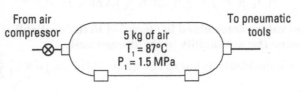

From air
compressor

5 kg of air
$T_1 = 87°C$
$P_1 = 1.5$ MPa

To pneumatic
tools

The stored availability (A_{stor}) is calculated using the following equation for an open system with transient flow:

$$A_{stor} = (m_1 - m_0)[(u_1 - u_0) + P_0(v_1 - v_0) - T_0(s_1 - s_0)]$$

In this equation, m_1 is the original mass of air in the tank, and m_0 is the mass of air in the tank at the dead state. Although mass flows out of the tank in this process, you use internal energy here because mass isn't flowing at the initial condition and at the dead state. After the air inside the tank reaches the dead state, you can't remove any additional mass for doing work. You can find the stored availability in the tank by following these steps:

1. **Find the internal energy u_0 and entropy s_0 of the air at the dead state of 25 degrees Celsius (298 Kelvin) and 100 kilopascals from Table A-1 in the appendix.**

   $u_0 = 213$ kJ/kg, and $s_0 = 6.863$ kJ/kg · K

2. **Find the internal energy u_1 and entropy s_1^o of the air at 87 degrees Celsius (360 Kelvin) and 1.5 megapascals from Table A-1 in the appendix.**

   $u_1 = 257.6$ kJ/kg, and $s_1^o = 7.053$ kJ/kg · K

3. **Find the entropy of the air s_1 in the tank at 1.5 megapascals, using the following equation.**

   This equation is introduced in Chapter 8.

   $$s_1 = s_1^o - R\ln\left(\frac{P_1}{P_0}\right) = 7.053 \text{ kJ/kg} \cdot \text{K} - 0.287 \text{ kJ/kg} \cdot \text{K}\ln\left(\frac{1.5\text{ MPa}}{0.1\text{ MPa}}\right)$$

   $$= 6.276 \text{ kJ/kg} \cdot \text{K}$$

4. **Calculate the mass of air in the tank at the dead state using the ideal-gas law relationship, $PV = mRT$.**

   The total volume of the tank is fixed, so $V = (m_1 RT_1/P_1) = (m_0 RT_0/P_0)$. You can rearrange this equation to solve for the dead state mass, m_0.

   $$m_0 = m_1 \left(\frac{T_1}{T_0} \right) \left(\frac{P_0}{P_1} \right) = 5 \text{ kg} \left(\frac{360 \text{ K}}{298 \text{ K}} \right) \left(\frac{0.1 \text{ MPa}}{1.5 \text{ MPa}} \right) = 0.40 \text{ kg}$$

5. **Insert the values determined in Steps 1–4 in the availability equation to determine the availability of the storage tank.**

   You can use the ideal-gas law to substitute $P_0 (v_1 - v_0) = RP_0 \left(\dfrac{T_1}{P_1} - \dfrac{T_0}{P_0} \right)$.

   $$A_{stor} = (5.0 - 0.4) \text{ kg} \left[(257.6 - 213) \text{ kJ/kg} + \left(0.287 \frac{\text{kJ}}{\text{kg} \cdot \text{K}} \right) (100 \text{ kPa}) \left(\frac{360 \text{ K}}{1,500 \text{ kPa}} - \frac{298 \text{ K}}{100 \text{ kPa}} \right) \right]$$

   $$+ (5.0 - 0.4) \text{ kg} \left[298 \text{ K} (6.276 - 6.863) \text{ kJ/kg} \cdot \text{K} \right]$$

   $$A_{stor} = 648 \text{ kJ}$$

Not all the energy in the tank can be used to produce work. For example, if the air in the tank is used to power pneumatic tools, the tools operate best above a certain pressure. The unusable availability below the tool's operating pressure is an irreversibility of the system.

Balancing the Availability of a System

Energy can move into or out of a system through a work input or output, heat transfer, or mass transfer. If a fluid mass can flow through a system, such as a heat exchanger, a turbine, or a pump, the system is defined as an open system. If mass can't flow through a system, as in a piston/cylinder arrangement, the system is defined as a closed system. Thus work and heat transfer processes are the only means by which energy can be transferred in a closed system.

You can use the first law of thermodynamics to write an energy balance on a system. Because availability is a thermodynamic property related to energy, a balance equation for availability can also be written for a system. The availability balance on a system has the following four components:

✔ A_{in}: Energy availability can be transferred *into* the system by heat transfer, work, or mass flow.

✔ A_{out}: Availability can be transferred *from* the system by heat transfer, work, or mass flow.

✔ $A_{destroyed}$: Availability can be destroyed within a system by irreversibilities such as friction, heat transfer through a finite temperature difference, mixing, chemical reactions, and unrestrained expansion, among other ways.

✔ ΔA_{system}: Availability can be stored or released within a system, causing a change in availability of the system.

You write the availability balance of a system using the preceding components of availability with the following equation:

$$A_{in} - A_{out} - A_{destroyed} = \Delta A_{system}$$

The units for availability are the same as for energy: kilojoules in the SI system and British thermal units in the English system. The uppercase variable A indicates that the property includes mass (extensive form). You use the lowercase form of the variable a for the intensive form (on a per-unit-mass basis). The availability balance equation can be written on a rate basis using a dot above the variable ($\dot{A}$) as usual with other rate equations, as shown here:

$$\dot{A}_{in} - \dot{A}_{out} - \dot{A}_{destroyed} = \frac{dA_{system}}{dt}$$

The following sections describe each of the components of availability in detail.

Transferring availability using work processes

Availability can be transferred into (A_{in}) or out of (A_{out}) a system using work. Typically, this transfer is accomplished by shaft or boundary work, but other forms of work exist (see Chapter 4). In a reversible work process, the availability transferred using work (A_{work}) equals the useful work (W_{useful}). If the process involves boundary work, the work on the surroundings (W_{surr}) must be subtracted from the actual work (W_{act}) to determine the useful work, because some of the work is used to push the atmosphere out of the way. The availability transferred to a system by boundary work is found using this equation:

$$A_{work} = W_{useful} = W_{act} - W_{surr}$$

In a process that begins at State 1 and ends at State 2, the *surroundings work* (W_{surr}) equals the atmospheric pressure (P_0) times the change in volume of the system ($V_2 - V_1$), as shown in this equation:

$$W_{surr} = P_0(V_2 - V_1)$$

For an open system, a reversible work process doesn't have to work against the surroundings; however, flow work is accounted for by enthalpy, so the useful work (W_{useful}) equals the actual work (W_{act}). In this case, the availability transfer by work (A_{work}) is found by this equation:

$$A_{work} = W_{useful} = W_{act}$$

The availability of the fluid increases when work is input to a system, as in a pump or a compressor, and the availability of a fluid decreases when work is done by the system, as in a turbine or an expansion stroke of a piston.

Transferring availability with heat transfer processes

You can move availability into or out of a system using heat transfer. In a heat transfer process, only a portion of the energy in a heat transfer process at a system boundary temperature (T) can be transferred as availability. The amount of energy that doesn't transfer in as availability becomes an irreversibility in the system. This irreversibility occurs because not all heat transfer to a system can be converted into useful work. Some heat must be rejected to the ambient environment surrounding the system.

In theory, it's possible to use a heat engine to produce useful work from the heat that's rejected by a system to the environment. But even a heat engine has to reject some heat, so you can never completely get out of rejecting heat by producing useful work. In practice, replacing a heat rejection process with a heat engine usually isn't cost-effective if the availability is relatively low compared to the dead state.

The portion of availability transferred in a system by heat, A_{heat}, at a boundary temperature T in an environment at T_0 is calculated with this equation:

$$A_{heat} = \left(1 - \frac{T_0}{T}\right)Q$$

Availability and heat move in or out of a system in the same direction as long as the temperature of the boundary is above the dead state temperature. If the system boundary temperature is below the dead state temperature, the availability transfers in the opposite direction of heat transfer.

Transferring availability with mass flow

Mass flows through an open system, so it can move energy, entropy, and availability into or out of the system. The rate of total availability (A_{mass})

that's transferred into or out of a system is calculated by the mass flow rate ($\dot{m}$) and the flow availability (a_f), as shown in this equation: $\dot{A}_{mass} = \dot{m}a_f$.

For a closed system, there is no mass flow so there's no transfer of availability by mass transfer, but the total availability (A) in the system is calculated using the mass of the system (m) and the specific availability (a), as shown by the equation $A = ma$.

Understanding the Decrease in Availability Principle

The first law of thermodynamics states that energy cannot be created or destroyed but can only change form. This means that energy is a conserved property. In any system, the energy that goes into a system equals the energy that leaves and accumulates in the system. Energy may enter as a form of heat and leave as a form of work, as it does in a heat engine. Mass has similar qualities: The mass entering a system equals the mass leaving and accumulating in the system, even if it changes phase from a liquid to a gas.

Entropy and availability are not conserved properties like energy and mass. In Chapter 8, I discuss the increase in entropy principle, whereby every process in a system causes entropy to increase from the perspective of the universe. Locally, you may cause entropy to decrease in a system by lowering its temperature or raising its pressure. But the result of entropy decreasing within a system is a proportionately greater increase in the entropy of the system's surroundings. The net effect of entropy changes to the system and its surroundings is that entropy always increases.

The nature of availability is that it diminishes as it's used to do work or provide heat. The inlet conditions of a thermodynamic process require high-quality energy relative to the availability at the outlet conditions, meaning the inlet conditions must have sufficient availability to perform the process. At the outlet of a work process, the availability has decreased such that it can't do as much work as it could before the process.

You may locally increase the availability of energy by adding heat, work, or mass into a process, but the availability of the source of heat, work, or mass decreases by supplying energy to the process. Globally, the decrease in availability is similar to the global increase in entropy. Eventually, the universe will have no availability for performing useful work. Irreversibility caused by heat transfer, friction, mixing, and so forth destroys availability and generates entropy. The destruction of availability, $A_{destroyed}$, is proportional to entropy generation, S_{gen}, as shown in this equation: $A_{destroyed} = T_0 S_{gen}$.

Figuring Out Reversible Work and Irreversibility

Reversible work is a thermodynamic concept that defines the best theoretical work process. A reversible work process in a system can proceed in both the forward and reverse directions, such that the system and the surrounding environment are returned to their initial state at the end of the reverse process.

Reversible work is accomplished by an ideal heat engine, as shown in Figure 9-4, that receives energy from a high-temperature thermal energy reservoir at T_H and rejects waste heat to a low-temperature reservoir at T_L. The ideal heat engine provides reversible work between two energy reservoirs. You calculate the reversible work using the heat input and the thermal efficiency of an ideal heat engine, as shown in this equation:

$$w_{rev} = \eta_{th,rev} \cdot q_{in}$$

Thermal efficiency, introduced in Chapter 7, is a measure of a heat engine's work output compared to heat input. For an ideal reversible heat engine operating between two thermal energy reservoirs, the maximum possible thermal efficiency is defined by this equation:

$$\eta_{th,rev} = 1 - \frac{T_L}{T_H}$$

Irreversibility (*I*) is a thermodynamic concept that's quantified by the following statements:

✔ Irreversibility is the difference between the work output or input of a reversible process and an actual work process ($W_{rev} - W_{act}$).

✔ Irreversibility is equivalent to the availability destroyed ($A_{destroy}$) in a system during a process.

✔ Irreversibility is proportional to the entropy a system generates (S_{gen}) during a process.

Because many different processes generate irreversibility, it can be calculated several ways. You can summarize the three basic statements about irreversibility stated in the preceding list like this:

$$I = A_{destroy} = T_0 S_{gen} = (W_{rev} - W_{act})$$

For a closed system, the irreversibility is calculated with this equation:

$$I = T_0(\Delta S_{sys} - S_{in} + S_{out})$$

In this equation, the entropy in (S_{in}) and the entropy out (S_{out}) terms are due to heat transfer across the system boundaries, such that the irreversibility can be calculated using this equation:

$$I = T_0\left[\left(S_2 - S_1\right)_{sys} - \frac{Q_{in}}{T_{in}} + \frac{Q_{out}}{T_{out}}\right]$$

For a steady-state open system, the rate of irreversibility in the system is calculated using the mass flow rates for each inlet and outlet, as shown in this equation:

$$\dot{I} = T_0\left[\sum_{out}\dot{m}s - \sum_{in}\dot{m}s - \frac{\dot{Q}_{in}}{T_{in}} + \frac{\dot{Q}_{out}}{T_{out}}\right]$$

Irreversibility comes from many different sources, such as friction, chemical reactions, mixing of fluids, expansion and compression of fluids, and heat transfer through a finite temperature difference. If you try to reverse an irreversible process, the system or the surroundings won't return to their initial state at the end of the reverse process. Friction, mixing, fluid expansion and compression, chemical reactions, and heat loss dissipate energy, and you can't recover that energy by reversing the process.

Irreversibility is always a positive quantity. Use the convention that work out of a system is positive and work into a system is negative to ensure the irreversibility is always positive. If the irreversibility of a device is calculated to be negative, the device isn't possible because it violates the second law of thermodynamics.

The following example shows you how to calculate the reversible work, useful work, and irreversibility for a heat engine.

Figure 9-4 shows a heat engine operating between two thermal energy reservoirs. A heat input of 1,310 kJ/kg comes from a high-temperature reservoir at 2,116 Kelvin. The heat engine rejects heat to the low-temperature reservoir at 293 Kelvin. The net work out of the engine is 665 kJ/kg.

Follow these steps to calculate the reversible work, useful work, and irreversibility of the heat engine:

1. **Find the maximum thermal efficiency of a heat engine operating between the two thermal energy reservoirs.**

$$\eta_{th,rev} = 1 - \frac{293\ K}{2,116\ K} = 0.861$$

2. **Calculate the reversible work (w_{rev}) using the ideal reversible heat engine efficiency $\eta_{th,rev}$ and the heat input q_{in}.**

$$w_{rev} = (0.861)1,310\ kJ/kg = 1,129\ kJ/kg$$

3. **Calculate the irreversibility of the heat engine by determining the difference between the reversible work and the useful work.**

The useful work equals the net work output ($w_u = w_{net}$) because no net work on the surroundings takes place in a cycle.

$$i = w_{rev} - w_u = (1{,}129 - 665) \text{ kJ/kg} = 464 \text{ kJ/kg}$$

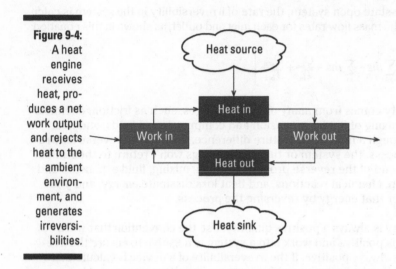

Figure 9-4:
A heat engine receives heat, produces a net work output and rejects heat to the ambient environment, and generates irreversibilities.

Calculating the Second-Law Efficiency of a System

No real work or heat transfer process performs as well as an idealized reversible work or heat transfer process. You can compare a real process to a reversible process to see how well the energy resources are used for the process by calculating the second-law efficiency. The *second-law efficiency* can be defined as a measure of the performance of an actual device or process compared to the performance of a reversible device or process.

For a heat engine, the measure of performance is typically the thermal efficiency of the system. I discuss how to calculate the thermal efficiency of a heat engine in Chapter 7. You can calculate the second-law efficiency by dividing the thermal efficiency of an actual heat engine by the efficiency of a reversible heat engine. Here's what the equation looks like:

$$\eta_{II} = \frac{\eta_{th}}{\eta_{th,rev}}$$

For a refrigeration cycle, you use the coefficient of performance of the system to find the second-law efficiency. I discuss the coefficient of performance of a refrigeration cycle in Chapter 7. You can calculate the second-law efficiency by dividing the coefficient of performance of an actual refrigeration cycle by the coefficient of performance of a reversible refrigeration cycle. The equation looks like this:

$$\eta_{II} = \frac{COP}{COP_{rev}}$$

Suppose you're designing a steam power plant that has a thermal efficiency of 35 percent. Furthermore, the power plant design has a maximum steam temperature of 850 degrees Celsius, and you must choose between two different heat sources for the plant. Figure 9-5 shows that one heat source is at 900 degrees Celsius and the other is at 1,500 degrees Celsius. Which is the best energy source? The heat sink temperature is at 25 degrees Celsius. You can calculate the second-law efficiency to figure out which resource is the better choice.

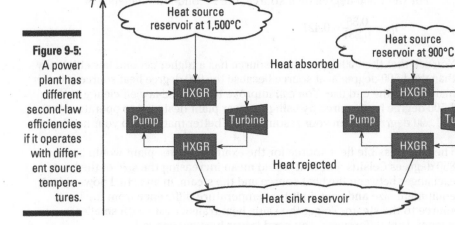

Figure 9-5:
A power plant has different second-law efficiencies if it operates with different source temperatures.

Follow these steps to calculate the second-law efficiency of the power plant:

1. **Calculate the efficiency** $\eta_{th,rev}$ **of a reversible heat engine for both heat-source temperatures.** Convert temperature from degrees Celsius to absolute temperature (Kelvin). For the 900-degrees-Celsius heat source, calculate the Carnot efficiency as follows:

$$\eta_{th,rev} = 1 - \frac{298 \text{ K}}{1,173 \text{ K}} = 0.746$$

For the 1,500-degrees-Celsius heat source, the ideal heat engine efficiency is

$$\eta_{th,rev} = 1 - \frac{298 \text{ K}}{1,773 \text{ K}} = 0.831$$

As expected with a Carnot heat engine (see Chapter 10), the higher-temperature heat source gives you a higher thermal efficiency. Calculating the second-law efficiency tells you how well your heat source is matched with the end use.

The efficiency of an ideal-heat engine increases as the temperature of the heat source increases.

2. **Calculate the second-law efficiency of the power plant for both heat sources.**

   For the 900-degree heat source, you calculate the second-law efficiency as follows:

   $$\eta_{II} = \frac{0.35}{0.746} = 0.469$$

   For the 1,500-degree heat source, the second-law efficiency is

   $$\eta_{II} = \frac{0.35}{0.831} = 0.421$$

You see that the 900-degree heat source has a higher second-law efficiency than the 1,500-degree heat source because the 900-degree heat source is better matched to the end use. You can improve the second-law efficiency of the 1,500-degree heat source by using a power plant designed to operate closer to 1,500 degrees. Then your resources are better matched to your needs.

The best possible heat source for the example power plant would be at 850 degrees Celsius, but that would mean increasing the size of the heat exchanger between the heat source and the steam. In an actual power plant, equipment size and cost limit the temperature difference from the heat source to the steam. To get the same heat transfer rate with smaller temperature differences, you need larger heat exchangers.

The temperature of the heat source for any heat engine should be as close as possible to the highest fluid temperature in the engine to maximize the effectiveness of the energy source. I discuss in detail the analysis of a steam power plant using the Rankine cycle heat engine in Chapter 12.

Part III

Planes, Trains, and Automobiles: Making Heat Work for You

The 5th Wave   By Rich Tennant

THERMODYNAMICS:

The act of continuous adjustment of a thermostat by two people of disparate body temperatures.

In this part . . .

You put together the engine that you tore apart in Part II. The first and second laws of thermodynamics are applied to whole systems that operate in cycles, like reciprocating engines, gas turbine engines, power plants, and, yes, your humble refrigerator. You can figure out how much energy you put in, how much work you get out, how efficient systems are, and more fun stuff like that.

Chapter 10

Working with Carnot and Brayton Cycles

. . . .

In This Chapter

▶ Getting acquainted with Carnot's ideal engine

▶ Examining the simple Brayton cycle for gas turbine engines

▶ Modifying the Brayton cycle to improve its performance

▶ Putting wings on the Brayton cycle with jet engines

. . . .

*1*f your air conditioner is running on a scorching summer afternoon, everyone else's is probably running at the same time. All those air conditioners put a big load on the electric utility. During the hottest part of the day, the power company has to generate a lot more power quickly to react to all those air conditioners turning on. The power company turns on something called a *gas turbine generator*. A gas turbine is an engine in which hot compressed air flows through a turbine to produce work. The turbine then turns a large electric generator to power your air conditioner (and everyone else's).

Gas turbine engines aren't just found in power plants operated by your electric utility company. If you've ever flown on a jet, you've been transported by a gas turbine engine called a turbojet or turbofan. A turbojet produces thrust in the form of a jet of high-velocity exhaust. Other airplanes may be powered by a gas turbine engine called a turboprop engine, which uses a turbine to turn a propeller. The turboprop engine also produces some exhaust thrust, though not as much as the propeller.

In this chapter you find out about a theoretically perfect engine: the Carnot cycle. I say *theoretically* because this engine is physically impossible to build, but it's fun to know what the maximum possible efficiency of an engine could be. That way, you'll know if someone tries to pull the wool over your eyes by telling you they can sell you an engine that's 100-percent efficient. I also tell you how to analyze the gas turbine engine using a thermodynamic model of the actual system called the Brayton cycle.

Analyzing the Ideal Heat Engine: The Carnot Cycle

Every heat engine in thermodynamics produces work from some sort of heat input. The heat input usually comes from burning a fuel in a combustion chamber, although some heat engines can use solar power for a heat input. A heat input to an engine is modeled in thermodynamics as if it comes from some infinitely large energy reservoir, so you don't have to worry about complications of the combustion process as part of your thermodynamic analysis. (I discuss the combustion process and show how you analyze it in Chapter 16.)

For a heat engine to operate, it must also be able to reject heat out of the cycle. At first, the idea of an engine rejecting heat may seem odd, because it means the engine isn't using all the heat you put into it. And that means the engine isn't 100-percent efficient. Isn't there some way to make an engine where all the heat that goes into it is converted to useful work and no heat needs to be rejected? I can answer that question with one word: No! It would violate the second law of thermodynamics, as discussed in Chapter 7.

The Carnot cycle models the best possible theoretical heat engine. In the Carnot cycle, heat is input (Q_H) from a high-temperature reservoir at T_H, a net work ($W_{net, out}$) is extracted, and heat is rejected (Q_L) to a low-temperature reservoir at T_L. Net work means there is a work input (W_{in}) to the cycle that's subtracted from the total work output (W_{out}) of the cycle.

The ideal Carnot cycle heat engine has some unique features or operating assumptions:

✔ It's completely frictionless. If you could make such an engine and you wanted to just spin it with your finger, it would spin perpetually.

✔ The heat input to and heat rejection from the cycle is accomplished without any temperature difference between the energy reservoirs and the engine itself. That means the hot side of the engine is at the same temperature as the high-temperature reservoir, and the cold side of the engine is at the temperature of the low-temperature reservoir.

In real life, achieving all of these features is impossible. Every engine has friction, all work creates entropy, and all heat transfer occurs with a temperature difference between two objects, which also creates entropy.

The combination of these special features means the Carnot cycle is totally reversible. (Chapter 9 tells you more about reversible and irreversible cycles.) A totally reversible cycle is one that can operate in reverse without additional input from outside the cycle. This concept is difficult to grasp. It's easier to think of a process as being irreversible. Friction is an irreversible process. Friction between parts generates heat. But if you try to make that process operate in reverse by adding heat to the parts and hoping they move by themselves, you're out of luck. The only way a process can be reversible is to not have any friction in the first place.

Examining the four processes in a Carnot cycle

The Carnot cycle is the gold standard of all heat engines. The performance of all heat engines can be compared to the Carnot cycle. Anyone who claims to have a heat engine with greater efficiency than a Carnot cycle is making a false claim. Although an engine based on the Carnot cycle is impossible to build, Figure 10-1 shows a sketch of the four processes in an open system Carnot engine. The Carnot engine can also be developed as a closed system using a piston-cylinder arrangement. Figure 10-2 shows the Carnot cycle mapped on a *T-s* diagram. I discuss these kinds of property diagrams in Chapter 3. Here's how the cycle works:

1. **Isothermal heat addition.**

   Heat is added from the high-temperature energy reservoir at T_H to the working fluid, which may be a gas or a vapor. Because adding heat to a gas increases its temperature, the gas has to be expanded in a turbine to keep the temperature constant. As a result of the expansion process, the turbine gets work out of the process at the same time. In reality, adding heat to a gas while it's expanding in a turbine isn't feasible.

   Figure 10-1 shows this process as the isothermal turbine. Figure 10-2 shows this as the process between States 1 and 2 on the *T-s* diagram.

2. **Isentropic work output.**

   The high-pressure, high-temperature gas expands in the isentropic turbine in Figure 10-1. No friction occurs in either the fluid or the rotating turbine. The temperature and pressure of the gas decrease in the process. Figure 10-2 shows this process between States 2 and 3 on the *T-s* diagram.

   Isentropic means that the process is reversible and adiabatic (that is, no heat transfer occurs through the system boundary).

3. **Isothermal heat rejection.**

   Heat is removed from the relatively low-temperature, low-pressure gas and rejected to the low-temperature thermal energy sink at T_L. Because removing heat from a gas decreases its temperature, the gas has to be compressed to keep the temperature constant by removing heat at the same rate at which work is being done. A compressor isn't a very good heat transfer device, so removing all the heat generated by compression in the same process is impractical.

Figure 10-1 shows this process as the isothermal compressor. Figure 10-2 shows this as the process between States 3 and 4 on the *T-s* diagram.

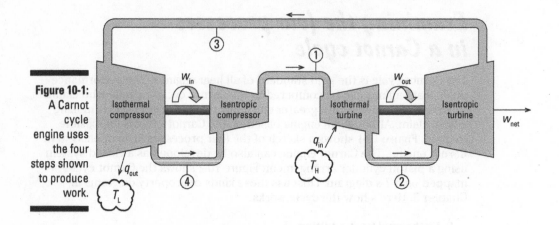

Figure 10-1:
A Carnot cycle engine uses the four steps shown to produce work.

4. Isentropic work input.

The low-pressure, low-temperature gas is compressed in the isentropic compressor in Figure 10-1. As in the turbine, no friction occurs in either the fluid or the rotating compressor. The temperature and pressure of the gas increase in the process. Figure 10-2 shows this process between States 4 and 1 on the *T-s* diagram.

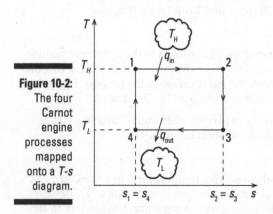

Figure 10-2:
The four Carnot engine processes mapped onto a *T-s* diagram.

Calculating Carnot efficiency

Thermal efficiency (η_{th}) is a measure of how effective an engine — or any power cycle — is at using the heat energy input (Q_{in}) for producing work (W_{net}). (I introduce thermal efficiency in Chapter 7). Mathematically, this is expressed as

$$\eta_{th} = \frac{W_{net}}{Q_{in}}$$

You can use the first law of thermodynamics to determine the net work for a cycle from the difference between the heat input and the heat output:

$$W_{net} = Q_{in} - Q_{out}$$

The second law of thermodynamics tells you that the heat input and the heat output to a reversible cycle result in a change in entropy ($s_2 - s_1$), as shown in these equations:

$$Q_{in,\,rev} = T_H(s_2 - s_1)$$
$$Q_{out,\,rev} = T_L(s_3 - s_4)$$

Note that for the isentropic processes in the Carnot cycle, $s_1 = s_4$ and $s_2 = s_3$. If you make substitutions for W_{net}, Q_{in}, and Q_{out} in the efficiency equation, the changes in entropy terms cancel each other out and you can calculate the efficiency (as shown in Chapter 7) using only the temperatures of the high- and low-thermal energy reservoirs:

$$\eta_{th,\,rev} = 1 - \frac{Q_{out}}{Q_{in}} = 1 - \frac{T_L}{T_H}$$

The low-temperature thermal energy reservoir, T_L, is usually the temperature of the ambient conditions in the air or whatever medium is being used for a heat sink, such as a river, a lake, the ocean, or the ground. The high-temperature thermal energy reservoir, T_H, is the temperature of whatever is being used as a heat source, such as a combustion process for a heat engine.

As an example, you can calculate the maximum possible (Carnot) efficiency for a heat engine where the high-temperature reservoir is at 1,400 Kelvin and the low-temperature reservoir is at 293 Kelvin. The Carnot efficiency of an engine operating between these two energy reservoirs is calculated as follows:

$$\eta_{th,\,rev} = 1 - \frac{293\ K}{1,400\ K} = 79.0\%$$

Working with the Ideal Gas Turbine Engine: The Brayton Cycle

Gas turbine engines provide power at electric utilities and on board ships, and they propel aircraft. Gas turbine engines can be started quickly, and they produce large amounts of power in a relatively small volume or space.

They're modeled thermodynamically as a Brayton cycle in honor of George Brayton, who developed the first internal combustion engine that featured a continuous combustion process (as opposed to periodic combustion in reciprocating engines; see Chapter 11). Figure 10-3 shows a cross-sectional view of a gas turbine engine. Fresh air is drawn into a compressor, which pressurizes the air. The air is heated as it flows through a combustion chamber; then it flows through a turbine where work is extracted from the air. The air then exhausts from the turbine at atmospheric pressure.

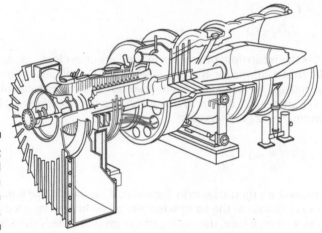

Figure 10-3:
A sectional
view of a
gas turbine
engine.

Because the air flows through the engine in a single pass, meaning it doesn't recirculate, the Brayton cycle is an open loop. Thermodynamic analysis of the engine closes the loop by replacing the exhaust process with a heat exchanger in which the air is returned to its initial condition prior to entering the compressor section. In reality, the atmosphere acts as a giant heat exchanger for this process.

Examining the four processes in a Brayton cycle

The Brayton cycle is modeled as an open system because energy, entropy, and mass flow through the system boundaries. Figure 10-4 shows a sketch of the four processes in a Brayton engine. Figure 10-5 shows the Brayton cycle mapped on *T-s* and *P-v* diagrams. I discuss these kinds of property diagrams in Chapter 3. Here's how the cycle works:

1. **Isentropic compression. (Work in.)**

   Fresh air enters the compressor of the engine at State 1. State 1 is defined by the ambient temperature and pressure of the air prior to entering the compressor. The air is compressed isentropically to State 2. As a result of compression, the temperature of the air increases with the pressure.

   Figure 10-4 shows this process as the compressor. Figure 10-5 shows this as the process between States 1 and 2 on the *T-s* and *P-v* diagrams.

   The amount of air compression that the compressor section provides is called the *pressure ratio* (r_p). Typical gas turbine engines compress the air to 5–30 times the inlet air pressure. Engine efficiency increases with the pressure ratio and generally ranges from about 40 to 60 percent.

2. **Isobaric heat input.**

   The high-pressure gas is heated in a constant-pressure process. This heat addition is the result of a continuous combustion process in a combustion chamber physically located between the compressor and the turbine sections of the engine.

   Figure 10-4 shows this as a heat exchanger that inputs heat to the air in the cycle. Figure 10-5 shows this process between States 2 and 3 on the *T-s* and *P-v* diagrams.

3. **Isentropic expansion. (Work out.)**

   The high-pressure and -temperature gas is expanded isentropically through a turbine to extract work from the cycle. As the gas decreases in pressure, its temperature also decreases. The turbine is often connected to the compressor on a single shaft so that the turbine is mechanically coupled to provide power to operate the compressor.

   Figure 10-4 shows this process as the turbine. Figure 10-5 shows this as the process between States 3 and 4 on the *T-s* and *P-v* diagrams.

A tremendous amount of work is required to compress a gas to high pressure. Consequently, much of the power the turbine produces is used by the compressor. The *back work ratio* (r_{bw}) is the ratio of compressor work to turbine work. In jet aircraft engines, all the turbine work is used to turn the compressor, electric generators, and hydraulic pumps. In a power plant gas turbine, more than half the turbine work is consumed by the compressor.

4. Isobaric heat rejection.

In the ideal Brayton cycle engine, the exhaust gas leaving the turbine is cooled in a constant-pressure process by a heat exchanger between the low-pressure side of the turbine and the compressor. Cooling the exhaust gas removes the energy from State 4 so that it reaches State 1 again to complete the four-process cycle in a closed loop.

Figure 10-4 shows the low-temperature, constant-pressure heat exchanger. Figure 10-5 shows the process between States 4 and 1 on the *T-s* and *P-v* diagrams.

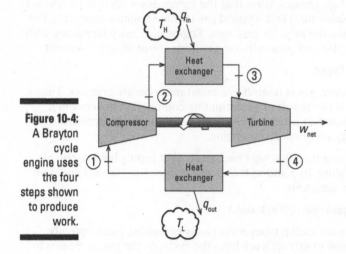

Figure 10-4: A Brayton cycle engine uses the four steps shown to produce work.

In a real gas turbine engine, the heat is rejected to the atmosphere, so the engine operates in an open loop. The atmosphere is the only physical heat exchanger between the turbine and the compressor. Furthermore, a significant amount of heat usually remains in the exhaust leaving the turbine, but the pressure can't be decreased any further to take advantage of this energy. In the section on regenerators, I discuss how to capture some of this energy to improve the efficiency of the cycle.

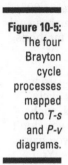

Figure 10-5:
The four
Brayton
cycle
processes
mapped
onto *T-s*
and *P-v*
diagrams.

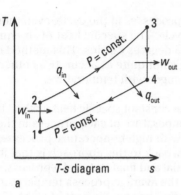

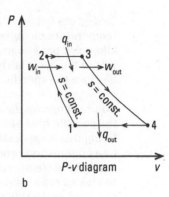

Analyzing the Brayton cycle

You can use either of two methods to analyze the Brayton cycle with the first law of thermodynamics and determine the amount of work and heat transfer that occurs in each process:

✔ **The constant specific heat method:** For this method, you assume that the specific heat of air remains constant in each process in the cycle. The analysis isn't as exact as the variable specific heat method, but you don't need to look up thermodynamic properties in a table.

✔ **The variable specific heat method:** This method takes into consideration that specific heat changes with temperature, and the method uses Table A-1 in the appendix. This method is more accurate than the constant specific heat method, but finding the values can require a bit of tedious interpolation of the tables. Many textbooks come with a computer program that enables you to easily determine thermodynamic properties accurately (without tedious interpolation).

Constant specific heat method

Analyzing the Brayton cycle with the first law of thermodynamics means finding the amount of heat transferred into and out of the cycle and finding the amount of work put into and produced by the cycle. Ultimately, it leads to finding out how efficient the engine is at converting energy from the fuel into useable power.

There are a couple ways to analyze a Brayton cycle using the constant specific heat assumption. The easiest way uses what's known as the *cold-air standard assumption,* which means that you use the specific heat of air at

25 degrees Celsius throughout all processes in the cycle, even though the air temperature changes during the cycle. The specific heat of air equals 1.005 kilojoules per kilogram-Kelvin at 25 degrees Celsius. This method makes analysis of the cycle fairly quick, but it's not quite as accurate as other methods because the specific heat of air changes with temperature.

You can improve the accuracy of the constant specific heat method by using the specific heat at the average temperature of each process in the cycle. Doing this is especially important with high-temperature processes, such as a combustion process. One drawback to this approach is that if you don't know the air temperature at the initial and final states of a process, you have to make a reasonable estimate for the average process temperature. You can often make this estimate by using the specific heat of air at 25 degrees Celsius and computing the unknown temperature of a process. Then you can use this temperature value to adjust your estimate for the specific heat by taking the average temperature of the process.

As an example, you can analyze a simple Brayton cycle using the cold-air standard assumptions by following this example with these specifications:

- ✔ The air temperature entering the compressor section of the engine is at 295 Kelvin, and the atmospheric pressure is 101 kilopascals.
- ✔ The pressure ratio (r_p) of the compressor is 10.
- ✔ The air temperature exiting the combustion chamber is 1,400 Kelvin.

The energy equations use constant pressure specific heat (c_p) in place of the change in enthalpy for each heat transfer and work process. I discuss enthalpy and specific heat in Chapter 2. The energy equation for each process in the Brayton cycle is written as follows:

- ✔ **For combustion, heat in:** $q_{in} = (h_3 - h_2) = c_p(T_3 - T_2)$
- ✔ **For exhaust, heat out:** $q_{out} = (h_4 - h_1) = c_p(T_4 - T_1)$
- ✔ **For compression, work in:** $w_{in} = (h_2 - h_1) = c_p(T_2 - T_1)$
- ✔ **For expansion, work out:** $w_{out} = (h_3 - h_4) = c_p(T_3 - T_4)$

You need to find the temperature at the endpoints of each process to determine the amount of energy associated with the process. The ambient air temperature is used at State 1 at the beginning of the compression process. In an isentropic compression process, the temperature at State 2 can be found from State 1 using an isentropic process equation for an ideal gas (introduced in Chapter 5), as shown here:

$$\frac{T_2}{T_1} = \left(\frac{P_2}{P_1}\right)^{\frac{k-1}{k}}$$

Notice the fraction (P_2/P_1) in this equation equals the pressure ratio (r_p) of the engine. The variable k is the ratio of specific heats, which has a value of $k = 1.4$ for air at 25 degrees Celsius.

These steps help you analyze the Brayton cycle engine using constant specific heat:

1. **Find the temperature at the exit of the compressor (T_2), as follows:**

$$T_2 = T_1 \left(\frac{P_1}{P_2}\right)^{\frac{k-1}{k}} = (295\ \text{K})\left(\frac{10}{1}\right)^{\frac{1.4-1}{1.4}} = 570\ \text{K}$$

Always use absolute temperatures when working with ideal-gas law equations.

2. **Find the temperature for State 4.**

Finding the temperature at State 4 is just like finding the temperature at State 2, but instead of isentropic compression, you have isentropic expansion from States 3 to 4.

$$T_4 = T_3 \left(\frac{P_4}{P_3}\right)^{\frac{k-1}{k}} = (1{,}400\ \text{K})\left(\frac{1}{10}\right)^{\frac{1.4-1}{1.4}} = 725\ \text{K}$$

3. **Find the amount of energy exchanged in each process using the energy equations:**

$q_{in} = 1.005\ \text{kJ/kg}(1{,}400\ \text{K} - 570\ \text{K}) = 834\ \text{kJ/kg}$

$q_{out} = 1.005\ \text{kJ/kg}(725\ \text{K} - 295\ \text{K}) = 432\ \text{kJ/kg}$

$w_{in} = 1.005\ \text{kJ/kg}(570\ \text{K} - 295\ \text{K}) = 276\ \text{kJ/kg}$

$w_{out} = 1.005\ \text{kJ/kg}(1{,}400\ \text{K} - 725\ \text{K}) = 678\ \text{kJ/kg}$

4. **Check your calculations to make sure the net heat transfer into the cycle ($q_{in} - q_{out}$) equals the net work out ($w_{out} - w_{in}$) of the cycle:**

$q_{in} - q_{out} = w_{out} - w_{in}$ or $834 - 432 = 678 - 276 = 399\ \text{kJ/kg}$

The specific heat of the air isn't actually constant for each process in the cycle, but assuming that it is enables you to quickly calculate the amount of work and heat transfer going on in the engine without consulting thermodynamic property tables.

You can see the effect of how specific heat changes with temperature by using the specific heat at the average temperature for the combustion process. The average temperature for the combustion process looks like this

$$T_{ave} = 0.5(1{,}400 + 570)\ \text{K} = 985\ \text{K}$$

You can find the specific heat of air around 1,000 Kelvin in Table A-1 of the appendix:

$$c_p = 1.142 \text{ kJ/kg} \cdot \text{K}$$

Then the heat input to the engine is as follows:

$$q_{in} = 1.142 \text{ kJ/kg}(1,400 \text{ K} - 570 \text{ K}) = 948 \text{ kJ/kg}$$

The result — 948 kJ/kg — is considerably higher than the 834 kJ/kg estimate you get by using the cold-air standard assumption.

Variable specific heat method

The constant specific heat method becomes inaccurate when temperatures in the cycle get pretty high, as they do in the heat-input process. The specific heat of air increases with temperature, so accounting for this change improves the accuracy of your analysis. Calculating more accurate values for work and heat transfer isn't overly difficult, but before you do it, you need to look up relative pressure and enthalpy properties for air. You can find the ideal-gas properties for air in Table A-1 of the appendix. Chapter 5 discusses how to use enthalpy properties, and Chapter 8 discusses relative pressure properties for an ideal-gas isentropic process.

If you look at Table A-1 in the appendix, you see that you need to know only the air temperature to find the enthalpy and the relative pressure. No units or dimensions are associated with the relative pressure property of air.

Don't be alarmed if the values for relative pressure you find in your textbook are different from the ones in this book's appendix. Different books use a different reference point for relative pressure, but it all works out in the end as long as you're consistent and use properties from the same source.

Follow these steps to analyze the Brayton cycle to account for the variation in specific heat with temperature:

1. **Look up the temperature T_1 (295 K) in Table A-1 in the appendix to find the enthalpy and relative pressure:**

   $$h_1 = 295.6 \text{ kJ/kg}$$

   $$P_{r1} = 1.055$$

   You will need to do some interpolation on the table because this temperature isn't listed. (I show you how to interpolate a table in Chapter 3.)

2. **Find the relative pressure, enthalpy, and temperature properties of the air at the end of compression (State 2) by using the engine pressure ratio (r_p) and the relative pressure at State 1:**

   $$r_p = \frac{P_2}{P_1} = \frac{P_{r2}}{P_{r1}}$$

The compression is an isentropic process, as Figure 10-5 shows. Chapter 8 gives you more details on isentropic processes.

3. **Use this equation to find the relative pressure at State 2:**

$$P_{r2} = P_{r1} \frac{P_2}{P_1} = (1.055)(10) = 10.55$$

4. **Use the relative pressure to find the temperature and enthalpy of the air at State 2.**

Interpolate Table A-1 in the appendix to find the temperature and enthalpy of the air at State 2:

$T_2 = 560$ K

$h_2 = 566$ kJ/kg

5. **Use the temperature at State 3 ($T_3 = 1,400$ K) to find the enthalpy and relative pressure from Table A-1 in the appendix:**

$h_3 = 1,515$ kJ/kg

$P_{r3} = 361.8$

The third process is isentropic expansion, and you can find the relative pressure at State 4 by using the pressure ratio:

$$P_{r4} = P_{r3} \frac{P_4}{P_3} = (361.8)\left(\frac{1}{10}\right) = 36.18$$

6. **Find the corresponding temperature and enthalpy by interpolating Table A-1 in the appendix:**

$T_4 = 785$ K

$h_4 = 806$ kJ/kg

7. **Use the enthalpy at each state to find the amount of energy exchanged in each process.**

Use the energy equations from the "Constant specific heat method" section, as follows:

$q_{in} = (1,515 - 566)$ kJ/kg $= 949$ kJ/kg

$q_{out} = (806 - 296)$ kJ/kg $= 510$ kJ/kg

$w_{in} = (566 - 296)$ kJ/kg $= 270$ kJ/kg

$w_{out} = (1,515 - 806)$ kJ/kg $= 709$ kJ/kg

8. **Check your calculations to make sure the net heat transfer ($q_{in} - q_{out}$) to the cycle equals the net work out ($w_{out} - w_{in}$) of the cycle:**

$q_{in} - q_{out} = w_{out} - w_{in}$ or $949 - 510 = 709 - 270 = 439$ kJ/kg

9. **Calculate the back work ratio (r_{bw}) of the engine:**

$$r_{bw} = \frac{w_{in}}{w_{out}} = \frac{270 \text{ kJ/kg}}{709 \text{ kJ/kg}} = 38\%$$

Comparing the results of the two analysis methods

The results you get from analyzing the same engine by using constant specific heat and variable specific heat differ in a few important ways, as summarized in Table 10-1.

Table 10-1	Comparing Results of Brayton Cycle Analyses	
State	**Constant c_p**	**Variable c_p**
1	295 K	295 K
2	570 K	560 K
3	1,400 K	1,400 K
4	725 K	785 K
Process	**E kJ/kg**	**E kJ/kg**
q_{in}	834	949
q_{out}	435	510
w_{in}	276	270
w_{out}	675	709

✓ **The difference in temperatures that you see between the two methods of analysis illustrates the amount of error you get when using the constant specific heat method.** The largest error in this example is the turbine exit temperature T_4, because the turbine inlet temperature was fixed at 1,400 Kelvin.

✓ **The amount of work going into the cycle during compression is about the same for both methods.** Because the temperature change is small, you're pretty safe in assuming that specific heat remains constant during compression.

The pressure ratio for this example is only 10:1. For engines with higher pressure ratios, the final compression temperature is higher, which means that assuming constant specific heat becomes less accurate.

✓ **The amount of work output you calculate when you assume constant specific heat is less than the work you calculate when you account for variable specific heat.** This difference occurs because the specific heat increases with temperature, meaning that at high temperatures, it takes more heat to make the same temperature rise compared to low temperatures.

✔ **The amount of heat input for the variable specific heat case is greater than that for the constant specific heat case.** The specific heat increases with temperature, so with a fixed combustion temperature like the one in this example, it takes more energy to get to 1,400 Kelvin with variable specific heat. The error in the heat input is the largest one in the analysis because it has the largest temperature change in the cycle.

✔ **The calculation shows that the amount of heat exhausted when you assume constant specific heat is less than the amount exhausted when you account for variable specific heat.** Although variable specific heat gives you more work output and rejects less heat, it requires more heat input for a fixed combustion temperature. Therefore, the engine efficiency is lower than when you account for the variation in specific heat.

Determining Brayton cycle efficiency

Finding efficiency just means figuring out how much useful work you get out of an engine compared to how much energy the engine uses to get that work done. Because an engine uses part of its work in the compressor section, you have to subtract the work into the engine from the work out of the engine to get the net work (w_{net}) from the engine.

You can calculate the efficiency of any heat engine using the following equation:

$$\eta_{th} = \frac{w_{net}}{q_{in}}$$

For the Brayton cycle analysis using the constant specific heat assumption, the efficiency is as follows:

$$\eta_{th} = \frac{w_{net}}{q_{in}} = \frac{q_{in} - q_{out}}{q_{in}} = \frac{(834 - 435) \text{ kJ/kg}}{834 \text{ kJ/kg}} = 48\%$$

For the Brayton cycle analysis using the variable specific heat assumption, you calculate the efficiency like this:

$$\eta_{th} = \frac{w_{net}}{q_{in}} = \frac{q_{in} - q_{out}}{q_{in}} = \frac{(949 - 510) \text{ kJ/kg}}{949 \text{ kJ/kg}} = 46\%$$

The constant specific heat method overestimates the efficiency of the cycle because it underestimates the amount of heat input to the cycle.

For the constant specific heat assumption, you can also compute the efficiency from the pressure ratio (r_p) and the ratio of specific heats (k) for air by using this equation:

$$\eta_{th} = 1 - r_p^{\frac{(1-k)}{k}} = 1 - \left(\frac{10}{1}\right)^{\frac{(1-1.4)}{1.4}} = 48\%$$

Calculating Brayton cycle irreversibility

When you examine a cycle using the second law of thermodynamics, you can figure out where the inefficiencies come from. Inefficiency arises from friction generated by parts and fluids that are moving. Inefficiency also occurs in heat transfer processes. Friction and heat losses in a cycle cause *irreversibility*. You find more on reversibility and irreversibility in Chapter 9. When you calculate the irreversibility of a process or a cycle, you are quantifying the change in entropy of the system. The following equation (introduced in Chapter 9) is used to calculate the irreversibility per unit mass of any thermodynamic process for an open system with a single inlet and a single exit:

$$i = T_0 \left[(s_{out} - s_{in}) - \frac{q_{in}}{T_{in}} + \frac{q_{out}}{T_{out}} \right]$$

You need to know the following points to use this equation:

- ✔ T_0 is the ambient temperature of the environment that the system operates in. For the Brayton cycle example I discuss in previous sections, you assume it's operating outside, where the temperature is 295 Kelvin.

- ✔ s_{out} is the entropy of the air at the end of a process. You need to know both the temperature and pressure of the air at the end of the process to find the entropy.

- ✔ s_{in} is the entropy of the air at the beginning of a process.

- ✔ q_{in} is the amount of heat transferred from the high-temperature reservoir to the system in the combustion process. q_{out} is the amount of heat transferred to the low-temperature reservoir from the system in the exhaust process.

- ✔ T_{in} is the temperature of the high-temperature thermal energy reservoir that transfers heat to the system. T_{out} is the temperature of the low-temperature thermal energy reservoir that the system rejects heat to. You usually use the maximum temperature of the cycle for the heat input reservoir temperature. The heat rejection reservoir temperature is usually the ambient air temperature.

You can use the preceding equation to determine the irreversibilities associated with the Brayton cycle. Isentropic processes are reversible, so the irreversibility or change in entropy ($s_{out} - s_{in}$) for the compressor and turbine is zero. For the heat transfer processes, you need to calculate the change in entropy ($s_{out} - s_{in}$) with the following equation:

$$s_{out} - s_{in} = s_{out}^0 - s_{in}^0 + R \ln \left(\frac{P_{out}}{P_{in}} \right)$$

As I describe in Chapter 8, the temperature and pressure of a gas determine its entropy. The s_{out}^0 and s_{in}^0 entropy terms are a function of the air temperature at the beginning and end of a process, respectively. You can find these values in Table A-1 in the appendix. The ideal-gas constant R, for air, is 0.287 kilojoule per kilogram-Kelvin. You can find out more about it in Chapter 3.

The following steps enable you to calculate the irreversibilities in the cycle using results from the variable specific heat method:

1. **Find the change in entropy for the combustion process.**

   $$s_3 - s_2 = s_3^0 - s_2^0 = (8.529 - 7.501) \text{ kJ/kg} \cdot \text{K} = 1.028 \text{ kJ/kg} \cdot \text{K}$$

   Because the combustion process occurs at constant pressure, the pressure ratio term in the change in entropy equation drops out. You can find values for s_2^0 and s_3^0 at temperatures T_2 (560 K) and T_3 (1,400 K) by interpolating Table A-1 in the appendix.

2. **Find the change in entropy for the exhaust process.**

   The change in entropy for the exhaust process is equal to that for the combustion process but opposite in sign. The following equation describes the change in entropy for both heat transfer processes:

   $$s_1 - s_4 = -(s_3 - s_2)$$

3. **Compute the irreversibility of the combustion process using absolute temperatures for the ambient temperature and the maximum cycle temperature:**

   $$i_{23} = T_0 \left[(s_3 - s_2) - \frac{q_{in}}{T_H} \right] = (295 \text{ K}) \left(1.028 \frac{\text{kJ}}{\text{kg} \cdot \text{K}} - \frac{949 \text{ kJ/kg}}{1,400 \text{ K}} \right) = 103.3 \text{ kJ/kg}$$

4. **Find the irreversibility of the exhaust process by using the ambient air temperature for the low-temperature reservoir:**

   $$i_{41} = T_0 \left[(s_1 - s_4) + \frac{q_{out}}{T_L} \right] = (295 \text{ K}) \left(-1.028 \frac{\text{kJ}}{\text{kg} \cdot \text{K}} + \frac{510 \text{ kJ/kg}}{295 \text{ K}} \right) = 206.7 \text{ kJ/kg}$$

Theoretically, the irreversibilities become smaller when the heat transfer processes occur at a constant temperature. In that instance, the cycle becomes the Carnot cycle.

Improving the Brayton Cycle with Regeneration

The exhaust in gas turbine engines often has a significant amount of thermal energy remaining after it has fully expanded to atmospheric pressure. Some of this energy can be used to preheat the air leaving the compressor before it enters the combustion chamber. Preheating the air reduces the amount of fuel that needs to be burned in the combustion chamber. A heat exchanger called a *regenerator* takes waste heat from the exhaust and preheats the combustion air.

Figure 10-6 shows the Brayton cycle modified with a regenerator. Heat from the exhaust enters the regenerator between States 4 and 6 in the figure, and the regenerator transfers heat to the combustion air supply between States 2 and 5.

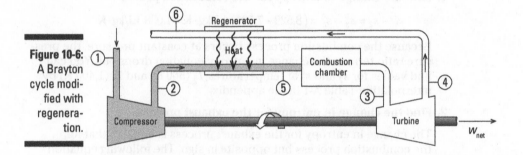

Figure 10-6:
A Brayton cycle modified with regeneration.

Figure 10-7 shows the regenerator heat transfer process on the *T-s* diagram of the Brayton cycle. The maximum possible temperature for State 5 is equal to the temperature at State 4 and is labeled 5'. Some heat losses normally occur in real systems, so State 5 is at a lower temperature than 5'. A regenerator is most effective when the difference between the compressor outlet temperature at State 2 and the turbine outlet temperature at State 4 is significant. These two temperatures are far apart for engines with a pressure ratio of less than 15. When the pressure ratio exceeds 15, the compressor outlet temperature (State 2) is very high and may actually be higher than the turbine exhaust temperature (State 4). A regenerator isn't beneficial under those conditions.

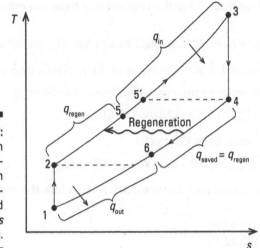

Figure 10-7:
The Brayton
engine pro-
cesses with
regenera-
tion mapped
onto a *T-s*
diagram.

The amount of heat transferred by the regenerator is determined by
this equation:

$$q_{regen,act} = h_5 - h_2 = h_4 - h_6$$

The maximum possible amount of heat transfer is calculated like this

$$q_{regen,max} = h_4 - h_2$$

The *effectiveness* (ε) of a regenerator compares the actual amount of heat
transferred in the regenerator to the maximum possible amount and is
calculated with this equation:

$$\varepsilon = \frac{h_5 - h_2}{h_4 - h_2}$$

If you add a regenerator with an effectiveness of 90 percent to the Brayton
cycle example from the previous section, you can see how much the efficiency of the cycle improves. Follow these steps in addition to the ones
in the previous "Variable specific heat method" section on analyzing the
Brayton cycle:

1. **Find the enthalpy at State 5 for the regenerator from the effectiveness equations:**

$$h_5 = \varepsilon(h_4 - h_2) + h_2 = 0.9(806 - 566) \text{ kJ/kg} + 566 \text{ kJ/kg} = 782 \text{ kJ/kg}$$

The enthalpy at States 1–4 is the same as in the previous example.

2. **Calculate the heat input to the cycle from State 5 to State 3:**

$$q_{in} = h_3 - h_5 = (1,515 - 782) \text{ kJ/kg} = 733 \text{ kJ/kg}$$

3. **Calculate the net work from the cycle:**

$$w_{net} = w_{turb} + w_{comp} = (709 - 270) \text{ kJ/kg} = 439 \text{ kJ/kg}$$

4. **Substitute the heat input and the net work results into the efficiency equation:**

$$\eta_{th} = \frac{w_{net}}{q_{in}} = \frac{439 \text{ kJ/kg}}{733 \text{ kJ/kg}} = 60\%$$

Adding a regenerator makes quite an improvement over the efficiency of the simple Brayton cycle engine. This addition brings the system closer to the performance of a Carnot cycle.

Adding Intercooling and Reheating to the Brayton Cycle

The amount of work that can be extracted from a gas turbine engine can be increased by cooling the gas during the compression process and heating the gas during the expansion process. Adding several nearly isentropic processes in steps makes the Brayton cycle operate more like a Carnot cycle, which has isothermal heat input and heat rejection. Heating or cooling air in a turbine or a compressor isn't very effective, so turbines and compressors are broken up into multiple sections called *stages*. Heat exchangers heat or cool the air between the stages.

Looking at how intercooling and reheating affect the Brayton cycle

Intercooling is the process whereby a heat exchanger is used to remove heat between stages of compression. For two-stage compression, you have low- and high-pressure compressors. *Reheating* is the process whereby additional fuel is burned in combustion chambers between different turbine stages. For two-stage expansion, you have high- and low-pressure turbines. Figure 10-8 shows the Brayton cycle with regeneration, modified with two stages of compression and expansion with intercooling and reheating.

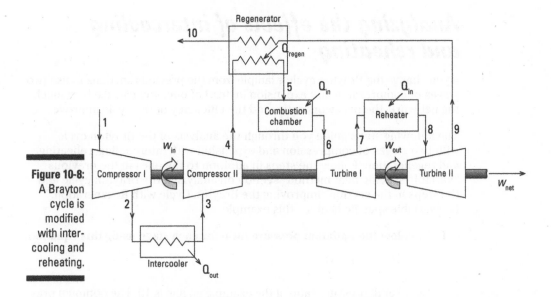

Figure 10-8:
A Brayton
cycle is
modified
with inter-
cooling and
reheating.

Figure 10-9 shows the modified cycle mapped onto a *T-s* diagram, where ten
thermodynamic states now need to be analyzed for the cycle.

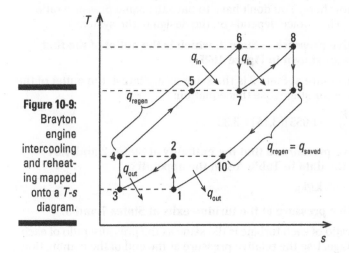

Figure 10-9:
Brayton
engine
intercooling
and reheat-
ing mapped
onto a *T-s*
diagram.

An additional benefit of intercooling and reheating is that the pressure ratio
of each stage is relatively small compared to the pressure ratio of a single-
stage system. Thus, the difference between the turbine exhaust temperature
and the compressor outlet temperature is significant. A regenerator can capture the waste heat from the exhaust and preheat the air entering the first-
stage combustion chamber.

Analyzing the effects of intercooling and reheating

If you change the Brayton cycle example from the previous sections to use two stages of compression and expansion instead of one, you can see how much the net work, the backwork ratio, and the efficiency of the cycle improve.

The following steps guide you through the analysis of the Brayton cycle with two stages of compression and expansion with intercooling, reheating, and regeneration. Follow the steps in addition to the ones in the previous "Variable specific heat method" section on analyzing the Brayton cycle and the steps in the section "Improving the Brayton Cycle with Regeneration." Use variable specific heat for this example.

1. **Calculate the optimum pressure ratio for each stage using this equation:**

$$\frac{P_2}{P_1} = \sqrt{r_p} = \sqrt{10} = 3.16$$

The overall pressure ratio of the example engine is 10. The optimum pressure ratio for each stage is the square root of the overall pressure ratio. The optimum pressure ratio calculated as shown here provides for the minimum work input to the compressors and the maximum work output of the turbines. (The pressure ratio between stages can be split up differently than I show here. You don't have to have the same pressure ratio for each stage. The choice depends on the design of the system.)

2. **Find the relative pressure at State 2, which is the outlet of the first-stage compressor shown in Figure 10-9.**

The relative pressure at State 2 is the same as at State 4, the outlet of the second compressor stage. Use this equation:

$$P_{r2} = P_{r1}\frac{P_2}{P_1} = (1.055)(3.16) = 3.33$$

3. **Use the relative pressure to find the enthalpy at States 2 and 4 by interpolating the data in Table A-1 in the appendix:**

$$h_2 = h_4 = 409 \text{ kJ/kg}$$

4. **Find the relative pressure at the turbine exits at States 7 and 9.**

The pressure ratio of each turbine is the same as the pressure ratio of each compressor stage. Use the relative pressure at the end of the combustion process P_{r3} from the Brayton cycle example in the "Variable specific heat method" section on analyzing the Brayton cycle. The equation looks like this:

$$P_{r7} = P_{r6}\frac{P_7}{P_6} = (361.8)\left(\frac{1}{3.16}\right) = 114.5$$

5. **Use the relative pressure to find the enthalpy at States 7 and 9 from Table A-1 in the appendix, using interpolation:**

$$h_7 = h_9 = 1,108 \text{ kJ/kg}$$

6. **Calculate the total compressor work.**

   Because both compressors are identical, the work is twice that of a single stage. Here's the equation:

   $$w_{comp} = 2(h_2 - h_1) = 2(409 - 296) \text{ kJ/kg} = 226 \text{ kJ/kg}$$

7. **Calculate the turbine work:**

   $$w_{turb} = 2(h_6 - h_7) = 2(1{,}515 - 1{,}108) \text{ kJ/kg} = 814 \text{ kJ/kg}$$

8. **Find the heat input.**

   First determine the amount of energy recovered from the exhaust in the regenerator. The effectiveness of the regenerator is calculated by this equation, using the states shown in Figure 10-9:

   $$\varepsilon = \frac{h_5 - h_2}{h_9 - h_2}$$

9. **Rearrange the equation to solve for the enthalpy at State 5:**

   $$h_5 = \varepsilon(h_9 - h_2) + h_2 = 0.9(1{,}108 - 409) + 409 \text{ kJ/kg} = 1{,}038 \text{ kJ/kg}$$

10. **Calculate the total heat input to the cycle.**

    The total heat input to the engine combines the primary heating prior to the first-stage turbine and the reheating between the stages. Note the heat input to both sections is not identical. The calculations look like this:

    $$q_{in} = q_{prime} + q_{reheat} = (h_6 - h_5) + (h_8 - h_7)$$
    $$q_{in} = (1{,}515 - 1{,}038) + (1{,}515 - 1{,}108) \text{ kJ/kg} = 884 \text{ kJ/kg}$$

11. **Use the net work and the total heat input to calculate the efficiency of the cycle:**

    $$\eta_{th} = \frac{w_{net}}{q_{in}} = \frac{w_{turb} - w_{comp}}{q_{in}} = \frac{(814 - 226) \text{ kJ/kg}}{884 \text{ kJ/kg}} = 67\%$$

12. **Calculate the backwork ratio using the compressor work and the turbine work:**

    $$r_{bw} = \frac{w_{comp}}{w_{turb}} = \frac{226 \text{ kJ/kg}}{814 \text{ kJ/kg}} = 28\%$$

The efficiency often increases by using intercooling between the compressor stages and reheating between the turbines. If you add additional stages of compression, intercooling, expansion, and reheating to the T-s diagram shown in Figure 10-9, the cycle begins to approximate the Carnot cycle T-s diagram shown in Figure 10-2. However, building more than two or three stages is usually too expensive. You can see that the backwork ratio also decreases as a result of the intercooling and reheating. The gain in efficiency is enhanced by using a regenerator.

Deviating from Ideal Behavior: Actual Brayton Cycle Performance

As marvelous as gas turbine engines are, they deviate somewhat from the ideal Brayton cycle assumptions. Friction occurs within the compressor and the turbine, thereby generating irreversibilities. The air flowing through the engine also has friction, which causes pressure drops within the flow passages of the engine. As a result, a real compressor requires more work input than an ideal compressor, and a real turbine puts out less work than an ideal turbine. The departure from ideal behavior of a turbine or a compressor is quantified by the isentropic efficiency. You calculate the efficiency of a turbine by dividing the actual turbine work by the ideal turbine work, as shown in this equation:

$$\eta_T = \frac{w_{actual}}{w_{ideal}} = \frac{h_3 - h_{4a}}{h_3 - h_{4s}}$$

For the efficiency of a compressor, you divide the work of an ideal compressor by the work of the actual compressor, as shown in this equation:

$$\eta_C = \frac{w_{ideal}}{w_{actual}} = \frac{h_{2s} - h_1}{h_{2a} - h_1}$$

Figure 10-10 shows the how the actual gas turbine cycle deviates from the ideal Brayton cycle. The states labeled 2a and 4a are the actual states at the compressor and turbine exits, respectively. The states labeled 2s and 4s are the exit states for the isentropic compressor and turbine.

You can analyze a Brayton cycle using isentropic efficiency to model a real gas turbine engine. Suppose the basic Brayton cycle engine I discuss in the section "Analyzing the Brayton cycle using two methods" has a compressor with an isentropic efficiency of 88 percent and a turbine with an isentropic efficiency of 92 percent. You can analyze the cycle and determine its efficiency by following these steps:

1. **Using results from the variable specific heat method, determine the actual compressor outlet enthalpy at State 2a with the isentropic efficiency equation:**

   $$h_{2a} = \frac{(h_{2s} - h_1)}{\eta_C} + h_1 = \frac{(566 - 296)\,\text{kJ/kg}}{0.88} + 296 \ \text{kJ/kg} = 603 \ \text{kJ/kg}$$

2. **Find the enthalpy of the actual turbine exit at State 4a:**

   $$h_{4a} = h_3 - \eta_T(h_3 - h_{4s}) = 1{,}515 - 0.92(1{,}515 - 806) \ \text{kJ/kg} = 863 \ \text{kJ/kg}$$

3. **Calculate the actual heat input to the engine using the actual enthalpy values:**

$$q_{in} = (h_3 - h_{2a}) = (1{,}515 - 603)\,\text{kJ/kg} = 912\ \text{kJ/kg}$$

4. **Calculate the actual heat rejection to the environment with the actual enthalpy values:**

$$q_{out} = (h_{4a} - h_1) = (863 - 296)\,\text{kJ/kg} = 567\ \text{kJ/kg}$$

5. **Use the heat input and heat output to determine the cycle efficiency:**

$$\eta_{th} = \frac{w_{net}}{q_{in}} = \frac{(912 - 567)\ \text{kJ/kg}}{912\ \text{kJ/kg}} = 38\%$$

The efficiency of the actual cycle is significantly less than the ideal Brayton cycle efficiency that's calculated in the section "Analyzing the Brayton cycle using two methods."

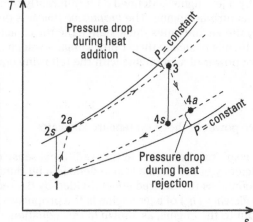

Figure 10-10:
An actual Brayton cycle mapped onto a *T-s* diagram.

Flying the Brayton Cycle in Jet Propulsion

While the first-class cabin passengers are sipping on champagne mimosas, you're stuck in the back of the plane where your view of the ground is blocked by the wing and a large jet engine, and you have nothing to do but wonder what's going on inside that jet engine. How can something that looks so sleek make so much power that it can propel a passenger plane at a speed of around 550 miles per hour or a fighter jet at a speed of up to 1,500 miles per hour or more? The modern jet engine is a marvel of engineering. In this section, I show you how to perform a thermodynamic analysis of this engine using the Brayton cycle.

The jet engine is a gas turbine engine specifically made for developing thrust instead of power. *Thrust* is a force that's produced by the difference in momentum of the air entering and leaving the engine. The air entering the engine has a much lower velocity and, hence, lower momentum than the air leaving the engine. Newton's second law of motion is used to determine the thrust force (F) produced by momentum, as shown in this equation:

$$F = \dot{m}\left(\mathbf{V}_{exit} - \mathbf{V}_{inlet}\right)$$

The momentum is a product of the mass flow rate ($\dot{m}$) of the air flowing through the engine and the difference between the inlet and exit velocities of the air ($\mathbf{V}_{exit} - \mathbf{V}_{inlet}$). The units of force in the SI system are newtons (N).

The power developed by a jet engine is defined a bit differently than the power of an ordinary gas turbine engine. The *propulsive power* is defined by the thrust produced by the engine times the distance the thrust acts on the engine per unit time, which is basically thrust times engine velocity. You can calculate the propulsive power of a jet engine with the following equation:

$$\dot{W}_P = F\mathbf{V}_{engine}$$

The units for propulsive power in the SI system are kilowatts.

Because a jet engine doesn't produce work in the traditional sense of a gas turbine engine, the efficiency of the engine is also defined a bit differently. The efficiency is still defined as the desired effect divided by the required input. The *propulsive efficiency* (η_p) of a jet engine is the propulsive power divided by the heat input to the engine, as shown in this equation:

$$\eta_p = \frac{\text{Propulsive power}}{\text{Heat input rate}} = \frac{\dot{W}_P}{\dot{Q}_{in}}$$

Seeing what happens in an ideal turbojet cycle

Because a jet engine is typically moving, you need to establish an inertial frame of reference for the engine in your analysis. If you fix the reference frame on the engine itself, the moving engine appears to be stationary to you as an observer. Then the air entering the engine appears to move at the relative velocity between the engine and the air.

Here's what happens in an ideal turbojet cycle. Figure 10-11 shows a diagram of the major components of a turbojet engine.

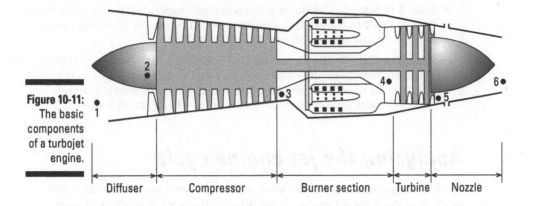

Figure 10-11: The basic components of a turbojet engine.

Diffuser   Compressor   Burner section   Turbine   Nozzle

Figure 10-12 shows the ideal turbojet cycle on a *T-s* diagram, with six states in the cycle.

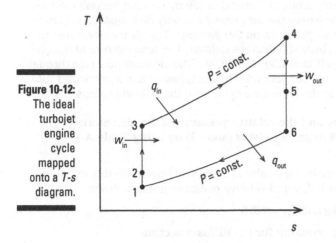

Figure 10-12: The ideal turbojet engine cycle mapped onto a *T-s* diagram.

✔ **State 1 to State 2:** When the jet engine is in flight, the incoming air has a significant amount of kinetic energy relative to the moving engine. The air slows down in the diffuser section and is compressed. The kinetic energy is converted to internal energy plus flow work so the enthalpy of the air increases isentropically in the diffuser from States 1 to 2.

✔ **State 2 to State 3:** The compressor section isentropically increases the pressure of the air from States 2 to 3 and usually has a pressure ratio ranging from 10 to 25. Military jet engines typically operate at the high end of the range.

✔ **State 3 to State 4:** Fuel is mixed with the air and burned in the combustion chamber, increasing the enthalpy of the air in an isobaric process from States 3 to 4.

✔ **State 4 to State 5:** The turbine isentropically expands the combustion air from States 4 to 5 and produces just enough power for the compressor, a small electric generator, and hydraulic pumps.

✔ **State 5 to State 6:** The air leaves the turbine at a high pressure (relative to the ambient pressure) and is accelerated to a high velocity in the nozzle in an isentropic process. The nozzle converts internal energy plus flow work (enthalpy) into kinetic energy from States 5 to 6.

Analyzing the jet engine cycle

The following example gives you an opportunity to perform a thermodynamic analysis of the turbojet engine. In addition to the normal analysis of determining the work of the compressor and turbine and the heat input, you can find the thrust, the propulsive power, and the propulsive efficiency of a turbojet engine.

Say your flight is cruising along at a speed of 250 meters per second (560 miles per hour) at an altitude where the air pressure is only 30 kilopascals. (That's about one third of the air pressure on the ground). The air temperature outside your window is a chilly –50 degrees Celsius. The temperature of the air entering the turbine section is 1,300 Kelvin, and the air mass flow rate through the engine is 60 kilograms per second. The compressor has a pressure ratio of 15:1. You can analyze the jet engine cycle with the following steps:

1. **Find the enthalpy and the relative pressure of the air entering the engine at –50 degrees Celsius (State 1) by using Table A-1 in the appendix.**

   You need to interpolate the table. I show you how to do this in Chapter 3. The enthalpy and relative pressure are as follows:

   $$h_1 = 223 \text{ kJ/kg and } P_{r1} = 0.396.$$

2. **Write the energy equation for the diffuser section.**

   The energy equation contains enthalpy and kinetic energy terms because the air has velocity associated with it from the moving engine. The equation looks like this:

   $$q_{12} - w_{12} = h_2 - h_1 + \frac{1}{2}\left(V_2^2 - V_1^2\right)$$

3. **Simplify the energy equation.**

   No heat transfer or work occurs in the diffuser, so the energy equation is a relationship only between enthalpy and kinetic energy. The velocity at State 1 is the speed of the engine relative to the wind and gives the kinetic energy at State 1. At State 2, the velocity of the air is zero because the air comes to rest at the end of the diffuser section and the V_2 term goes away. Here's the equation:

$$h_2 = h_1 + \frac{1}{2}\left(V_1^2\right)$$

4. **Calculate the enthalpy at State 2 with this equation:**

$$h_2 = 223 \text{ kJ/kg} + \frac{1}{2}\left(250 \text{ m/s}\right)^2\left(1 \text{ kJ/1,000 J}\right) = 254 \text{ kJ/kg}$$

5. **Find the temperature and the relative pressure of the air at State 2 from the enthalpy, by interpolating Table A-1 in the appendix.**

   You find that $T_2 = -19°C$ and $P_{r2} = 0.628$.

   You can see the air temperature rises quite a bit just in the diffuser section due to the conversion of kinetic energy into enthalpy.

6. **Find the relative pressure at State 3 using the relative pressure at State 2 and the pressure ratio of the compressor:**

$$\frac{P_{r3}}{P_{r2}} = \frac{P_3}{P_2} \text{ or } P_{r3} = 0.628(15) = 9.42$$

7. **Use the relative pressure at State 3 to find the enthalpy at the compressor exit by interpolating Table A-1 in the appendix.**

   You find that $h_3 = 547$ kJ/kg.

8. **Compute the compressor work:**

$$w_{comp} = h_3 - h_2 = \left(547 - 254\right) \text{ kJ/kg} = 293 \text{ kJ/kg}$$

9. **Find the enthalpy and relative pressure at State 4 from the combustion air temperature of 1,300 K in Table A-1 in the appendix.**

   You find that $h_4 = 1,396$ kJ/kg and $P_{r4} = 265.8$.

10. **Calculate the heat input to the engine:**

$$q_{in} = h_4 - h_3 = \left(1,396 - 547\right) \text{ kJ/kg} = 849 \text{ kJ/kg.}$$

11. **Calculate the enthalpy at the turbine exit State 5. In the ideal engine, the turbine work is equal to the compressor work.**

   The calculation looks like this:
   $$w_{turb} = w_{comp} = h_4 - h_5$$

   or

   $$h_5 = \left(1,396 - 293\right) \text{ kJ/kg} = 1,103 \text{ kJ/kg}$$

12. **Find the relative pressure at the turbine exit State 5 by interpolating Table A-1 in the appendix.**

   You find that $P_{r5} = 112.8$.

13. **Calculate the pressure at State 5:**

$$\frac{P_5}{P_4} = \frac{P_{r5}}{P_{r4}} \text{ or } P_5 = P_4 \frac{P_{r5}}{P_{r4}} \text{ where } P_4 = P_3 = P_1 \frac{P_{r3}}{P_{r1}} \text{ so } P_5 = P_1\left(\frac{P_{r3}}{P_{r1}}\right)\left(\frac{P_{r5}}{P_{r4}}\right)$$

Then $P_5 = 30 \text{ kPa}\left(\frac{9.42}{0.396}\right)\left(\frac{112.8}{265.8}\right) = 303 \text{ kPa}$

14. **Find the relative pressure at State 6, the nozzle exit:**

$$\frac{P_{r6}}{P_{r5}} = \frac{P_6}{P_5} \text{ or } P_{r6} = 112.8\left(\frac{30 \text{ kPa}}{303 \text{ kPa}}\right) = 11.2$$

15. **Find the enthalpy at the nozzle exit State 6 using the relative pressure by interpolating Table A-1 in the appendix.**

You find that $h_6 = 576 \text{ kJ/kg}$.

16. **Write the energy equation for the nozzle to find the velocity of the air leaving the engine.**

The velocity of the air at the turbine exit is assumed to be approximately zero.

$$h_6 + \frac{1}{2}\mathbf{V}_6^2 = h_5 + \frac{1}{2}\mathbf{V}_5^2 \text{ or } \mathbf{V}_6 = \sqrt{2(h_5 - h_6)}$$

17. **Compute the velocity at the nozzle exit:**

$$V_6 = \sqrt{2(1{,}103 - 576) \text{ kJ/kg}(1{,}000 \text{ J/kJ})} = 1{,}027 \text{ m/s}$$

18. **Calculate the total heat rejected by the cycle:**

$$q_{out} = h_6 - h_1 = (576 - 223) \text{ kJ/kg} = 353 \text{ kJ/kg}$$

19. **Calculate the thrust produced by the engine using Newton's second law, with the mass flow rate of the air through the engine and the difference in air velocity entering and leaving the engine:**

$$F = \dot{m}(\mathbf{V}_{exit} - \mathbf{V}_{inlet}) = (60 \text{ kg/s})(1{,}027 - 250) \text{ m/s} = 46{,}620 \text{ N}$$

20. **Determine the propulsive power of the engine by multiplying the engine thrust (F) by the engine velocity (V$_{engine}$):**

$$\dot{W}_P = F\mathbf{V}_{engine} = (46.62 \text{ kN})(250 \text{ m/s}) = 11{,}660 \text{ kW}$$

21. **Calculate the propulsive efficiency of the engine, which is the amount of propulsive power out of the engine divided by the heat input to the engine:**

$$\eta_p = \frac{\dot{W}_P}{\dot{Q}_{in}} = \frac{\dot{W}_P}{\dot{m}q_{in}} = \frac{11{,}660 \text{ kW}}{(60 \text{ kg/s})(849 \text{ kJ/kg})} = 23\%$$

You can see that quite a bit of energy isn't used by the engine, and the waste heat is merely dumped into the air.

Chapter 11

Working with Otto and Diesel Cycles

---

*I*n simplest terms, you put fuel into your car, and then you use it to get where you want to go. How it does what it does and how efficiently it does it probably doesn't cross your mind while you're tooling around on a Sunday drive. But you're a budding thermodynamicist now; you're prepared to look much deeper, and this chapter is where you find out how much work the engine in your car or truck or boat gets out of the fuel it burns. I take you through the cycles that gasoline- and diesel-burning engines use, showing you some calculations that help you find out what you're getting out of your engine. You won't be able to fix your car after reading this chapter, but at least you'll have a better understanding of the ins and outs of your engine.

Understanding the Basics of Reciprocating Engines

Most cars, as well as many other land and water vehicles, have a reciprocating engine packed under the hood. A *reciprocating engine* is one that has pistons moving back and forth inside cylinders. I know that when you open the hood, what you see looks complicated — and it is. But four basic processes make these vehicles move. (As I tell you in Chapter 10, every heat engine operates with only four basic processes: work in, heat in, work out, heat out.) Whether the reciprocating engine is in a garden tractor, a top fuel dragster, a pickup truck, a train, or a boat, the same four processes produce useful power.

Reciprocating engines come in *two-stroke* versions or *four-stroke* versions. Figure 11-1 shows you how the four fundamental thermodynamic processes of a heat engine are accomplished in a four-stroke engine. I tell you more about the two-stroke variety in Chapter 18. A stroke means the piston moves from one end of the cylinder to the other end of the cylinder.

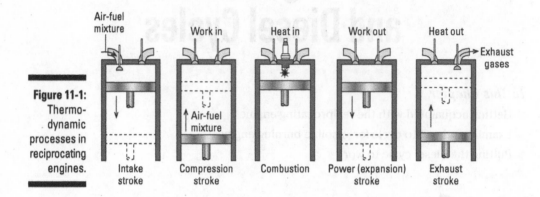

Figure 11-1: Thermo-dynamic processes in reciprocating engines.

Reciprocating engines are further divided into different kinds of thermodynamic cycles. In this chapter I discuss the Otto cycle and the diesel cycle heat engines. Many features of both cycles are similar. I spell out the differences between them as I describe how reciprocating engines work. Figure 11-2 shows the pressure-volume (*P-v*) diagram of the processes in an actual Otto cycle engine.

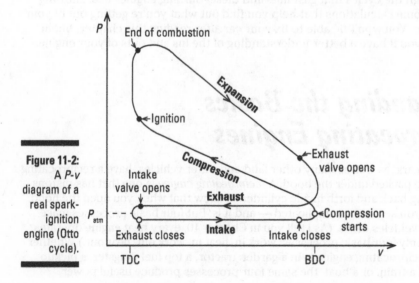

Figure 11-2: A *P-v* diagram of a real spark-ignition engine (Otto cycle).

Figure 11-3 shows the *P-v* and temperature-entropy (*T-s*) diagrams for the ideal Otto cycle.

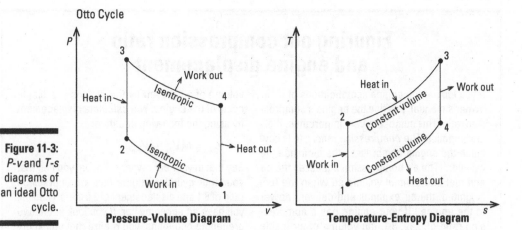

Figure 11-3:
P-v and *T-s* diagrams of an ideal Otto cycle.

Here are the steps involved in a four-stroke reciprocating engine, as shown in Figures 11-1, 11-2, and 11-3:

1. **Fresh air is drawn into the cylinder as the piston moves from top dead center (TDC) down to bottom dead center (BDC) in the cylinder.**

   Figure 11-2 shows that the pressure inside the cylinder is slightly below atmospheric pressure, which allows the air to fill the cylinder. In this figure, the maximum and minimum cylinder volumes are at BDC and TDC, respectively. A small amount of work is involved in this process, but it's usually ignored in a typical thermodynamic analysis of the engine. This process isn't part of the ideal Otto cycle and isn't shown in Figure 11-3. The ideal Otto cycle assumes the same air is used over and over and doesn't need to refresh the air between cycles.

2. **The piston compresses the air inside the cylinder from BDC to TDC, so work goes into the air. (Work in.)**

   The source of work for compressing the air comes from the work output from another piston that's in its power stroke, unless the source is a single-cylinder engine. For a single-cylinder engine, the compression work input comes from the energy stored in the momentum of a flywheel connected to the engine.

   In Figure 11-3, the lines from States 1 to 2 on the diagrams show this process. The compression process in the ideal engine is *reversible* (frictionless) and *adiabatic* (transfers no heat to or from the cylinder). Because the compression is reversible and adiabatic, the process is *isentropic*, meaning that entropy is constant, as shown by the vertical line from States 1 to 2 on the *T-s* diagram.

3. **Fuel is burned in the cylinder, putting heat into the air. (Heat in.)**

Figuring out compression ratio and engine displacement

If you look at the engine specifications of a car owner's manual, you find the engine's compression ratio. This number is very important in thermodynamics. The *compression ratio, r,* tells how much the engine compresses the air inside the cylinder. (The air compression squeezes the air and fuel into a small volume so when the fuel is ignited, the air expands and delivers power to the engine.) If your engine has a compression ratio of, say, 8:1, the volume of air inside the cylinder when the piston is at the bottom is 8 times larger than the volume when the piston is at the top of the cylinder. These two volumes are defined as V_{max} and V_{min}. You calculate the compression ratio by using the following equation:

$$r = \frac{V_{max}}{V_{min}}$$

You use these two cylinder volumes, along with the number of cylinders in the engine, *N*, to calculate the engine size or *displacement*—the

volume of air moved by the pistons in a single cycle of the engine. You calculate displacement by using the following equation:

$$V_{Disp} = N(V_{max} - V_{min})$$

Say that in your car owner's manual you find that your four-cylinder engine has a compression ratio of 8:1 and displacement of 2.0 liters. I can tell you have an economy car. If you look at the two preceding equations, you realize that you know everything except for V_{max} and V_{min}. With these two equations, you can find the values for the two unknowns — V_{max} and V_{min} — for each cylinder by using basic algebra:

$$8 = \frac{V_{max}}{V_{min}}$$

$$(2.0 \text{ liters}) = (4 \text{ cyl})(V_{max} - V_{min})$$

So, after doing a little math, you can find that $V_{max} = 0.571$ liter and $V_{min} = 0.0714$ liter.

In an Otto cycle engine, a spark initiates the combustion process, which is treated as a constant-volume process, as shown in Figure 11-2. The combustion process is completed so quickly the piston doesn't move very much. In a diesel cycle engine, the fuel spontaneously combusts because the air is compressed so much that the temperature is above the fuel flash temperature.

The addition of heat to the air causes the temperature, pressure, and entropy to increase, as shown in the lines from States 2 to 3 in the *P-v* and *T-s* diagrams of Figure 11-3.

4. **The hot gas expands and pushes the piston down from TDC to BDC. (Work out.)**

In the Otto cycle, the combustion process is completed before the piston moves down, as shown in Figure 11-2. In the Diesel cycle, the combustion process continues during a portion of the piston stroke, thereby doing work at the same time.

As the gas expands, the volume increases while both the pressure and the temperature decrease. This process is shown by the line from States 3 to 4 in Figure 11-3. Notice that this line is vertical in the *T-s* diagram. The expansion process is isentropic just like the compression process.

5. **Exhaust is pushed out of the cylinder from nearly BDC to TDC, and the excess heat leaves with it. (Heat out.)**

A small amount of work is used in this process, but, like the work involved in the intake stroke, it's usually disregarded in a typical thermodynamic analysis.

Even though the piston obviously moves, the exhaust process is modeled as a constant-volume process in the ideal Otto cycle, because the piston is at the bottom of the cylinder at the beginning of the exhaust process, and it's back at the bottom at the start of the compression process. Thus, the cylinder volume at the start of the exhaust process is the same as the volume at the start of compression, which makes it appear to be a constant-volume process from States 4 to 1 in the ideal Otto cycle engine shown in Figure 11-3.

Working with the Ideal Spark Ignition Engine: The Otto Cycle

In 1876 a German fellow by the name of Nikolaus Otto built the first four-stroke spark ignition reciprocating engine, and the *Otto cycle* was born. If you can come up with a new kind of engine, you may get it named after yourself and be famous one day.

The Otto cycle is an idealized model of a spark-ignition engine, where the combustion and exhaust processes are modeled as heat transfer processes connected to thermal energy reservoirs. The Otto cycle is a *closed system* because mass neither enters nor leaves the cycle during the heat transfer or work processes.

When you analyze the Otto cycle with the first law of thermodynamics, you find the amount of heat transferred into and out of the air during the cycle and the amount of work put into and produced by the air during the cycle. Ultimately, the analysis leads to finding out how efficient an engine is at converting energy from the fuel into moving the car. The following sections give you a detailed overview of how you can analyze the cycle using the first law of thermodynamics and then determine its efficiency and irreversibility using the second law of thermodynamics.

Analyzing the Otto cycle

You can analyze an Otto cycle with the first law of thermodynamics using either of two methods in this section to determine the amount of work and heat transfer that occurs in each process.

- ✔ **The constant specific heat method:** For this method, you assume the specific heat of air remains constant in each process in the cycle. The analysis isn't as exact as the variable specific heat method, but you don't need to look up thermodynamic properties in a table.

- ✔ **The variable specific heat method**. This method takes into consideration that specific heat changes with temperature and uses values from Table A-1 in the appendix. This method is more accurate but finding the values can require a bit of tedious interpolation of the tables. Many textbooks come with a computer program that enables you to easily determine thermodynamic properties accurately (without tedious interpolation).

Constant specific heat method: Otto cycle

There are a couple ways to analyze an Otto cycle using the constant specific heat assumption. The easiest way uses what's known as the *cold-air standard assumption*. This assumption means that you use the specific heat of air at 25 degrees Celsius throughout all processes in the cycle, even though the air temperature changes during the cycle. The constant volume specific heat of air equals 0.719 kilojoule per kilogram-Kelvin at 25 degrees Celsius. This method makes analysis of the cycle fairly quick, but it isn't quite as accurate as other methods because the specific heat of air changes with temperature.

You can improve the accuracy of the constant specific heat method by using the constant volume specific heat at the average temperature of each process in the cycle. Doing this is especially important with high-temperature processes, such as a combustion process. One drawback to this approach is that if you don't know the air temperature at the initial and final states of a process, you have to make a reasonable estimate for the average process temperature. You can often make this estimate by using specific heat of air at 25 degrees Celsius and computing the unknown temperature of a process. Then you can use this temperature value to adjust your estimate for the specific heat by taking the average temperature of the process.

Here's an example of an Otto cycle analysis using the first law of thermodynamics. In this analysis, I show you how to compute the heat rejected by the engine, the work input, and the work output of the engine, using the following specifications:

- ✔ The engine has a compression ratio (r) of 8.

- ✔ The air temperature entering the engine is 20 degrees Celsius.

- ✔ The fuel combustion process provides a heat input (q_{in}) to the engine of 1,310 kJ/kg of air.

The energy equations use constant volume specific heat (c_v) in place of the change in internal energy for each heat transfer and work process. Chapter 5 gives more details on the energy analysis of closed systems — there, I show that for an ideal gas, the change in internal energy (u) is related to the change in temperature (T), using the constant volume specific heat (c_v). The following energy equations apply to each of the four processes in the Otto cycle heat engine:

✔ **For combustion, heat in:** $q_{in} = (u_3 - u_2) = c_v(T_3 - T_2)$

✔ **For exhaust, heat out:** $q_{out} = (u_4 - u_1) = c_v(T_4 - T_1)$

✔ **For compression, work in:** $w_{in} = (u_2 - u_1) = c_v(T_2 - T_1)$

✔ **For expansion, work out:** $w_{out} = (u_3 - u_4) = c_v(T_3 - T_4)$

You need to find the temperature at the endpoint of each process to determine the amount of energy associated with the processes. The ambient air temperature is used as State 1 at the beginning of the compression process. In the isentropic compression process, the temperature at State 2 can be found from State 1 using an isentropic process equation (see Chapter 5), as shown here:

$$T_2 = T_1\left(\frac{V_1}{V_2}\right)^{k-1} = T_1 \, r^{k-1}$$

Notice the fraction (V_1/V_2) in this equation is simply the compression ratio of the engine. The variable k is the ratio of specific heats, which has a value of $k = 1.4$ for air at 25 degrees Celsius.

Follow these steps to help you analyze the Otto cycle engine:

1. **Convert your temperature from Celsius to Kelvin.**

   Chapter 2 can help. Doing so here makes $T_1 = 20°C = 293$ K.

2. **Find the temperature at the end of the compression stroke (T_2), as follows:**

   $$T_2 = 293 \text{ K}(8)^{1.4-1} = 673 \text{ K}$$

   When working with ideal-gas law equations, you should always use absolute temperatures.

3. **To find the temperature at State 3, rearrange the energy equation for q_{in} from earlier in this section.**

   $$T_3 = \frac{q_{in}}{c_v} + T_2 = \frac{1,310 \text{ kJ/kg}}{0.719 \text{ kJ/kg}\cdot\text{K}} + 673 \text{ K} = 2,495 \text{ K}$$

4. **Find the temperature for State 4, using the isentropic process equation for the expansion process.**

   $$T_4 = T_3\left(\frac{V_3}{V_4}\right)^{k-1} = 2,495 \text{ K}(1/8)^{1.4-1} = 1,086 \text{ K}$$

Finding the temperature at State 4 is just like finding the temperature at State 2, but instead of isentropic compression, you have isentropic expansion from State 3 to State 4.

5. **Now that you know the temperature at each state, find the amount of energy exchanged in each process, using the energy equations earlier in this section.**

$$q_{out} = 0.719 \text{ kJ/kg}(1{,}086 \text{ K} - 293 \text{ K}) = 570 \text{ kJ/kg}$$

$$w_{in} = 0.719 \text{ kJ/kg}(673 \text{ K} - 293 \text{ K}) = 273 \text{ kJ/kg}$$

$$w_{out} = 0.719 \text{ kJ/kg}(2{,}495 \text{ K} - 1{,}086 \text{ K}) = 1{,}013 \text{ kJ/kg}$$

6. **Check your calculations to make sure the net heat transfer ($q_{in} - q_{out}$) to the cycle equals the net work out ($w_{out} - w_{in}$) of the cycle.**

$$q_{in} - q_{out} = w_{out} - w_{in}$$

$$(1{,}310 - 570) \text{ kJ/kg} = (1{,}013 - 273) \text{ kJ/kg} = 740 \text{ kJ/kg}$$

Good job; the energy of the cycle is balanced!

The specific heat of the air is not actually constant for each process in the cycle, but making that assumption enables you to quickly estimate the amount of work and heat transfer going on in the engine without consulting thermodynamic property tables.

Variable specific heat method: Otto cycle

The constant specific heat method gets inaccurate when temperatures in the cycle get pretty high, as they do in the heat input process. The specific heat of air increases with temperature, so accounting for this change improves the accuracy of your analysis. Calculating more accurate values for work and heat transfer isn't overly difficult, but before you can do so, you need to look up the *relative volume* and *internal energy* properties for air. You can find the ideal-gas properties of air in the appendix. Chapter 5 discusses how to use internal energy properties, and Chapter 8 discusses relative volume properties for an ideal-gas isentropic process. The following example shows you how to figure out accurate values for work and heat transfer.

If you look at Table A-1 in the appendix, you see that you need to know only the air temperature to find the internal energy and the relative volume. The relative volume property of air has no units or dimensions associated with it. Don't be alarmed if the values for relative volume you find in your textbook are different from the ones you see here. Different books use a different reference point for relative volume, but it all works out in the end as long as you're consistent and use values from a single source.

Follow these steps to analyze the Otto cycle using the first law of thermo-dynamics to find the energy associated with the heat transfer and work processes. This method accounts for the variation in specific heat with temperature.

1. **Determine the internal energy and relative volume of the air at State 1 for $T_1 = 20°C$ or 293 K, using Table A-1 in the appendix.**

   $u_1 = 209$ kJ/kg and $v_{r1} = 190.2$

   You'll need to do some interpolation on the table because the tempera-ture you're looking for isn't listed. (I show you how to interpolate a table in Chapter 3.)

2. **Find the relative volume, internal energy, and temperature properties of the air at the end of compression (State 2) by using the engine com-pression ratio and the relative volume at State 1.**

   $$r = \frac{v_1}{v_2} = \frac{v_{r1}}{v_{r2}}$$

   The compression stroke is an isentropic process, as Figure 11-2 shows. Chapter 8 gives you the scoop on isentropic processes.

3. **Use this equation to find the relative volume at State 2.**

   $$\frac{v_{r2}}{v_{r1}} = \frac{v_2}{v_1} = \frac{1}{8}, \text{ so } v_{r2} = \frac{190.2}{8} = 23.8$$

4. **Use the relative volume to find the temperature and internal energy of the air at State 2.**

   Interpolate Table A-1 in the appendix, and you find that the temperature and internal energy of the air at State 2 are as follows:

   $T_2 = 385°C$ (or 658 K)

   $u_2 = 480$ kJ/kg

   After compression is finished, the combustion process adds 1,310 kJ/kg of heat.

5. **Use the energy equation on this process to find the internal energy at State 3 as you did in the previous analysis of the Otto cycle combus-tion process.**

   This time you use only the internal energy — you don't use the constant volume specific heat.

   $u_3 = q_{in} + u_2 = (1,310 + 480)$ kJ/kg $= 1,790$ kJ/kg

6. Look up this value of u_3 in Table A-1 of the appendix to find the temperature and relative volume at State 3 by interpolation:

   $T_3 = 1,843°C$ (or $2,116$ K)

   $v_{r3} = 0.665$

7. **Find the relative volume (v_{r4}) at State 4 using the compression ratio.**

   The expansion process is isentropic, so you can use the isentropic process equation to find (v_{r4}).

   $$r = \frac{v_4}{v_3} = \frac{v_{r4}}{v_{r3}} = 8$$

   $v_{r4} = 8(0.665) = 5.32.$

8. **Look up the corresponding temperature and internal energy for v_{r4} by interpolating Table A-1 of the appendix:**

   $T_4 = 837°C$ (or $1,110$ K)

   $u_4 = 854$ kJ/kg

9. **Use the internal energy at each state to find the amount of energy exchanged in each process, using the energy equations in the earlier "Constant specific heat method: Otto cycle" section.**

   $q_{out} = (854 - 209)$ kJ/kg $= 645$ kJ/kg

   $w_{in} = (480 - 209)$ kJ/kg $= 271$ kJ/kg

   $w_{out} = (1,790 - 854)$ kJ/kg $= 936$ kJ/kg

10. **Check your calculations to make sure the net heat transfer ($q_{in} - q_{out}$) to the cycle equals the net work out ($w_{out} - w_{in}$) of the cycle.**

    $q_{in} - q_{out} = w_{out} - w_{in}$

    $(1,310 - 645)$ kJ/kg $= (936 - 271)$ kJ/kg $= 665$ kJ/kg

    Yes, it's balanced again!

Comparing the results of the two analysis methods

The results you get from analyzing the same engine by using constant specific heat and variable specific heat differ in a few important ways, as summarized in Table 11-1.

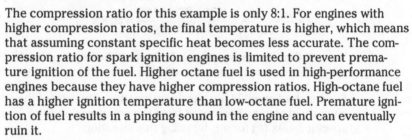

Table 11-1 Comparing Results of the Otto Cycle Analyses

State	Constant c_v	Variable c_v
1	293 K	293 K
2	673 K	658 K
3	2,495 K	2,116 K
4	1,086 K	1,100 K
Process	E kJ/kg	E kJ/kg
q_{in} 2 → 3	1,310	1,310
q_{out} 4 → 1	570	645
w_{in} 1 → 2	273	271
w_{out} 3 → 4	1,013	936

✔ **The amount of work going into the cycle during the compression stroke is about the same for both methods.** Because the temperature change is small, you're pretty safe assuming that specific heat remains constant during compression.

The compression ratio for this example is only 8:1. For engines with higher compression ratios, the final temperature is higher, which means that assuming constant specific heat becomes less accurate. The compression ratio for spark ignition engines is limited to prevent premature ignition of the fuel. Higher octane fuel is used in high-performance engines because they have higher compression ratios. High-octane fuel has a higher ignition temperature than low-octane fuel. Premature ignition of fuel results in a pinging sound in the engine and can eventually ruin it.

✔ **The amount of work you calculate for the power stroke when you assume specific heat is constant is more than the work you calculate when you account for variable specific heat.** This is because the specific heat increases with temperature, meaning that at high temperatures, it takes more heat to make the same temperature rise compared to low temperatures.

✔ **The calculation shows the amount of heat exhausted when you assume constant specific heat is less than when you account for variable specific heat.** Assuming constant specific heat gives you more work output and less heat rejected than when you account for the variation in specific heat.

Calculating Otto cycle efficiency

Finding efficiency just means figuring out how much useful work you get out of an engine compared to how much energy the engine uses to get that work done. Because an engine uses part of its work to compress the air inside the cylinders, you have to subtract the work into (w_{in}) the engine from the work out (w_{out}) of the engine to get the net work (w_{net}) from the engine.

You calculate the efficiency of any heat engine by using the following equation:

$$\eta_{th} = \frac{w_{net}}{q_{in}}$$

For the Otto cycle analysis using the constant specific heat assumption, the efficiency is

$$\eta_{th} = \frac{w_{out} - w_{in}}{q_{in}} = \frac{(1{,}013 - 273) \text{ kJ/kg}}{1{,}310 \text{ kJ/kg}} = 56.5\%$$

For the Otto cycle analysis using the variable specific heat assumption, the efficiency is

$$\eta_{th} = \frac{w_{out} - w_{in}}{q_{in}} = \frac{(936 - 271) \text{ kJ/kg}}{1{,}310 \text{ kJ/kg}} = 50.8\%$$

You see that the constant specific heat method overestimates the efficiency of the cycle because it overestimates the amount of work output from the cycle.

For constant specific heat assumption, the efficiency of the Otto cycle heat engine can also be computed from the compression ratio (r) and the ratio of specific heats (k) for air, by using this equation:

$$\eta_{th} = 1 - r^{(1-k)} = 1 - 8^{(1-1.4)} = 56.5\%$$

In this equation, you can see that the efficiency of the Otto cycle engine is proportional to the compression ratio. So, the higher the compression ratio, the higher the efficiency of the engine.

Calculating Otto cycle irreversibility

No engine is 100 percent efficient. But what does that mean, and can you get better performance somehow? When you examine a cycle using the second law of thermodynamics, you figure out where the inefficiencies come from, which may make you wonder about ways to make them more efficient. In a word, inefficiency arises from friction generated by parts and fluids that are moving. Inefficiency also occurs in heat transfer processes. Friction and heat losses in a cycle cause *irreversibility*. I discuss irreversibility in detail in

Chapter 9. The following equation is used to calculate the irreversibility of any thermodynamic process in a closed system:

$$i = T_0 \left[\left(s_{\text{final}} - s_{\text{initial}} \right) - \frac{q_{\text{in}}}{T_{\text{in}}} + \frac{q_{\text{out}}}{T_{\text{out}}} \right]$$

Before you use this equation, you need to know the following points about each of its terms:

- T_0 is the ambient or dead state temperature of the environment that the process operates in. For this Otto cycle engine, the ambient temperature is 20 degrees Celsius.

- s_{final} is the entropy of the air at the end of a process. You need to know both the temperature and pressure of the air at the end of the process to find the entropy.

- s_{initial} is the entropy of the air at the beginning of a process.

- q_{in} is the amount of heat transferred from the high-temperature reservoir to the system during the combustion process. q_{out} is the amount of heat transferred to the low-temperature reservoir from the system during the exhaust process.

- T_{in} is the temperature of the high-temperature thermal energy reservoir that transfers heat to the system. T_{out} is the temperature of the low-temperature thermal energy reservoir that the system rejects heat to. You usually use the maximum temperature of the cycle for the heat input reservoir temperature. The heat rejection reservoir temperature is usually the ambient air temperature.

You can use this equation to determine the irreversibilities associated with the Otto cycle. Isentropic processes are reversible, so the irreversibility for the compression $(s_2 - s_1)$ and expansion $(s_4 - s_3)$ strokes is zero. For the heat transfer processes, you need to calculate the change in entropy $(s_{\text{final}} - s_{\text{initial}})$ with the following equation:

$$s_{\text{final}} - s_{\text{initial}} = s^0_{\text{final}} - s^0_{\text{initial}} - R\ln\left(\frac{P_{\text{final}}}{P_{\text{initial}}} \right)$$

As I describe in Chapter 8, the temperature and pressure of a gas determine its entropy. The s^0_{initial} and s^0_{final} entropy terms are a function of the air temperature at the beginning and end of a process, respectively. You can find these values in Table A-1 of the appendix. The ideal gas constant R for air is 0.287 kilojoules per kilogram-Kelvin. You can find out more about it in Chapter 3.

You can find the pressure at each state in the cycle by using the ideal-gas law (see Chapter 3). Suppose your car is at some elevation above sea level, and the ambient air pressure is 95 kilopascals. This defines the pressure at State 1: (P_1). You also need the compression ratio of the engine and the temperature at the beginning and the end of the compression process from the example in the previous section.

The following steps help you calculate the irreversibilities in the cycle:

1. **Find the pressure at the end of compression with the ideal-gas law arranged in the following form:**

$$\frac{P_2 v_2}{T_2} = \frac{P_1 v_1}{T_1} \text{ or } P_2 = P_1 \left(\frac{v_1}{v_2}\right)\left(\frac{T_2}{T_1}\right) = (95 \text{ kPa})(8)\left(\frac{673 \text{ K}}{293 \text{ K}}\right) = 1{,}746 \text{ kPa}$$

2. **Calculate the pressure at the end of the combustion process by using the ideal-gas law:**

$$P_3 = P_2 \left(\frac{v_2}{v_3}\right)\left(\frac{T_3}{T_2}\right) = (1{,}746 \text{ kPa})(1)\left(\frac{2{,}510 \text{ K}}{673 \text{ K}}\right) = 6{,}512 \text{ kPa}$$

Remember that combustion for the Otto cycle is a constant-volume process, so the volume ratio v_2/v_3 is 1.

3. **Find the change in entropy for the combustion process using the following equation:**

$$s_3 - s_2 = s_3^0 - s_2^0 - R\ln\left(\frac{P_3}{P_2}\right)$$

You can find values for s_2^0 and s_3^0 at temperatures T_2 and T_3 by interpolation in Table A-1 of the appendix.

4. **Plug in the values for s_2^0 and s_3^0 and the pressure at States 2 and 3 to get the result for the change in entropy for the combustion process:**

$$s_3 - s_2 = (9.253 - 7.698) \text{ kJ/kg} \cdot \text{K} - (0.287 \text{ kJ/kg} \cdot \text{K})\ln\left(\frac{6{,}512 \text{ kPa}}{1{,}746 \text{ kPa}}\right)$$

$$= 1.777 \text{ kJ/kg} \cdot \text{K}$$

You may find it strange that the change in entropy for the exhaust process is equal to that of the combustion process but opposite in sign. Ideal compression and expansion processes are isentropic, so the entropy at State 1 equals that at State 2. The same reasoning applies for States 3 and 4. The following equation describes the change in entropy for both heat transfer processes:

$$s_1 - s_4 = -(s_3 - s_2)$$

5. **Compute the irreversibility of the combustion process using absolute temperatures for the ambient temperature and the maximum cycle temperature.**

$$i_{23} = T_0\left[(s_3 - s_2) - \frac{q_{in}}{T_H}\right] = (293 \text{ K})\left(1.777\frac{\text{kJ}}{\text{kg} \cdot \text{K}} - \frac{1{,}310 \text{ kJ/kg}}{2{,}510 \text{ K}}\right) = 192 \text{ kJ/kg}$$

6. **Find the irreversibility of the exhaust process by using the ambient air temperature for the low-temperature reservoir.**

$$i_{41} = T_0\left[(s_1 - s_4) + \frac{q_{out}}{T_L}\right] = (293 \text{ K})\left(-1.777\frac{\text{kJ}}{\text{kg} \cdot \text{K}} + \frac{645 \text{ kJ/kg}}{293 \text{ K}}\right) = 300 \text{ kJ/kg}$$

Theoretically, the irreversibilities become smaller when the heat transfer processes occur at a constant temperature. In that instance, T_2 equals T_3 and the cycle becomes the Carnot cycle, which I discuss in Chapter 10. Increasing the compression ratio makes the temperature at State 2 (refer to the *T-s* diagram of Figure 11-2) closer to the temperature at State 3. The heat input is reduced to keep the temperature at State 3 fixed. In the same manner, it makes the temperature at State 4 closer to the temperature at State 1. Increasing the compression ratio gets the engine closer to the ideal Carnot cycle by increasing its efficiency.

Determining the mean effective pressure of reciprocating engines

Imagine you crack open your car owner's manual to the engine specifications and find that your 2.0-liter engine can produce 90 horsepower at 6,000 revolutions per minute (rpm) (or, if you're thinking in the metric system, 67 kilowatts [kW]). You can use the *mean effective pressure* (MEP) to compare the performance of your engine to your neighbor's engine. Whoever has the higher MEP has the better performing (that is, more powerful) engine but probably spends more money on gas. One way to determine how effective your engine is at producing power is to calculate the MEP, also known as the *average pressure,* inside the cylinder during the power stroke.

You calculate MEP by using the power produced by the engine, W_{net}, at a given engine speed, s (don't confuse this with entropy), and the volumetric flow rate, $\dot{V}$, of air moving through the engine. You find the volumetric flow rate by multiplying one half of the engine speed by the displacement volume. You use one half because every other revolution in each cylinder intakes air; thus, there are 3,000 intake strokes in each cylinder per minute in your engine when it's operating at 6,000 rpm.

You find volumetric flow rate by using the following equation:

$$\dot{V} = \frac{s}{2}V_{Disp} = \left(\frac{1}{2}\right)\left(\frac{6{,}000 \text{ rev}}{\text{min}}\right)\left(\frac{1 \text{ min}}{60 \text{ sec}}\right)(2.0 \text{ L})\left(\frac{\text{m}^3}{1{,}000 \text{ L}}\right) = 0.1 \frac{\text{m}^3}{\text{sec}}$$

You need to convert the engine speed from revolutions per minute to revolutions per second and the displacement from liters to cubic meters.

You determine mean effective pressure by dividing the engine power by the air flow rate through the engine, as follows:

$$MEP = \frac{\dot{W}_{net}}{\dot{V}}$$

Then you can substitute the values for the power and the volumetric flow rate to calculate the MEP:

$$MEP = \frac{67 \text{ kJ/sec}}{0.1 \text{ m}^3/\text{sec}} = 670 \text{ kPa}$$

When you calculate the MEP, make sure you use the right units in the equation by using cubic meters for volume and seconds for time. The way a pressure unit like kPa is related to units of power (kJ/s) and volumetric flow rate (m³/s) isn't obvious, so the following unit conversions may be helpful to you: kW = 1 kJ/s and 1 kPa = 1 kJ/m³.

Working with the Ideal Compression Ignition Engine: The Diesel Cycle

The red warning lights are blinking and a bell is ringing as you pull up to a railroad crossing. Suddenly you hear the train horn blowing and the train whizzes by right in front of you. You begin to wonder, can I do a thermodynamic analysis of the locomotive engine? You bet!

Your basic freight train uses a very large diesel engine to turn an electric generator. The electric generator powers electric motors that move the wheels under the engine. A train doesn't use a mechanical transmission like a car or a truck to get power to the wheels because a lot of torque is needed to get the train moving from a standing start. An electric motor is the best way to have high torque at low speed. In the upcoming sections, you dig into the pistons and cylinders of the diesel engine to understand thermodynamically how it does its job.

Examining the four processes in a diesel cycle

The diesel cycle is just like all other heat engines and requires the four basic cycle processes: work in, heat in, work out, and heat out. The main difference between the diesel cycle and the Otto cycle is that you assume that the combustion process in the diesel cycle occurs at a constant pressure instead of a constant volume.

In the diesel cycle, only air is compressed in the cylinder and fuel is injected as the piston approaches the top of the cylinder. The compression ratio of the diesel cycle is higher than that of the Otto cycle; the air temperature at the end of compression in the diesel cycle is hot enough to spontaneously ignite the fuel when it's injected into the cylinder. Diesel fuel burns more slowly than gasoline, so it adds heat to the air while the piston moves down the cylinder. A constant-pressure combustion process more accurately represents what happens in a diesel cycle.

Take a look at the four thermodynamic processes for the diesel cycle in the pressure-volume and temperature-entropy diagrams of Figure 11-4. At first glance, the processes look the same as the Otto cycle. However, you can spot some differences. Here's what's happening:

- From State 1 to State 2, you see isentropic compression. This is where the work goes into the cycle.

- From State 2 to State 3 comes constant-pressure heat input. The combustion process in a diesel cycle lasts longer than that of the Otto cycle so the piston moves down during combustion, and the expanding gas does work in the process. This process is the only difference from the Otto cycle.

✔ Additional work comes out of the cycle during isentropic expansion from State 3 to State 4.

✔ Finally, heat comes out of the cycle during the constant volume exhaust process from State 4 back to State 1.

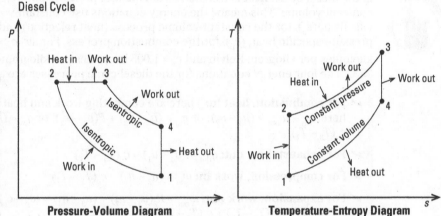

Figure 11-4:
Pressure-volume and temperature-entropy diagrams for the diesel cycle.

Analyzing the Diesel cycle

Just like the Otto cycle section, you can use either of two methods in this section to analyze the diesel cycle to determine the amount of work and heat transfer that occurs in each process. The constant specific heat method is quicker than the variable specific heat method if you don't have computer software to look up thermodynamic properties, but it's not as accurate. Software properties are available with most textbooks and make this old chore a breeze.

Constant specific heat method: Diesel cycle

You can use the steps for analyzing the Otto cycle (see the earlier section "Constant specific heat method: Otto cycle") to analyze the diesel cycle. Remember that the diesel cycle forms a closed system and, like the Otto cycle, operates in a piston-cylinder mechanism. Here's an example using the following specifications from a diesel locomotive engine:

✔ The ambient conditions for the air are 23 degrees Celsius and 100 kilopascals pressure.

✔ The heat input to the engine from the diesel fuel is 1,230 kilojoules per kilogram of air.

✔ The compression ratio for a typical diesel locomotive is 15:1.

✔ The diesel engine includes 16 cylinders, and each cylinder has a displacement volume of 15.7 liters.

✔ The engine produces 6,250 horsepower (4,663 kW) at 1,050 rpm.

You don't really need to know the last two pieces of information to do a first law analysis of the engine. They're just fun facts for you to marvel over.

In the diesel cycle, heat is transferred at constant pressure and rejected at constant volume. This means the energy equations use constant-volume specific heat, (c_v), for the constant-volume process (heat rejection) and constant-pressure specific heat, (c_p), for the combustion process. For air, $c_v = 0.719$ kilojoules per kilogram-Kelvin and $c_p = 1.005$ kilojoules per kilogram-Kelvin at 25°C. The four energy equations for the diesel cycle processes are as follows:

✔ **For combustion, heat in:** There are competing work and heat processes here: $q_{in} - w_{out} = (u_3 - u_2)$, or $q_{in} = (u_3 - u_2) + P(v_3 - v_2)$, or $q_{in} = (h_3 - h_2) = c_p(T_3 - T_2)$.

✔ **For exhaust, heat out:** $q_{out} = (u_4 - u_1) = c_v(T_4 - T_1)$

✔ **For compression, work in:** $w_{in} = (u_2 - u_1) = c_v(T_2 - T_1)$

✔ **For expansion, work out:** $w_{out} = P_2(v_3 - v_2) + (u_3 - u_4) = (c_p - c_v)(T_3 - T_2) + c_v(T_3 - T_4) = R(T_3 - T_2) + c_v(T_3 - T_4)$

You probably noticed that the energy equation for the expansion process for the diesel cycle is more complicated than the one for the Otto cycle. This is because the expansion process has two distinct phases:

✔ In the first phase, work is coming out of the engine during the combustion process because the piston is moving, so this is represented by the term $P_2(v_3 - v_2)$. You don't need to know the pressure at State 2 or the volumes at States 2 and 3; you can find the work done during combustion using the following substitute expression, which can be derived from the ideal-gas law:

$$Pv = RT \text{ and } R = c_p - c_v \text{ so } P_2(v_3 - v_2) = (c_p - c_v)(T_3 - T_2)$$

✔ In the second phase, the hot air is expanding and the work done by the piston is represented by the term $(u_3 - u_4) = c_v(T_3 - T_4)$.

✔ The *cutoff ratio* (r_c) defines the position of the piston (or the volume in the cylinder) where the combustion process stops during the expansion stroke. The cutoff ratio can be calculated from the specific volumes at States 2 and 3, as shown in this equation:

$$r_c = \frac{v_3}{v_2}$$

You can now find the temperature at each step of the diesel cycle to determine the amount of heat and work going in and out of the system. The ambient conditions you chose are used to describe State 1 and find the temperature and pressure at State 2. The following steps show you how to analyze the diesel cycle using the constant specific heat assumption:

1. **Recall the ratio of specific heats, $k = 1.4$ for air, and use the ideal-gas isentropic equations in Chapter 5 to find the temperature T_2 and pressure P_2 at the end of the compression process.**

   Remember to use absolute temperatures, so $T_1 = 23°C = 296$ K.

   $$T_2 = T_1 \left(\frac{V_1}{V_2} \right)^{k-1} = 296 \text{ K} (15)^{1.4-1} = 874 \text{ K}$$

2. **Find T_3 by rearranging the energy equation for q_{in}.**

   In a diesel cycle, the heat addition occurs with constant pressure.

   $$T_3 = \frac{q_{in}}{c_p} + T_2 = \frac{1,230 \text{ kJ/kg}}{1.005 \text{ kJ/kg} \cdot \text{K}} + 874 \text{ K} = 2,097 \text{ K}$$

3. **Solve for the conditions at State 4 using the same equations that were used to solve for State 2 in Step 1.**

   But remember that the engine is now undergoing an isentropic expansion instead of an isentropic compression.

   $$T_4 = T_3 \left(\frac{V_3}{V_4} \right)^{k-1}$$

4. **Use the ideal-gas law to determine v_3.**

   $$\left(\frac{P_2 v_2}{T_2} \right) = \left(\frac{P_3 v_3}{T_3} \right) \quad \text{or} \quad v_3 = v_2 \frac{T_3}{T_2}$$

5. **Make a quick substitution for v_3, and you have an equation to find the temperature at State 4.**

   $$T_4 = T_3 \left(\frac{v_2 \cdot T_3}{v_4 \cdot T_2} \right)^{k-1} = 2,097 \text{ K} \left(\frac{1 \cdot 2,097 \text{ K}}{15 \cdot 874 \text{ K}} \right)^{1.4-1} = 1,007 \text{ K}$$

6. **Now that you know the temperature at each state, find the amount of energy exchanged in each process, using the diesel cycle energy equations.**

   $q_{out} = 0.719 \text{ kJ/kg} \cdot \text{K} (1,007 \text{ K} - 296 \text{ K}) = 511 \text{ kJ/kg}$

   $w_{in} = 0.719 \text{ kJ/kg} \cdot \text{K} (874 \text{ K} - 296 \text{ K}) = 416 \text{ kJ/kg}$

   $w_{out} = (1.005 - 0.719) \text{ kJ/kg} \cdot \text{K} (2,097 - 874) \text{ K} + 0.719 \text{ kJ/kg} \cdot \text{K} (2,097 - 1,007) \text{ K} = 1,135 \text{ kJ/kg}$

7. **Check your calculations to make sure the net heat transfer ($q_{in} - q_{out}$) to the cycle equals the net work ($w_{out} - w_{in}$) out of the cycle.**

$$q_{in} - q_{out} = w_{out} - w_{in}$$

$$(1{,}230 - 511) \text{ kJ/kg} = (1{,}135 - 416) \text{ kJ/kg} = 719 \text{ kJ/kg}$$

Variable specific heat method: Diesel cycle

As I discuss in the previous section "Variable specific heat method: Otto cycle," the constant specific heat method gets inaccurate at high temperatures because specific heat of air increases with temperature. The following method accounts for the variation of specific heat with temperature and gives more accurate results.

1. **Use Table A-1 in the appendix to find the internal energy and relative volume for State 1 at 23°C or 296 K**

    $u_1 = 212 \text{ kJ/kg}$ and $v_{r1} = 185.4$

2. **Find the relative volume, internal energy, and temperature properties of the air at the end of compression (State 2) using the engine compression ratio and the relative volume at State 1.**

    Use the same expression you used in the Otto cycle to find the relative volume at State 2:

    $$\frac{v_{r2}}{v_{r1}} = \frac{v_2}{v_1} = \frac{1}{15}, \text{ so } v_{r2} = \frac{185.4}{15} = 12.4$$

3. **Use the value for v_{r2} to find the temperature, internal energy, and enthalpy of the air at State 2 in Table A-1.**

    You need both the internal energy and the enthalpy at State 2 because $u_1 - u_2$ is used in the energy equation for the compression process and $h_3 - h_2$ is used in the combustion process energy equation.

    $T_2 = 561°C$ (or 834 K), $u_2 = 620 \text{ kJ/kg}$, and $h_2 = 859 \text{ kJ/kg}$

4. **Use the energy equation on the combustion process to find the enthalpy at State 3.**

    This is different than what you do in the Otto cycle where you find the internal energy at State 3. After compression is finished, the combustion process adds 1,230 kilojoules per kilogram of heat.

    $$h_3 = q_{in} + h_2 = (1{,}230 + 859) \text{ kJ/kg} = 2{,}089 \text{ kJ/kg}$$

5. **Look up this value of h_3 in Table A-1 and you find by interpolation that the temperature and relative volume at State 3 are as follows:**

   $T_3 = 1{,}596°C$ (or 1,869 K) and $v_{r3} = 1.007$

   The third process is isentropic expansion, and finding the relative volume at State 4 is a bit trickier in the diesel cycle than it is in the Otto cycle because you don't know the cylinder volume at State 3. But you can help yourself to the ideal-gas law and find what you need without even knowing the volume at State 3. Here's how you do it. The combustion process is constant pressure, so the volume and temperature at States 2 and 3 are related to each other by the ideal-gas law:

   $$\frac{v_3}{v_2} = \frac{T_3}{T_2}$$

6. **Rearrange this equation to solve for v_3 and substitute it into the equation for isentropic expansion of the air, like this:**

   $$\frac{v_3}{v_4} = \frac{v_{r3}}{v_{r4}} = \left(\frac{v_2}{v_4}\right)\left(\frac{v_3}{v_2}\right) = \left(\frac{v_2}{v_4}\right)\left(\frac{T_3}{T_2}\right)$$

7. **Then rearrange this equation to solve for v_{r4}:**

   $$v_{r4} = v_{r3}\left(\frac{v_4}{v_2}\right)\left(\frac{T_2}{T_3}\right) = 1.007(15)\left(\frac{834\ K}{1{,}869\ K}\right) = 6.74$$

8. **Now you can look up the corresponding temperature and internal energy by interpolating Table A-1.**

   $T_4 = 753°C$ (or 1,026 K) and $u_4 = 781$ kJ/kg

9. **Find the amount of energy exchanged in each process using the energy equations with the energy properties at each state.**

   $q_{out} = (781 - 212)$ kJ/kg $= 569$ kJ/kg

   $w_{in} = (620 - 212)$ kJ/kg $= 408$ kJ/kg

10. **Manipulate the energy equation for the work output, because you don't know what the pressure is at the combustion process.**

    The pressure term $P_2(v_3 - v_2)$ can be replaced by the difference between the enthalpy and internal energy changes, as shown here:

    $w_{out} = P_2(v_3 - v_2) + (u_3 - u_4) = (h_3 - h_2) - (u_3 - u_2) + (u_3 - u_4)$

    The internal energy at State 3 cancels out, so the resulting energy equation allows you to find the work output, like this:

    $w_{out} = (h_3 - h_2) + (u_2 - u_4)$

    $w_{out} = (2{,}089 - 859) + (620 - 781)$ kJ/kg $= 1{,}069$ kJ/kg

11. **Check your calculations to make sure the net heat transfer ($q_{in} - q_{out}$) to the cycle equals the net work ($w_{out} - w_{in}$) out of the cycle.**

$$q_{in} - q_{out} = w_{out} - w_{in}$$

$$(1{,}230 - 569) \text{ kJ/kg} = (1{,}069 - 408) \text{ kJ/kg} = 661 \text{ kJ/kg}$$

Calculating diesel cycle efficiency

Finding efficiency just means figuring out how much useful work you get out of an engine compared to the amount of energy the engine uses to get that work done. Because a diesel engine uses part of its work to compress the air inside the cylinders, you have to subtract the work into (w_{in}) the engine from the work out (w_{out}) of the engine to get the net work (w_{net}) from the engine.

You can calculate the efficiency of any heat engine using the following equation:

$$\eta_{th} = \frac{w_{net}}{q_{in}}$$

For the diesel cycle analysis using the constant specific heat assumption, the efficiency is as follows:

$$\eta_{th} = \frac{w_{out} - w_{in}}{q_{in}} = \frac{(1{,}135 - 416) \text{ kJ/kg}}{1{,}230 \text{ kJ/kg}} = 58.5\%$$

For the diesel cycle analysis using the variable specific heat assumption, the efficiency is

$$\eta_{th} = \frac{w_{out} - w_{in}}{q_{in}} = \frac{(1{,}069 - 408) \text{ kJ/kg}}{1{,}230 \text{ kJ/kg}} = 53.7\%$$

You see that the constant specific heat method overestimates the efficiency of the cycle because it overestimates the amount of work output from the cycle.

For constant specific heat assumption, the efficiency of the Diesel cycle heat engine can also be computed from the compression ratio (r), the cutoff ratio (r_c), and the ratio of specific heats (k) for air by using this equation:

$$\eta_{th} = 1 - r^{(1-k)} \left[\frac{r_c^k - 1}{k(r_c - 1)} \right]$$

You can find the cutoff ratio (r_c) for the diesel engine by recognizing the combustion process is at constant pressure, so the volume ratios (v_2 and v_3) are related to the temperatures (T_2 and T_3) using the ideal-gas law, as shown in this equation:

$$r_c = \frac{v_3}{v_2} = \frac{T_3}{T_2} = \frac{2{,}097 \text{ K}}{874 \text{ K}} = 2.40$$

Now you can calculate the efficiency of the diesel engine using the constant specific heat assumption, as follows:

$$\eta_{th} = 1 - r^{(1-k)} \left[\frac{r_c^k - 1}{k(r_c - 1)} \right] = 1 - 15^{(1-1.4)} \left[\frac{2.40^{1.4} - 1}{1.4(2.40 - 1)} \right] = 58.4\%$$

This value is very close to the efficiency calculated for the cycle using the previous net work and heat input method for the constant specific heat assumption.

Calculating diesel cycle irreversibility

You can calculate the irreversibilities of the diesel cycle in exactly the same way as you do for the Otto cycle. This means the irreversibility of the compression and expansion process is zero, so you only have to calculate it for the heat transfer processes. In the earlier section "Calculating Otto cycle irreversibility," I show you how to determine the change in entropy for the heat transfer processes when you account for the variation in specific heat during the cycle. Here, I show you how to find the change in entropy when you assume the specific heat is constant for the cycle. This method is easier to calculate than the method used for calculating when specific heat varies, but it's not quite as accurate. The changes in entropy for the combustion process and the exhaust process are equal to each other but opposite in direction or sign. In a real engine, the change in entropy for both processes isn't equal in magnitude because the compression and expansion processes aren't isentropic.

You can determine the change in entropy for a constant-pressure process with this equation:

$$s_3 - s_2 = c_p \ln\left(\frac{T_3}{T_2} \right) = (1.005 \text{ kJ/kg} \cdot \text{K}) \ln\left(\frac{2{,}097 \text{ K}}{874 \text{ K}} \right) = 0.880 \text{ kJ/kg} \cdot \text{K}$$

You already know the specific heat and the temperature at the beginning and end of the combustion process, so this is simple.

You should recognize the following equations as those that you use in the Otto cycle analysis. You just have to plug in the numbers that you found for the diesel cycle.

$$i_{23} = T_0\left[(s_3 - s_2) - \frac{q_{in}}{T_H}\right] = (296\ K)\left(0.880\frac{kJ}{kg \cdot K} - \frac{1,230\ kJ/kg}{2,097\ K}\right) = 87\ kJ/kg$$

$$i_{41} = T_0\left[(s_1 - s_4) + \frac{q_{out}}{T_L}\right] = (296\ K)\left(-0.880\frac{kJ}{kg \cdot K} + \frac{511\ kJ/kg}{296\ K}\right) = 251\ kJ/kg$$

You can see that the irreversibility is higher for the exhaust process. A lot of energy is still left in the exhaust because it's a hot gas. Unfortunately, it can't easily be put to work, but maybe you can think of some clever ideas for utilizing this energy source.

Chapter 12

Working with Rankine Cycles

You're standing by the window looking out across the menacing sky. Rain pelts the glass as lightning strikes and thunder rolls. Suddenly, the power goes out. You reach for matches and candles.

Until you're without it, you probably take power for granted; you don't think about where it comes from, how it's made, or how it gets to your house. Electric power is generated from many sources. Some comes from renewable sources such as wind or solar or geothermal energy. Most electricity comes from fossil fuels where power plants burn coal, oil, or natural gas to produce electricity. Nuclear power is also used to generate a fair amount of electricity.

The modern electric power plant is complex and filled with equipment such as boilers, condensers, pumps, turbines, and generators. This equipment is connected by a maze of piping, plumbing, and wiring. Despite the complexity of a power plant, the thermodynamic analysis of the basic power cycle is relatively straightforward. In this chapter, I discuss how to analyze a heat engine called the Rankine cycle, which uses water as the working fluid to generate power with a steam turbine. The steam turbine then turns a generator to produce electricity for your home — as long as a problem with the transmission lines isn't encountered during a powerful storm.

The analysis of a Rankine cycle is the same regardless of the kind of fuel used (including nuclear power) to provide the heat input. I start the chapter with the most simple and basic form of the Rankine cycle. Then I discuss several modifications that are commonly made to the cycle to improve its overall efficiency and power generation capability. Finally, I discuss how a real Rankine cycle system deviates from the ideal thermodynamic analysis.

Understanding the Basics of the Rankine Cycle

When most people hear about steam engines, they think of old train engines that huff and puff a disappearing trail of steam while belching out clouds of black smoke. This kind of engine uses high-pressure steam that does work by expanding the steam against a piston in a cylinder. Steam engines may seem old-fashioned, but they're still used every day in a different form to generate power.

Figure 12-1 shows a diagram of the four components in the Rankine cycle. Here's how the components contribute to the cycle:

- ✔ **Pump:** The pump pressurizes the water between States 1 and 2 in Figure 12-1 and sends it through the boiler to turn it into steam.

- ✔ **Steam generator:** The steam generator (or boiler) consists of a firebox where fuel burns. You see it between States 2 and 3 in Figure 12-1. The walls of the firebox are lined with tubes filled with high-pressure water. As the water boils in the tubes, it turns into steam and flows into a heat exchanger located in the hottest part of the boiler. The steam becomes superheated, meaning it's heated above its saturation temperature.

- ✔ **Turbine:** A *turbine* is a mechanical device that spins on an axle like a fan with many blades. Instead of using a piston/cylinder, the modern power plant uses a turbine to extract work from the steam as shown between States 3 and 4 in Figure 12-1. High-pressure gas or steam enters a turbine and flows through a series of blades. The blades react to the flowing gas (using Newton's first law of motion . . . from physics) and spin the turbine, doing work in the process.

- ✔ **Condenser:** After the steam goes through the turbine in the Rankine cycle, it's condensed back into water. The steam condensation process takes place in a heat exchanger that circulates cool water or air over a series of steam-filled tubes. (The cool water may come from a nearby lake, river, or ocean because large power plants require a lot of cooling water; the air just comes from the atmosphere.) This heat exchanger is called a *condenser;* you see it between States 4 and 1 in Figure 12-1. The pressure of the steam inside the condenser is usually less than atmospheric pressure, which is 101 kilopascals.

The *saturation temperature* is the boiling point of water for a given pressure. For example, when you boil water on a stove, it's at atmospheric pressure and the boiling point normally is 100 degrees Celsius. At higher pressures, like those in a Rankine cycle boiler, the saturation temperature is much higher than 100 degrees Celsius. You can find the saturation temperature in Tables A-3 and A-4 in the appendix.

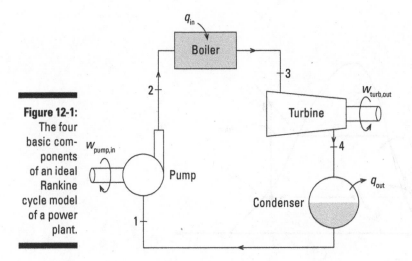

Figure 12-1:
The four
basic com-
ponents
of an ideal
Rankine
cycle model
of a power
plant.

The steam turbine used in the Rankine cycle is very similar to the gas turbine engines used in the Brayton cycle. In Chapter 10 I discuss how to analyze gas turbine engines, which are used on aircraft as jet engines or in power plants to generate electric power. In gas turbines, the working fluid is air, and it's exhausted into the atmosphere after leaving the turbine rather than circulating around a cycle as in the Rankine engine.

The amount of electricity that consumers use varies quite a bit on a daily basis. During the day, the demand for electricity is high; at night, when most people are sleeping, the demand is low. An electric power plant may use both the Rankine cycle engine and the Brayton cycle engine to produce electricity. In the Rankine cycle, turning a boiler on and off very quickly isn't possible, so this cycle is used to provide what is known as a base electric load to the utility company. In contrast, the Brayton cycle is able to start up quickly (think how quickly a jet engine can start), so it's used to meet the peak electric load for a utility system.

Examining the Four Processes of the Rankine Cycle

In this section I describe how the Rankine cycle works. I show you the thermodynamic path taken by each of the four processes.

You can understand how a Rankine cycle works by drawing it on a temperature-entropy (*T-s*) diagram, as shown in Figure 12-2. The *T-s* diagram shows how the temperature and entropy change for each process. Chapter 3 tells you more about these kinds of diagrams.

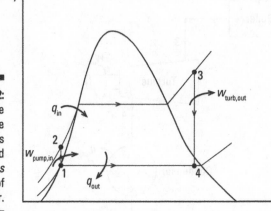

Figure 12-2:
The Rankine cycle processes mapped onto a T-s diagram of water.

The inverted u-shaped line you see in Figure 12-2 is called the *liquid-vapor dome*. (I discuss the features of the liquid-vapor dome in detail in Chapter 3.) It represents the liquid to vapor transition region in the phase diagram. The portion of the line on the left side of the peak is 100 percent liquid, and the portion on the right side of the peak is 100 percent vapor. The area beneath the liquid-vapor line is a mixture of liquid and vapor. The area to the left of the liquid line is called compressed liquid, while the area to the right of the vapor line is called superheated vapor. At temperatures and pressures above the peak (374 degrees Celsius and 22 megapascals), the fluid is neither liquid nor vapor but is called a *supercritical fluid*. The peak of the liquid-vapor dome is called the *critical point*.

The four processes of the Rankine cycle shown in Figure 12-2 are as follows:

1. Work in, isentropic compression.

Water is pressurized by isentropic compression in a pump to the operating pressure of the steam generator. The line from State 1 to State 2 on the T-s diagram shows this process.

The compression process in the ideal pump is reversible (frictionless) and adiabatic (transfers no heat to or from the pump). Because it's reversible and adiabatic, the process is *isentropic*, meaning that entropy is constant, as shown by the vertical line from States 1 to 2 on the T-s diagram. Later in the chapter, in the section "Deviating from Ideal Behavior: Actual Rankine Cycle Performance," I show you how to account for deviations from the isentropic process. The outlet pressure can be as high as 24 megapascals in a supercritical boiler, although not every steam generator operates at a pressure this high.

2. Heat in at constant pressure.

The steam generator (or boiler) adds heat to the water to produce high-temperature steam. The line from State 2 to State 3 on the *T-s* diagram shows this process.

The water temperature increases from State 2 as heat is added until it reaches the boiling point. Then the temperature remains constant until the water becomes a vapor.

Finally, additional heat causes the vapor temperature to continue rising to State 3. Because water boils into a vapor only if the operating pressure is less than the critical pressure of 22 megapascals, this steam generator is called a *subcritical steam generator.* Boiling doesn't take place above the critical pressure, and the fluid simply increases in temperature without a phase change (see Chapter 3), meaning the steam generator is a *supercritical steam generator.* In the ideal steam generator, pressure remains constant, and the process is therefore *isobaric.*

3. Work out, isentropic expansion.

The high-temperature and high-pressure steam has a lot of energy when it leaves the steam generator. Work is extracted from the steam by isentropic expansion in a turbine. The vertical line from State 3 to State 4 on the *T-s* diagram shows this process.

As steam flows through the turbine, the pressure and temperature decrease so the specific volume of the steam increases. As a result of this volumetric increase, the flow passages and the turbine blades increase in size with every successive stage in the turbine. The steam can become a liquid-vapor mixture in the final stages of the turbine.

In practice, the minimum quality of the liquid-vapor mixture leaving the turbine is 90 percent. The quality of the steam is a measure of moisture content (liquid water) in the steam. This means the steam mixture must contain less than 10 percent liquid water by mass; otherwise, water droplets impacting the turbine blades at high speed can cause pitting and erosion on the blades. The lowercase letter x usually represents the thermodynamic property of steam quality (discussed in Chapter 3).

4. Heat out at constant pressure.

As the steam leaves the turbine, it condenses back into liquid water in a constant pressure process from State 4 to State 1, as shown in the *T-s* diagram. The water is condensed into a liquid because pressurizing a liquid takes a lot less work than pressurizing a gas.

At the exit of the condenser, the water is often a saturated liquid, which means it has a liquid-vapor quality of 0.0. Sometimes it may be a subcooled liquid because it has been cooled below its saturation temperature. Chapter 3 discusses the concept of saturation temperature and pressure for phase-change processes.

Analyzing the Cycle Using Steam Tables

In the thermodynamic analysis of any Rankine cycle, you need to find thermodynamic properties at each state, using the steam tables in the appendix. You will probably need to do interpolation to find the temperature, enthalpy, or entropy values for most of the states in the cycle. I discuss how to interpolate a thermodynamic table in Chapter 3. For each state in a thermodynamic cycle, you need two independent properties, such as temperature and specific volume, to determine all the other thermodynamic properties, like enthalpy, internal energy, or entropy of a state. I discuss this idea, called the *state postulate,* in Chapter 2.

Imagine a cycle with turbine inlet conditions (State 3) at 10 megapascals, 800 degrees Celsius. The condenser pressure (at State 1 or 4) is 10 kilopascals. The steam mass flow rate through the system is 60 kilograms per second. In your analysis, you can find the pump work, turbine work, heat input, heat rejected, and steam quality at the turbine exit. In the following sections, I show you how to calculate the efficiency and the irreversibility of the cycle. I also use these same given conditions for analyzing the modifications to the Rankine cycle in the sections "Improving the Rankine Cycle with Reheat" and "Improving the Rankine Cycle with Regeneration."

You can analyze this simple Rankine cycle by following these steps:

1. **Find the thermodynamic properties at State 1, the pump inlet.**

   At State 1, the pressure (P_1) is 10 kilopascals, and the water is a saturated liquid. This means the quality (x_1) is 0.0. Use these two independent thermodynamic properties to find the following properties from Table A-4 in the appendix by interpolation: h_1 = 191.8 kJ/kg, s_1 = 0.6492 kJ/kg · K, and v_1 = 0.00101 m^3/kg.

2. **Find the thermodynamic properties at State 2, the boiler inlet.**

   At State 2, the pressure (P_2) is 10 megapascals and the entropy (s_2) is equal to the entropy at State 1 (s_1) because the pump is an isentropic process. In this state, the water is a *compressed liquid.* You only need to find the enthalpy at State 2 (h_2) from the compressed liquid table for water, because the specific volume (v_2) at State 2 is nearly the same as it is at State 1 (water is nearly incompressible). By interpolating Table A-2 in the appendix, you find that h_2 = 201.9 kJ/kg.

3. **Find the thermodynamic properties at State 3, the turbine inlet.**

   At State 3, the pressure (P_3) is 10 megapascals, and the temperature (T_3) is 800 degrees Celsius. Under these conditions, the water is a superheated vapor. Use these two properties to find the enthalpy (h_3) and entropy (s_3) from Table A-5 in the appendix. By interpolation, you find that h_3 = 4,115 kJ/kg and s_3 = 7.408 kJ/kg · K.

4. **Find the thermodynamic properties at State 4, the condenser inlet.**

At State 4, the pressure (P_4) is 10 kilopascals, and the entropy (s_4) is equal to the entropy at State 3 (s_3) because the turbine is an isentropic process. At the condenser inlet, the water is usually a liquid-vapor mixture, and, in this problem, you know the saturation pressure. Use these two properties to find the enthalpy (h_4) at State 4 from Table A-4 in the appendix. By interpolating the table, you find that h_4 = 2,348 kJ/kg.

5. **Find the steam quality at State 4, the condenser inlet.**

At State 4, the pressure (P_4) is 10 kilopascals, and the enthalpy (h_4) is 2,248 kilojoules/kilogram. You first find the enthalpy of both the saturated vapor (h_g) and the saturated liquid (h_f) at State 4. Use Table A-4 in the appendix to find h_f and h_g: h_f = 191.8 kJ/kg and h_g = 2,585 kJ/kg.

Compute the quality at State 4 using the following equation, as shown in Chapter 3:

$$x_4 = \frac{h_4 - h_f}{h_g - h_f} = \frac{(2{,}348 - 191.8) \text{ kJ/kg}}{(2{,}585 - 191.8) \text{ kJ/kg}} = 0.901$$

6. **Find the pump power.**

You can determine the pump power (rate of doing work) by using either of two methods. Because the pump compresses a liquid, the specific volume (v_1) doesn't change very much because liquid water is a nearly incompressible substance. You can use the following equations to calculate the pump power:

$$\dot{W}_{in} = \dot{m} v_1 (P_2 - P_1)$$

$$\dot{W}_{in} = \left(60 \ \frac{\text{kg}}{\text{s}}\right)\left(0.00101 \ \frac{\text{m}^3}{\text{kg}}\right)(10{,}000 - 10) \text{ kPa}\left(\frac{1 \text{ MW}}{1{,}000 \text{ kJ/s}}\right) = 0.61 \text{ MW}$$

Or, you can compute the power from the change in enthalpy between States 1 and 2, if these are known values, as they are in this example. The result is the same.

$$\dot{W}_{in} = \dot{m}(h_2 - h_1) = \left(60 \ \frac{\text{kg}}{\text{s}}\right)(201.9 - 191.8) \ \frac{\text{kJ}}{\text{kg}}\left(\frac{1 \text{ MW}}{1{,}000 \text{ kJ/s}}\right) = 0.61 \text{ MW}$$

7. **Find the turbine power.**

The turbine power is calculated from the change in enthalpy from States 3 to 4:

$$\dot{W}_{out} = \dot{m}(h_3 - h_4) = \left(60 \ \frac{\text{kg}}{\text{s}}\right)(4{,}115 - 2{,}348) \ \frac{\text{kJ}}{\text{kg}}\left(\frac{1 \text{ MW}}{1{,}000 \text{ kJ/s}}\right) = 106 \text{ MW}$$

8. **Find the heat input from the steam generator.**

The heat input is calculated from the change in enthalpy from States 2 to 3:

$$\dot{Q}_{in} = \dot{m}(h_3 - h_2) = \left(60 \ \frac{\text{kg}}{\text{s}}\right)(4{,}115 - 201.9) \ \frac{\text{kJ}}{\text{kg}}\left(\frac{1 \text{ MW}}{1{,}000 \text{ kJ/s}}\right) = 235 \text{ MW}$$

9. **Find the heat rejected by the condenser.**

The heat rejected is calculated from the change in enthalpy from States 4 to 1:

$$\dot{Q}_{out} = \dot{m}(h_4 - h_1) = \left(60 \ \frac{kg}{s}\right)(2,348 - 191.8) \ \frac{kJ}{kg}\left(\frac{1 \ MW}{1,000 \ kJ/s}\right) = 129 \ MW$$

Calculating Rankine cycle efficiency

Finding efficiency just means figuring out how much useful work you get out of a cycle compared to how much energy you put in to get that work done. Because a Rankine cycle uses part of its work to compress the water, you have to subtract the power into the pump from the power out of the turbine to get the net rate of work ($\dot{W}_{net}$). You calculate the efficiency like this:

$$\eta_{th} = \frac{\dot{W}_{net}}{\dot{Q}_{in}} = \frac{\dot{W}_{out} - \dot{W}_{in}}{\dot{Q}_{in}} = \frac{(106 - 0.6) \ MW}{235 \ MW} = 45\%$$

Notice how small the work of the pump is compared to the output of the turbine.

You can determine the *back work ratio* for the Rankine cycle just as you did for the Brayton cycle in Chapter 10. The back work ratio compares the amount of work input required by the pump to the work output from the turbine. You calculate the back work ratio as shown in this equation:

$$r_{bw} = \frac{\dot{W}_{in}}{\dot{W}_{out}} = \frac{0.61 \ MW}{106 \ MW} = 0.6\%$$

The back work ratio for the Brayton cycle example in Chapter 10 is 39 percent. The amount of work per unit mass required to compress a gas is considerably more than the work required to compress a liquid.

Calculating Rankine cycle irreversibility

When you examine a cycle using the second law of thermodynamics, you figure out where the inefficiencies come from, which may make you think of ways to make them more efficient. Inefficiency arises from friction generated by parts and fluids that are moving. It also occurs in heat transfer processes. Friction in a fluid; heat transfer in a boiler and a condenser; and heat losses from piping, turbine housings, and boiler walls, among other places in a cycle, cause irreversibility. You find out about reversibility and irreversibility in Chapter 9. When you calculate the irreversibility of a process or a cycle, you're quantifying the change in entropy of the system. Use the following equation to calculate the irreversibility of any thermodynamic process:

$$\dot{I} = \dot{m}T_0\left[(s_{final} - s_{initial}) - \frac{q_{in}}{T_{in}} + \frac{q_{out}}{T_{out}}\right]$$

You need to know the following points to use this equation:

✔ T_0 is the ambient or dead state temperature of the environment that the process operates in. For this Rankine cycle example, the cycle operates where the ambient temperature is 25 degrees Celsius.

✔ s_{final} is the entropy of the water or steam at the end of a process.

✔ $s_{initial}$ is the entropy of the water or steam at the beginning of a process.

✔ q_{in} is the amount of heat transferred from the high-temperature reservoir to the system in the steam generator.

✔ q_{out} is the amount of heat transferred from the system to the low-temperature reservoir in the condenser.

✔ T_{in} is the temperature of the high-temperature thermal energy reservoir that transfers heat to the system. Usually, you use the maximum temperature of the cycle for the heat input reservoir temperature. For this Rankine cycle example, say that the combustion gases in the steam generator are at 850 degrees Celsius. That gives a difference of 50 degrees Celsius between the combustion gas and the steam exit temperature in the boiler, which is typical of real boilers.

✔ T_{out} is the temperature of the low-temperature thermal energy reservoir that the system rejects heat to. The heat rejection reservoir temperature is the sink temperature (coolant temperature). In this example, the coolant temperature is 20 degrees Celsius.

You can use the previous equation to determine the irreversibilities associated with the Rankine cycle. Isentropic processes are reversible, so the irreversibility or change in entropy ($s_{final} - s_{initial}$) for the pressurization of the water in the pump and expansion of the steam in the turbine is zero. To compute the irreversibility of the heat input from the steam generator, use absolute temperatures for the ambient environment (T_0) and the combustion gases (T_H) heating up the water in the steam generator. For this example, T_0 is 298 K and T_H is 1,123 K.

$$\dot{I}_{S.G.} = \dot{m}T_0\left[s_3 - s_2 - \frac{q_{in}}{T_H}\right]$$

$$\dot{I}_{S.G.} = \left(60\ \frac{kg}{s}\right)298\ K\left[(7.408 - 0.6492)\ \frac{kJ}{kg\cdot K} - \frac{3,913.1\ kJ/kg}{1,123\ K}\right]\left(\frac{1\ MW}{1,000\ kJ/s}\right)$$

$$= 58.5\ MW$$

Find the rate of irreversibility of the heat rejected by the condenser by using the sink temperature for the low-temperature reservoir where T_L is 293K:

$$\dot{i}_{cond} = \dot{m} T_0 \left[s_1 - s_4 + \frac{q_{out}}{T_L} \right]$$

$$\dot{i}_{cond} = \left(60\ \frac{kg}{s} \right) 298\ K \left[(0.6492 - 7.408)\frac{kJ}{kg \cdot K} + \frac{2{,}156.2\ kJ/kg}{293\ K} \right] \left(\frac{1\ MW}{1{,}000\ kJ/s} \right) = 10.7\ MW$$

The irreversibilities become smaller when the heat transfer processes occur with a smaller difference between the temperatures of the reservoir and the working fluid. You can see this by comparing the small condenser irreversibility to the large boiler irreversibility. Much of the irreversibility in the boiler occurs from the large temperature difference between the combustion gases and the water in the tubes.

Improving the Rankine Cycle with Reheat

The efficiency of the Rankine cycle can be improved by reducing the irreversibilities. The biggest gain in efficiency occurs by decreasing the temperature difference between the combustion gases and the steam in the boiler. One way to decrease this average temperature difference is to increase the average temperature of the steam in the boiler by using a higher boiler pressure. One consequence of increasing the boiler pressure is that the steam quality at the exit of the turbine decreases for a fixed maximum boiler temperature. You can increase the steam quality at the turbine exit by separating the turbine into multiple stages and reheating the steam between stages, as shown in Figure 12-3. Increasing the steam quality avoids having excessive moisture, which forms droplets, in the final stages of the turbine.

The Rankine cycle with reheat has a high-pressure and a low-pressure turbine, and an additional heat exchanger is put inside the steam generator to provide the reheat. The practical limit for the number of turbine stages is two for a subcritical steam generator (less than 22 megapascals of operating pressure) and three for a supercritical steam generator.

Figure 12-4 shows the *T-s* diagram of the Rankine cycle modified with reheat. The high-pressure turbine is the isentropic line between States 3 and 4. The steam is reheated at constant pressure from States 4 to 5. Usually the optimal pressure for the reheat is 25 percent of the maximum operating pressure. Then the low-pressure turbine expands the steam from States 5 to 6. Lastly, the condenser removes heat from the steam between State 6 and State 1. A three-stage turbine Rankine cycle would show one more reheat section and a mid-pressure turbine stage on the *T-s* diagram.

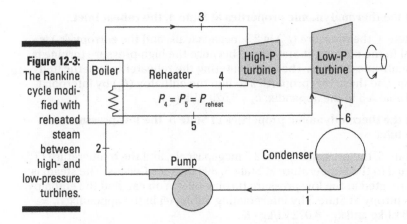

Figure 12-3:
The Rankine cycle modified with reheated steam between high- and low-pressure turbines.

You can see in Figure 12-4 that if the steam was expanded from State 3 all the way to the condenser pressure, the quality of the steam would be lower than the steam quality at State 6.

The upcoming steps use the same specifications as the example in the section "Analyzing the Cycle Using Steam Tables" but now with reheat pressure (State 4) at 2.5 megapascals. In your analysis, you can find the pump work, turbine work, heat input, heat rejected, steam quality at the turbine exit, and efficiency for comparison to the cycle without reheat. You can use already known properties for States 1, 2, and 3.

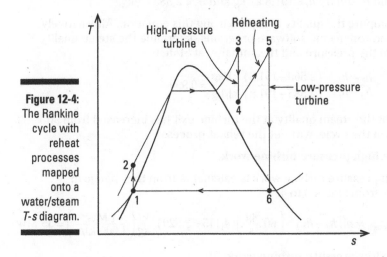

Figure 12-4:
The Rankine cycle with reheat processes mapped onto a water/steam *T-s* diagram.

Follow these steps to analyze a simple Rankine cycle modified with reheat:

1. **Find the thermodynamic properties at State 4, the reheat inlet.**

   At State 4, the pressure (P_4) is 2.5 megapascals, and the entropy (s_4) is equal to the entropy at State 3 (s_3) because the high-pressure turbine is an isentropic process. The water entering the reheater is superheated steam. Use these two properties to find the enthalpy (h_4) by interpolating Table A-5 in the appendix: h_4 = 3,529 kJ/kg.

2. **Find the thermodynamic properties at State 5, the low-pressure turbine inlet.**

   At State 5, the pressure (P_5) is 2.5 megapascals, and the temperature (T_5) is equal to the temperature at State 3, 800 degrees Celsius. The steam is superheated at the low-pressure turbine inlet. You can find the enthalpy and entropy at State 5 by interpolating Table A-5 in the appendix: h_5 = 4,148 kJ/kg and s_5 = 8.072 kJ/kg · K.

3. **Find the thermodynamic properties at State 6, the condenser inlet.**

   At State 6, the pressure (P_6) is 10 kilopascals, and the entropy (s_6) is equal to the entropy at State 5 (s_5) because the turbine is an isentropic process. The water is a liquid-vapor mixture at the turbine exit. Use these two properties to find the enthalpy (h_6) at State 6 in Table A-4 in the appendix: h_6 = 2,560 kJ/kg.

4. **Find the steam quality at State 6, the condenser inlet.**

   At State 6, the pressure (P_6) is 10 kilopascals, and the enthalpy (h_6) is 2,560 kilojoules/kilogram. First, find the enthalpy of the saturated vapor (h_g) and the saturated liquid (h_f) at State 6. Use Table A-4 in the appendix to find h_f and h_g: h_f = 191.8 kJ/kg and h_g = 2,585 kJ/kg.

   Then compute the quality at State 6 using this equation. Alternatively, most thermodynamic software programs determine the steam quality (x_6) from the pressure and the enthalpy at State 6:

   $$x_6 = \frac{h_6 - h_f}{h_g - h_f} = \frac{(2,560 - 191.8) \text{ kJ/kg}}{(2,585 - 191.8) \text{ kJ/kg}} = 0.990$$

   Note that the steam quality at the turbine exit has increased from 90 percent in the cycle, without the reheat process.

5. **Find the high-pressure turbine work.**

   The high-pressure turbine work is calculated from the change in enthalpy from States 3 to 4:

   $$\dot{W}_{\text{hi-turb}} = \dot{m}(h_3 - h_4) = \left(60 \ \frac{\text{kg}}{\text{s}}\right)(4,115 - 3,529) \ \frac{\text{kJ}}{\text{kg}}\left(\frac{1 \text{ MW}}{1,000 \text{ kJ/s}}\right) = 35 \text{ MW}$$

6. **Find the low-pressure turbine work.**

   The low-pressure turbine work is calculated from the change in enthalpy from States 5 to 6:

$$\dot{W}_{\text{lo-turb}} = \dot{m}(h_5 - h_6) = \left(60\ \frac{\text{kg}}{\text{s}}\right)(4{,}148 - 2{,}560)\ \frac{\text{kJ}}{\text{kg}}\left(\frac{1\ \text{MW}}{1{,}000\ \text{kJ/s}}\right) = 95\ \text{MW}$$

7. **Find the heat input to the steam generator.**

   The heat input includes both the first heating and the reheating of the steam. It's calculated from the change in enthalpy from States 2 to 3 and from States 4 to 5:

   $$\dot{Q}_{\text{in}} = \dot{Q}_{\text{boiler}} + \dot{Q}_{\text{reheat}}$$
   $$\dot{Q}_{\text{in}} = \dot{m}(h_3 - h_2 + h_5 - h_4)$$
   $$\dot{Q}_{\text{in}} = \left(60\ \frac{\text{kg}}{\text{s}}\right)(4{,}115 - 201.9 + 4{,}148 - 3{,}529)\ \frac{\text{kJ}}{\text{kg}}\left(\frac{1\ \text{MW}}{1{,}000\ \text{kJ/s}}\right) = 272\ \text{MW}$$

8. **Find the heat rejected by the condenser.**

   The heat rejected is calculated from the change in enthalpy from States 6 to 1:

   $$\dot{Q}_{\text{out}} = \dot{m}(h_6 - h_1) = \left(60\ \frac{\text{kg}}{\text{s}}\right)(2{,}560 - 191.8)\ \frac{\text{kJ}}{\text{kg}}\left(\frac{1\ \text{MW}}{1{,}000\ \text{kJ/s}}\right) = 142\ \text{MW}$$

9. **Find the efficiency of the cycle.**

   The efficiency is calculated from the net work of both turbines minus the pump work and the heat input to the cycle:

   $$\eta_{\text{th}} = \frac{\dot{W}_{\text{net}}}{\dot{Q}_{\text{in}}} = \frac{\dot{W}_{\text{hi-turb}} + \dot{W}_{\text{lo-turb}} - \dot{W}_{\text{in}}}{\dot{Q}_{\text{in}}} = \frac{(35 + 95 - 0.6)\ \text{MW}}{272\ \text{MW}} = 48\%$$

   Notice the efficiency of the Rankine cycle with reheat has increased compared to the case without reheat. The pump work remains the same, but the total net work from the cycle increases from 105 to 135 megawatts, while the heat input increases from 235 to 272 megawatts.

Improving the Rankine Cycle with Regeneration

Heating the water in a boiler generates a significant amount of irreversibility because the combustion gas temperature is so much hotter than the water temperature. In Chapter 8, I discuss how heat transfer between objects with a large temperature difference generates more irreversibility than heat transfer between objects with a small temperature difference. Irreversibility can be reduced by heating the water with a heat source that isn't as hot as the combustion gases.

The efficiency of a heat engine cycle increases as irreversibility of a cycle decreases.

The best source for a low-temperature heating system is the steam expanding in the turbine. Yes, it sounds like you're robbing Peter to pay Paul, but making some of the steam that could be doing work do heating instead brings significant gains in efficiency. Extracting steam from the turbine to preheat the boiler *feedwater* (the water that feeds the steam generator) is called *regeneration*.

Modern steam power plants use two kinds of regeneration systems, each called a feedwater heater (FWH):

✓ An *open feedwater heater* mixes the steam extracted from the turbine directly with the water leaving the condenser.

✓ A *closed feedwater heater* does not mix the steam from the turbine with the water. Instead, the steam warms the feedwater in a heat exchanger that keeps the two fluids separate.

Figure 12-5 shows a diagram of a Rankine cycle modified with an open FWH. In Figure 12-5, y is the fraction of steam drawn off the turbine that goes into the FWH, and $1 - y$ is the rest of the steam that goes to the condenser.

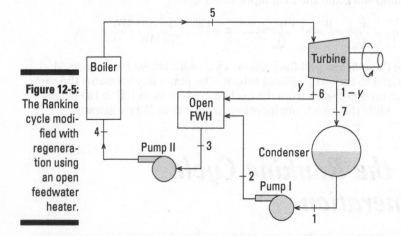

Figure 12-5:
The Rankine cycle modified with regeneration using an open feedwater heater.

Figure 12-6 shows the open FWH process on a *T-s* diagram of the Rankine cycle. The hot steam extracted from the turbine at State 6 in the diagram mixes with the cold feedwater at State 2 and warms the water up to State 3.

Because the steam extracted from the turbine is at a higher pressure than the water coming from the condenser at State 1, a low-pressure pump is inserted between the condenser and the FWH. This pump brings the pressure of the water at State 2 up to the pressure of the extracted steam at State 6. A second pump is used after the FWH to bring the pressure of the feedwater up to the steam generator operating pressure at State 4.

Figure 12-6:
The Rankine cycle with open feedwater heater regeneration mapped onto a water/steam *T-s* diagram.

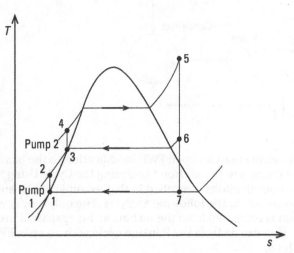

Figure 12-7 shows a diagram of a Rankine cycle modified with a closed FWH. The steam and the feedwater don't mix in the heat exchanger, so they can be at different pressures; therefore, you don't need an additional pump as in the open FWH system. The extracted steam from the turbine at State 5 is condensed in the FWH to State 6. The condensed water goes through a trap with a valve that throttles it to a lower pressure at State 7 so it can go back to the condenser and rejoin the feedwater. A *trap* is a container that collects water so it can drain through the throttling valve while trapping the steam in the FWH.

Each type of regenerator system has its own advantages and disadvantages. The open FWH has a simple fluid-mixing heat exchanger but requires an additional pumping stage. The closed FWH has a more complex heat exchanger, which is less effective than the fluid-mixing heat exchanger. Although it doesn't need an additional pump, it requires a trap, which is a fairly simple device. Most power plants have several stages of regeneration and use a combination of open and closed feedwater heaters.

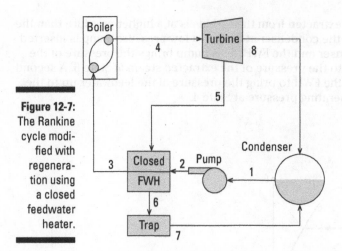

Figure 12-7:
The Rankine
cycle modi-
fied with
regenera-
tion using
a closed
feedwater
heater.

The following example uses an open FWH modification to the basic Rankine cycle example I discuss in the section "Analyzing the Cycle Using Steam Tables." All the specifications described in that example apply, and I call this the "baseline example" in the following analysis. The only new specification is that the steam is extracted from the turbine at 1-megapascal pressure for the open FWH. You can analyze the Rankine cycle with an open FWH by following these steps:

1. **Find the thermodynamic properties at State 2, the low-pressure pump outlet.**

   At State 2, the pressure (P_2) is 1 megapascal, and the entropy (s_2) is equal to the entropy at State 1 (s_1) because the low-pressure pump is an isentropic process. At the low-pressure pump outlet, the water is a compressed liquid. Use these two properties to find the enthalpy (h_2) from Table A-2 in the appendix. By interpolating the table, you find that h_2 = 192.8 kJ/kg.

   Note that the properties at State 1, the pump inlet, are the same as those in the baseline example.

2. **Find the thermodynamic properties at State 3, the high-pressure pump inlet.**

   At State 3, the pressure (P_3) is 1 megapascal, and the water is a saturated liquid, so the quality (x_3) is equal to 0. Use these two properties to find the enthalpy (h_3) and the entropy (s_3) by interpolating Table A-4 in the appendix: h_3 = 762.8 kJ/kg and s_3 = 2.139 kJ/kg · K.

3. **Find the thermodynamic properties at State 4, the steam generator inlet.**

At State 4, the pressure (P_4) is 10 megapascals, and the entropy (s_4) is equal to the entropy at State 3 (s_3) because the high-pressure pump is an isentropic process. The water is a compressed liquid at this state. You find the enthalpy at State 4 by interpolating Table A-2 in the appendix: $h_4 = 773.1$ kJ/kg.

Note that the properties at State 5, the turbine inlet, are the same as they are in the baseline example State 3. $h_5 = 4,115$ kJ/kg and $s_6 = 7.408$ kJ/kg · K.

4. **Find the thermodynamic properties at State 6, the steam extraction point.**

At State 6, the pressure (P_6) is 1 megapascal, and the entropy (s_6) is equal to the entropy at State 5 (s_5) because the turbine is an isentropic process. At this state, the water is superheated steam. Use these two properties to find the enthalpy at State 6 by interpolating the data in Table A-5 in the appendix: $h_6 = 3,226$ kJ/kg.

Note that the properties at State 7, the condenser inlet, are the same as they are in the baseline example State 4. $h_7 = 2,348$ kJ/kg.

5. **Find the mass flow rate of steam flowing through the FWH.**

The mass of steam and water flowing into the FWH equals the mass flowing out. Makes sense, doesn't it? The conservation of mass equation for the FWH is $\dot{m}_3 = \dot{m}_2 + \dot{m}_6$.

The energy flowing into the FWH equals the energy flowing out. Because this is an ideal FWH, it's perfectly insulated and doesn't lose heat to the air around it. The conservation of energy equation for the FWH is $\dot{m}_3 h_3 = \dot{m}_2 h_2 + \dot{m}_6 h_6$.

Combine these two equations and solve for the mass flow rate of steam flowing into the FWH:

$$\dot{m}_6 = \dot{m}_3 \left[\frac{h_3 - h_2}{h_6 - h_2} \right] = (60 \text{ kg/s}) \left[\frac{(762.8 - 192.8) \text{ kJ/kg}}{(3,226 - 192.8) \text{ kJ/kg}} \right] = 11.3 \text{ kg/s}$$

Then the mass flow rate of steam that flows through the rest of the turbine and into the condenser is as follows:

$$\dot{m}_2 = \dot{m}_3 - \dot{m}_6 = (60 - 11.3) \text{ kg/s} = 48.7 \text{ kg/s}$$

6. **Find the low-pressure pump work.**

The low-pressure pump work is calculated from the change in enthalpy from States 1 to 2 and the mass flow rate of water going through it.

$$\dot{W}_{\text{lo-pump}} = \dot{m}_1 (h_2 - h_1) = \left(48.7 \ \frac{\text{kg}}{\text{s}} \right) (192.8 - 191.8) \ \frac{\text{kJ}}{\text{kg}} \left(\frac{1 \text{ MW}}{1,000 \text{ kJ/s}} \right) = 0.05 \text{ MW}$$

7. Find the high-pressure pump work.

The high-pressure pump work is calculated from the change in enthalpy from States 3 to 4 and the mass flow rate of water going through it.

$$\dot{W}_{\text{hi-pump}} = \dot{m}_3 (h_4 - h_3) = \left(60.0 \, \frac{\text{kg}}{\text{s}}\right)(773.1 - 762.8) \, \frac{\text{kJ}}{\text{kg}}\left(\frac{1 \text{ MW}}{1{,}000 \text{ kJ/s}}\right) = 0.62 \text{ MW}$$

8. Find the turbine work.

The turbine work has two different mass flows through it because some steam is extracted at State 6. The total work is calculated using the appropriate mass flow rate through each part of the turbine.

$$\dot{W}_{\text{out}} = \dot{m}_5 (h_5 - h_6) + \dot{m}_7 (h_6 - h_7)$$

$$\dot{W}_{\text{out}} = \left[60 \, \frac{\text{kg}}{\text{s}}(4{,}115 - 3{,}226) \, \frac{\text{kJ}}{\text{kg}} + 48.7 \, \frac{\text{kg}}{\text{s}}(3{,}226 - 2{,}348) \, \frac{\text{kJ}}{\text{kg}} \right]\left(\frac{1 \text{ MW}}{1{,}000 \text{ kJ/s}}\right)$$

$$= 96.1 \text{ MW}$$

9. Find the net work.

The net work is the sum of the turbine work and the work of both pumps:

$$\dot{W}_{\text{net}} = \dot{W}_{\text{out}} - \dot{W}_{\text{lo-pump}} - \dot{W}_{\text{hi-pump}} = (96.1 - 0.05 - 0.62) \text{ MW} = 95.4 \text{ MW}$$

10. Find the heat input to the steam generator.

The heat input is calculated from the change in enthalpy from States 3 to 4 and the total mass flow rate of water.

$$\dot{Q}_{\text{in}} = \dot{m}_3 (h_5 - h_4) = 60 \, \frac{\text{kg}}{\text{s}}(4{,}115 - 773.1) \, \frac{\text{kJ}}{\text{kg}}\left(\frac{1 \text{ MW}}{1{,}000 \text{ kJ/s}}\right) = 201 \text{ MW}$$

11. Find the heat rejected by the condenser.

The heat rejected is calculated from the change in enthalpy from State 7 to State 1 and the mass flow rate of water through it.

$$\dot{Q}_{\text{out}} = \dot{m}_7 (h_7 - h_1) = 48.7 \, \frac{\text{kg}}{\text{s}}(2{,}348 - 191.8) \, \frac{\text{kJ}}{\text{kg}}\left(\frac{1 \text{ MW}}{1{,}000 \text{ kJ/s}}\right) = 105 \text{ MW}$$

Notice the heat rejected with the regenerator is less than the heat rejected by the baseline case, which was 129 MW.

12. Find the efficiency of the cycle.

The efficiency is calculated from the net work and the heat input to the cycle.

$$\eta_{\text{th}} = \frac{\dot{W}_{\text{net}}}{\dot{Q}_{\text{in}}} = \frac{95.4 \text{ MW}}{201 \text{ MW}} = 48\%$$

Notice the efficiency of the Rankine cycle with regeneration has increased compared to the baseline case without regeneration (45 percent). The amount of increase in efficiency is about the same as in the case with reheat (48 percent). Many power plants use both reheat and regeneration to maximize efficiency. Additional stages of regeneration can bring the actual efficiency of modern plants up to about 50 percent.

Deviating from Ideal Behavior: Actual Rankine Cycle Performance

Real power plants have several differences from the ideal Rankine cycle assumptions. The water and steam flowing through the piping have friction, which causes pressure drops within the steam generator and condenser. Friction also occurs within the pump and the turbine; this friction generates irreversibilities. Friction causes a real pump to use more work input than an ideal pump and a real turbine to put out less work than an ideal turbine. The departure from ideal behavior, as shown in Figure 12-8, is captured by the isentropic efficiencies of the turbine and the pump with these equations:

$$\eta_{turbine} = \frac{h_3 - h_{4a}}{h_3 - h_{4s}}$$

$$\eta_{pump} = \frac{h_1 - h_{2s}}{h_1 - h_{2a}}$$

Figure 12-8 shows the how the actual Rankine cycle deviates from the ideal Rankine cycle. The states labeled 2a and 4a are the actual states at the pump and turbine exits, respectively. Because friction generates entropy, the entropy of the actual states is greater than the entropy of the ideal states, so they're to the right of the ideal conditions in the diagram. The states labeled 2s and 4s are the exit states for the isentropic compressor and turbine. State 3 is the exit of the ideal steam generator, while State 3a is the exit of the actual steam generator.

Other deviations from the ideal Rankine cycle include pressure drops in the boiler and in the condenser. Figure 12-8 shows that the actual pressure at the boiler exit at State 3a is less than the pressure at State 3 for the ideal boiler exit. Similarly, the pressure drop in the condenser makes the inlet pressure of the actual condenser at State 4a higher than the inlet pressure of an ideal condenser at State 4s.

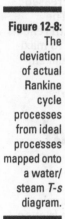

Figure 12-8:
The
deviation
of actual
Rankine
cycle
processes
from ideal
processes
mapped onto
a water/
steam *T-s*
diagram.

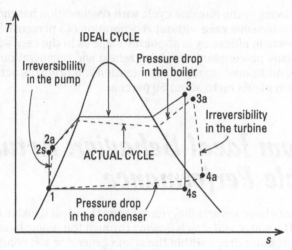

You can analyze a Rankine cycle using isentropic efficiency to model a real pump and turbine. You can also account for pressure drops in the steam generator and the condenser. Suppose the baseline Rankine cycle engine I discuss in the section "Analyzing the Rankine Cycle Using Steam Tables" has a pump with an isentropic efficiency of 95 percent and a turbine with an isentropic efficiency of 90 percent. Furthermore, the pressure drop in the steam generator causes the turbine inlet (State 3a) pressure to be 9.9 megapascals. And the inlet pressure to the condenser (State 4a) is 12 kilopascals because of its pressure drop. You can analyze the cycle and determine its efficiency by following these steps:

1. **Find the thermodynamic properties at State 2a, the pump outlet.**

   Use the isentropic efficiency equation for a pump to find the actual State 2.

   $$h_{2a} = h_1 - \frac{h_1 - h_{2s}}{n_{\text{pump}}} = 191.8 \ \frac{\text{kJ}}{\text{kg}} - \frac{(191.8 - 201.9) \ \text{kJ/kg}}{0.95} = 202.4 \ \text{kJ/kg}$$

   Note that the properties at State 1, the pump inlet, are the same as those in the baseline example.

2. **Find the thermodynamic properties at State 3a, the turbine inlet.**

   At State 3a, the pressure (P_3) is 9.9 megapascals, and the temperature is 800 degrees Celsius. At this state, the water is superheated steam. Use these two properties and interpolate to find the enthalpy (h_{3a}) and the entropy (s_{3a}), using Table A-5 in the appendix: h_{3a} = 4,115 kJ/kg and s_{3a} = 7.413 kJ/kg · K.

   Note that the enthalpy is the same as in the baseline example, but the entropy is a little higher.

3. Find the thermodynamic properties at State 4a, the condenser inlet.

At State 4a, the pressure (P_{4a}) is 12 kilopascals. First, find the enthalpy at State 4s. The entropy (s_{4s}) is equal to the entropy at State 3a (s_{3a}) for an isentropic turbine. You find the enthalpy at State 4s using pressure (P_{4a}) and entropy (s_{4s}) and interpolating Table A-4 in the appendix: $h_{4s} = 2,374$ kJ/kg.

Then use the isentropic efficiency equation for the turbine to find the actual State 4a:

$$h_{4a} = h_3 - \eta_{\text{turbine}}\left(h_3 - h_{4s}\right) = 4,115 \ \tfrac{\text{kJ}}{\text{kg}} - 0.9\left(4,115 - 2,374\right) \ \tfrac{\text{kJ}}{\text{kg}} = 2,548 \ \tfrac{\text{kJ}}{\text{kg}}$$

4. Find the steam quality at State 4a, the condenser inlet.

At State 4a, the pressure (P_{4a}) is 12 kilopascals, and the enthalpy (h_{4a}) is 2,548 kilojoules/kilogram. First, find the enthalpy of the saturated vapor (h_g) and the saturated liquid (h_f) at State 4a. Use Table A-4 in the appendix to find h_f and h_g: $h_f = 206.9$ kJ/kg and $h_g = 2,591$ kJ/kg.

Compute the quality at State 4a using this equation. Alternatively, most thermodynamic software programs determine the steam quality from the pressure and the enthalpy at 4a:

$$x_{4a} = \frac{h_{4a} - h_f}{h_g - h_f} = \frac{\left(2,548 - 206.9\right) \ \text{kJ/kg}}{\left(2,591 - 206.9\right) \ \text{kJ/kg}} = 0.982$$

5. Find the pump work.

To calculate the pump work, you use the change in enthalpy from States 1 to 2a and the mass flow rate of water going through it.

$$\dot{W}_{\text{in}} = \dot{m}\left(h_{2a} - h_1\right) = \left(60 \ \tfrac{\text{kg}}{\text{s}}\right)\left(202.4 - 191.8\right) \ \tfrac{\text{kJ}}{\text{kg}}\left(\frac{1 \ \text{MW}}{1,000 \ \text{kJ/s}}\right) = 0.64 \ \text{MW}$$

6. Find the turbine work.

You calculate the turbine work from the change in enthalpy from States 3 to 4a and the mass flow rate of water going through it.

$$\dot{W}_{\text{out}} = \dot{m}\left(h_3 - h_{4a}\right) = \left(60 \ \tfrac{\text{kg}}{\text{s}}\right)\left(4,115 - 2,548\right) \ \tfrac{\text{kJ}}{\text{kg}}\left(\frac{1 \ \text{MW}}{1,000 \ \text{kJ/s}}\right) = 94 \ \text{MW}$$

7. Find the heat input from the steam generator.

Use the change in enthalpy from State 2a to State 3 and the mass flow rate of water.

$$\dot{Q}_{\text{in}} = \dot{m}\left(h_3 - h_{2a}\right) = \left(60 \ \tfrac{\text{kg}}{\text{s}}\right)\left(4,115 - 202.4\right) \ \tfrac{\text{kJ}}{\text{kg}}\left(\frac{1 \ \text{MW}}{1,000 \ \text{kJ/s}}\right) = 235 \ \text{MW}$$

8. **Find the heat rejected by the condenser.**

   Calculate the heat rejected from the change in enthalpy from States 4a to 1 and the mass flow rate of water through it.

   $$\dot{Q}_{out} = \dot{m}(h_{4a} - h_1) = \left(60 \ \frac{kg}{s}\right)(2,548 - 191.8) \ \frac{kJ}{kg}\left(\frac{1 \ MW}{1,000 \ kJ/s}\right) = 141 \ MW$$

   Notice the heat rejected in the actual condenser is more than the heat rejected by the baseline case of 129 megawatts. This difference is because the real turbine can't extract as much work from the cycle as an ideal turbine.

9. **Find the efficiency of the cycle.**

   Use the net work and the heat input to the cycle.

   $$\eta_{th} = \frac{\dot{W}_{net}}{\dot{Q}_{in}} = \frac{\dot{W}_{out} - \dot{W}_{in}}{\dot{Q}_{in}} = \frac{(94 - 0.6) \ MW}{235 \ MW} = 39.7\%$$

   Notice the efficiency of the actual Rankine cycle is lower than the efficiency of the ideal baseline case (45 percent). This difference probably comes as no surprise to you, because friction and heat loss occur throughout the cycle.

Chapter 13

Cooling Off with Refrigeration Cycles

Air conditioning and its cousins, refrigerators and heat pumps, have changed many things over the last century. The population of the southeastern corner of the United States has blossomed thanks to air conditioning. Foods stay fresh longer and don't require salt preservation because of refrigeration. Houses can be heated relatively inexpensively in mild winter climates with heat pumps. Each of these appliances can be classified as refrigeration cycles (yes, even heat pumps).

In this chapter, you see how refrigeration cycles use work to move heat from one location to another. I cover two basic kinds of refrigeration cycles. One is called the *reverse Brayton cycle* because it operates much like a Brayton power cycle, except it uses a work input to move heat instead of a heat input to provide power. The reverse Brayton cycle uses a compressor, a turbine, and heat exchangers in its processes. The working fluid remains a gas through all processes in the cycle.

The other type of refrigeration cycle is called the *vapor compression system*. In this cycle, the working fluid changes between a liquid and a vapor at different points in the cycle, similar to the way the fluid changes phase in a Rankine cycle. However, calling the vapor compression cycle a reverse Rankine cycle isn't appropriate.

The thermodynamic analysis of a refrigeration cycle is very similar to the analysis of any of the heat engines I discuss in Chapters 10–12. Think of a refrigeration cycle as a heat engine operating in reverse. I discuss the fundamental concepts of heat and work in detail in Chapter 4.

Understanding the Basics of Refrigeration Cycles

Any machine that provides a cooling effect can be classified as a refrigeration cycle (the following sections in the chapter discuss the details of the different types of refrigeration cycles). Every refrigeration cycle operates between two energy reservoirs that are at different temperatures:

- **The low-temperature reservoir (T_L) is where you want the cooling.** In summer, the area you want to cool is the inside of your house via an air-conditioning system. Another area that needs to stay cool is the inside of your refrigerator. In order to cool a low-temperature reservoir, the refrigeration cycle must absorb energy from the reservoir. Heat absorbed from the low reservoir becomes the heat input to the cycle. In winter, a heat pump (which works opposite of an air conditioner) absorbs heat from the outdoors, so it basically cools the outdoors (although that's not its purpose).

- **The high-temperature reservoir (T_H) is where the refrigeration cycle rejects heat.** In the summer, an air conditioner expels heat to the outdoor environment. A refrigerator rejects heat to the surrounding kitchen. Heat rejected to the high reservoir becomes the heat output of the cycle. In winter, a heat pump rejects heat to the inside of your house to keep you warm.

At first, it may seem counterintuitive to absorb heat from a low-temperature reservoir and reject heat to a high-temperature reservoir. However, a refrigeration cycle absorbs heat from the low-temperature reservoir by a heat exchanger that's colder than the low-temperature reservoir. And the refrigeration cycle rejects heat to the high-temperature reservoir with a heat exchanger that's hotter than the high-temperature reservoir.

A refrigeration cycle requires a work input to move or pump heat from the low-temperature reservoir up to the high-temperature reservoir. The work is used to compress a gas, which basically turns the work into heat because the gas temperature increases with pressure. The heat rejected by a refrigeration cycle includes both the heat absorbed from the low-temperature reservoir and the heat generated by compressing the gas.

Chilling with the Reverse Brayton Cycle

When you fly in an airplane, the temperature outside is usually well below –40 degrees Celsius. Outside air is brought into the cabin to provide fresh air for the passengers and crew. You may think that it needs to be heated before it gets piped into those little nozzles above your head because it's so cold outside. Not so: The airplane is flying very fast, and the air has to slow down and pressurize before entering the cabin. All the kinetic energy of the incoming air and compression is therefore converted to heat. (I tell you about kinetic energy in Chapter 2.) In fact, the air warms up so much that it has to be cooled down to make it comfortable for everyone.

Airplanes don't use the same kind of air-conditioning system that you may be familiar with in your home. Instead, they use the *reverse Brayton cycle* to cool the air. Some thermodynamic textbooks may call the reverse Brayton cycle the *gas refrigeration cycle* because the fluid used in the cycle is air, a gas. In the section "Cooling with the Vapor-Compression Refrigerator," you see that the fluid used in the refrigeration cycle changes between a liquid and a vapor as it goes around the cycle.

Examining the four processes of the reverse Brayton cycle

Like many thermodynamic cycles, the reverse Brayton refrigeration cycle is made up of four processes, as shown in Figure 13-1. The cycle includes two heat transfer processes and two work processes to achieve cooling.

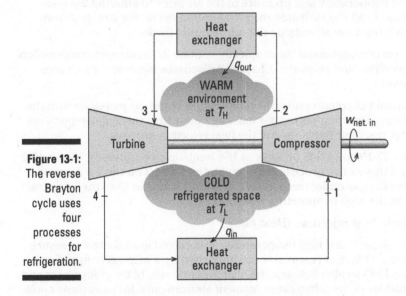

Figure 13-1: The reverse Brayton cycle uses four processes for refrigeration.

You can understand how the reverse Brayton cycle works by drawing the four processes on a temperature-entropy (*T-s*) diagram as shown in Figure 13-2. Chapter 3 tells you more about these kinds of diagrams. Chapters 8 and 9 tell you more about entropy.

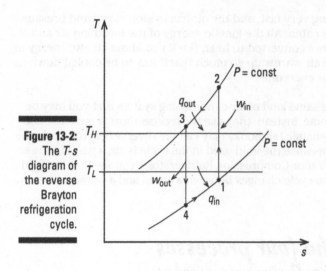

Figure 13-2:
The *T-s*
diagram of
the reverse
Brayton
refrigeration
cycle.

Here's how the reverse Brayton refrigeration cycle works:

1. **Isentropic compression. (Work in.)**

   Air enters the compressor of the refrigerator at State 1. State 1 is defined by the temperature and pressure of the air prior to entering the compressor. I call the air inside the refrigeration cycle "the gas" to distinguish it from the air outside the refrigeration cycle.

   The gas is compressed isentropically to State 2. (Isentropic compression is reversible and adiabatic. Chapter 8 discusses isentropic processes in detail.)

   As a result of compression, the temperature of the gas increases with the pressure. The compressed gas must be hotter than the high-temperature energy reservoir that's used in the heat rejection process.

   Figure 13-1 shows this process as the isentropic compressor. Figure 13-2 shows this as the process between States 1 and 2 on the *T-s* diagram. Notice the gas outlet temperature at State 2 is higher than the temperature of the high-temperature reservoir (T_H).

2. **Isobaric heat rejection. (Heat out.)**

   The high-pressure, high-temperature gas is cooled in a constant-pressure process. This heat removal is accomplished by a heat exchanger (see Figure 13-1) located between the high-pressure side of the compressor and the turbine of the refrigerator. Ambient air from outside the airplane cools the hot compressed gas inside the heat exchanger. The gas inside the heat exchanger is hotter than the air outside the airplane. Figure 13-2 shows this process between States 2 and 3 on the *T-s* diagram.

3. Isentropic expansion. (Work out.)

The high-pressure gas is expanded isentropically through a turbine to extract work from the cycle. As the gas decreases in pressure, its temperature also decreases. The turbine is connected to the compressor on a single shaft so that the turbine is mechanically coupled to provide some power for the compressor.

Figure 13-1 shows this process as the isentropic turbine. Figure 13-2 shows this as the process between States 3 and 4 on the *T-s* diagram.

4. Isobaric heat addition. (Heat in.)

The low-pressure, low-temperature gas is heated in a constant-pressure process. This heat addition is accomplished by a heat exchanger (see Figure 13-1) located between the low-pressure side of the compressor and the turbine of the refrigerator. Heat is not added to the system by a combustion process like a heat engine. Ambient air from inside the airplane warms the cold expanded gas inside the heat exchanger. The gas inside the heat exchanger is colder than the air inside the cabin of the airplane.

Figure 13-2 shows this process between States 4 and 1 on the *T-s* diagram.

Analyzing the cycle with constant specific heat

You can assume the specific heat of the gas within the refrigeration cycle remains constant. The temperature change of the gas in the cycle isn't very large, so you don't have to worry about how the specific heat changes with temperature as you do in a Brayton, Otto, or diesel cycle heat engine. In heat engines, the combustion process creates very high gas temperatures in the cycle, so the specific heat changes throughout the cycle, and using variable specific heat gives more accurate results in your analysis.

You can analyze an example of a simple reverse Brayton cycle given the following information and referring to Figures 13-1 and 13-2:

- ✔ The air temperature (T_1) and pressure (P_1) entering the compressor at State 1 are 288 Kelvin and 300 kilopascals, respectively.

- ✔ The pressure ratio (r_p) of the compressor is 3. The *pressure ratio* is the amount of air compression that the compressor provides.

- ✔ The temperature of the air entering the turbine at State 3 (T_3) is 323 Kelvin.

- ✔ The temperature of the warm environment (T_H) is 313 Kelvin.

- ✔ The temperature of the cold environment (T_L) is 293 Kelvin.

In the energy equations, you can use constant pressure specific heat (c_p) in place of the change in enthalpy for each heat transfer and work process, so you can just work with temperature directly without having to look up enthalpy values in a table. I discuss enthalpy and specific heat in Chapter 2. The energy equation for each process in the Brayton cycle is written as follows:

- **For heat in:** $q_{in} = (h_1 - h_4) = c_p(T_1 - T_4)$
- **For heat out:** $q_{out} = (h_2 - h_3) = c_p(T_2 - T_3)$
- **For compression, work in:** $w_{in} = (h_2 - h_1) = c_p(T_2 - T_1)$
- **For expansion, work out:** $w_{out} = (h_3 - h_4) = c_p(T_3 - T_4)$

You need to find the temperature at the endpoint of each process to determine the amount of energy associated with the processes. In the isentropic compression process, the temperature at State 2 can be found from State 1 using an isentropic process equation for an ideal gas (introduced in Chapter 5), as shown here:

$$\frac{T_2}{T_1} = \left(\frac{P_2}{P_1}\right)^{\frac{k-1}{k}}$$

Notice the fraction (P_2/P_1) in this equation is simply the pressure ratio (r_p) of the compressor. The variable k is the ratio of specific heats, which has a value of $k = 1.4$ for air at 25 degrees Celsius.

Follow these steps to analyze this reverse Brayton Cycle refrigerator example using constant specific heat:

1. **Find the temperature at the compressor exit (T_2) using the ideal gas relationships for isentropic processes:**

$$T_2 = T_1\left(\frac{P_2}{P_1}\right)^{\frac{k-1}{k}} = (288 \text{ K})\left(\frac{3}{1}\right)^{\frac{1.4-1}{1.4}} = 394 \text{ K}$$

When working with ideal-gas law equations, you always use absolute temperatures.

2. **Find the temperature at the turbine exit (T_4):**

$$T_4 = T_3\left(\frac{P_4}{P_3}\right)^{\frac{k-1}{k}} = (323 \text{ K})\left(\frac{1}{3}\right)^{\frac{1.4-1}{1.4}} = 236 \text{ K}$$

Finding the temperature at State 4 is just like finding the temperature at State 2, but instead of isentropic compression, you have isentropic expansion from States 3 to 4.

3. **Find the amount of energy exchanged in each process using the energy equations and the specific heat of air at 298 K from Table A-1 in the appendix:**

$$q_{in} = 1.005 \text{ kJ/kg} \cdot \text{K}(288 - 236) \text{ K} = 52 \text{ kJ/kg}$$

$$q_{out} = 1.005 \text{ kJ/kg} \cdot \text{K}(394 - 323) \text{ K} = 71 \text{ kJ/kg}$$

$$w_{in} = 1.005 \text{ kJ/kg} \cdot \text{K}(394 - 288) \text{ K} = 107 \text{ kJ/kg}$$

$$w_{out} = 1.005 \text{ kJ/kg} \cdot \text{K}(323 - 236) \text{ K} = 88 \text{ kJ/kg}$$

4. **Check your calculations to make sure the net heat transfer to the cycle equals the net work by the cycle:**

$$q_{in} - q_{out} = w_{out} - w_{in}$$

$$(52 - 71) \text{ kJ/kg} = (88 - 107) \text{ kJ/kg} = -19 \text{ kJ/kg}$$

The negative value means that more work goes into the cycle than comes out of the cycle because it's a refrigeration system, not a heat engine.

The amount of heat the refrigerator rejects to the hot reservoir equals the heat absorbed from the cold reservoir plus the net work input required by the compressor, which is a consequence of the conservation of energy.

Calculating the reverse Brayton cycle coefficient of performance

Any measure of performance compares what you want to what you provide. Finding the *coefficient of performance* (COP) for a cycle involves comparing how much heat you move from the low-temperature reservoir (q_{in}) to the net amount of work you put in (w_{net}). (I introduce the coefficient of performance in Chapter 7.) The COP isn't quite the same as efficiency, but it's similar in the fact that the higher the COP, the better the refrigerator is at moving heat for the least amount of work. You calculate the coefficient of performance like this:

$$\text{COP} = \frac{q_{in}}{w_{in} - w_{out}} = \frac{52 \text{ kJ/kg}}{(107 - 88) \text{ kJ/kg}} = 2.74$$

The best possible COP comes from a Carnot refrigerator, and it depends only on the temperatures of the high and low reservoirs (T_H and T_L), respectively. In general, the closer T_H and T_L are to each other, the higher the COP for both the Carnot refrigerator and an actual refrigeration cycle. In the Carnot refrigerator, the heat transfer is accomplished by constant-temperature processes, and no temperature difference exists between the reservoirs and their respective heat exchangers. In real refrigeration cycles, heat transfer requires a temperature difference between the reservoir and the gas inside the heat exchanger. This temperature difference in heat-transfer processes reduces the overall COP of real refrigeration cycles, and it generates irreversibility, which I discuss in the following section.

You calculate the Carnot COP for this cycle as follows:

$$\text{COP}_{\text{Carnot}} = \frac{T_L}{T_H - T_L} = \frac{313 \text{ K}}{(313 - 293) \text{ K}} = 15.7$$

The Carnot coefficient of performance is much higher than the COP of an actual cycle. Calculating the irreversibilities of the cycle in the following section shows you why the difference between the ideal Carnot refrigerator and the reverse Brayton refrigeration cycle is so great.

Calculating irreversibility for Brayton's refrigerator

When you examine a cycle using the second law of thermodynamics, you figure out where the inefficiencies come from, which may make you think of ways to make the cycle more efficient. Inefficiency arises from friction generated by parts and fluids that are moving. It also occurs in heat-transfer processes. Friction and heat losses in a cycle cause *irreversibility*. You find out about reversibility and irreversibility in Chapter 9.

When you calculate the irreversibility of a process or cycle, you're quantifying the change in entropy of the system. The following equation is used to calculate the irreversibility of any thermodynamic process:

$$i = T_0 \left[\left(s_{\text{final}} - s_{\text{initial}} \right) - \frac{q_{\text{in}}}{T_L} + \frac{q_{\text{out}}}{T_H} \right]$$

You need to know the following points to use this equation:

- T_0 is the ambient temperature of the environment that the cycle operates in. For refrigeration cycles, T_0 is the high-temperature reservoir, T_H.

- s_{final} is the entropy of the air at the end of a process. You need to know both the temperature and pressure of the air at the end of the process to find the entropy.

- s_{initial} is the entropy of the air at the beginning of a process.

- q_{in} is the amount of heat transferred from the low-temperature reservoir to the system.

✔ q_{out} is the amount of heat transferred to the high-temperature reservoir from the system.

✔ T_H is the temperature of the high-temperature thermal energy reservoir that the system rejects heat to.

✔ T_L is the temperature of the low-temperature thermal energy reservoir that transfers heat into the system.

You can use the equation previously given to determine the irreversibilities associated with the reverse Brayton cycle. Isentropic processes are reversible, so the irreversibility or change in entropy ($s_{final} - s_{initial}$) for the compressor and the turbine is zero. For the heat transfer processes, you calculate the change in entropy ($s_{final} - s_{initial}$) with the following equation:

$$s_{final} - s_{initial} = s^0_{final} - s^0_{initial} - R\ln\left(\frac{P_{final}}{P_{initial}}\right)$$

As I describe in Chapter 8, the temperature and pressure of a gas determine its entropy. The $s^0_{initial}$ and s^0_{final} entropy terms are a function of the air temperature at the beginning and end of a process, respectively. You can find these values in Table A-1 in the appendix. The ideal gas constant (R) for air is 0.287 kilojoules per kilogram-Kelvin. You can find out more about it in Chapter 3.

The following steps enable you to calculate the irreversibilities for the reverse Brayton refrigeration cycle example. Refer to Figures 13-1 and 13-2 for the diagrams of the cycle. Recall the high reservoir temperature is T_H = 313 Kelvin and the low reservoir temperature is T_L = 293 Kelvin.

1. **Find the change in entropy ($s_4 - s_1$) for the heat absorption process from States 4 to 1, as shown in Figure 13-2.**

$$s_{final} - s_{initial} = s^0_{final} - s^0_{initial} - R\ln\left(\frac{P_{final}}{P_{initial}}\right)$$

Because the heat absorption process occurs at constant pressure, the pressure term ($P_{final}/P_{initial}$) drops out. You can find values for s^0_1 and s^0_4 at temperatures T_1 = 288 Kelvin and T_4 = 236 Kelvin in Table A-1 of the appendix. The change in entropy is as follows:

$$s_1 - s_4 = s^0_1 - s^0_4 = (6.829 - 6.629)\ \text{kJ/kg} \cdot \text{K} = 0.20\ \text{kJ/kg} \cdot \text{K}$$

2. **Find the change in entropy ($s_2 - s_3$) for the heat rejection process from States 2 to 3.**

The change in entropy for the heat rejection process is equal to that of the heat absorption process but opposite in sign. The following equation describes the change in entropy for both heat transfer processes:

$$(s_1 - s_4) = -(s_3 - s_2)$$

3. **Compute the irreversibility of the heat absorption process (i_{41}), using absolute temperatures for the ambient temperature (T_0) and the low-temperature reservoir (T_L).**

$$i_{41} = T_0 \left[(s_1 - s_4) - \frac{q_{in}}{T_L} \right] = (313 \text{ K}) \left(0.20 \frac{\text{kJ}}{\text{kg} \cdot \text{K}} - \frac{52 \text{ kJ/kg}}{293 \text{ K}} \right) = 7.1 \text{ kJ/kg}$$

4. **Find the irreversibility of the heat-rejection process (i_{23}) by using the ambient air temperature (T_0) and the temperature of the high-temperature reservoir (T_H).**

$$i_{23} = T_0 \left[(s_3 - s_2) + \frac{q_{out}}{T_H} \right] = (313 \text{ K}) \left(-0.20 \frac{\text{kJ}}{\text{kg} \cdot \text{K}} + \frac{71 \text{ kJ/kg}}{313 \text{ K}} \right) = 8.4 \text{ kJ/kg}$$

The irreversibilities become smaller when the heat transfer processes occur with smaller temperature differences between the reservoir and the fluid in the system.

Cooling with the Vapor-Compression Refrigerator

Ever since movie houses opened their doors to entice passersby inside with cool air flowing out to the sidewalk, people have enjoyed comfort from air conditioning. The majority of today's air-conditioning systems use the vapor compression refrigeration cycle. Some very large commercial air-conditioning systems use an absorption cycle, which you can read more about in Chapter 18.

Warming up to refrigerants

A refrigerant is a fluid that meets several thermodynamic criteria for efficient use in a refrigeration cycle, such as boiling point, liquid and gas density, thermal conductivity, and heat capacity. The most commonly used refrigerant chemicals are fluorocarbons.

All refrigerants are assigned an R-number for identification by the American Society of Heating, Refrigeration, and Air Conditioning Engineers (ASHRAE). Refrigerant R-134a is one of the most popular refrigerants in use today; this book uses R-134a for example problems, and its thermodynamic properties appear in the appendix. R-12, which has the trade name "Freon," was once a very popular refrigerant — until it was found to deplete the earth's ozone layer. Many people still call refrigerants Freon, much like they may ask for a "Kleenex" when they want a facial tissue.

The vapor compression cycle uses a refrigerant that changes phase between liquid and vapor. In one part of the cycle, the refrigerant is a vapor and a compressor increases its pressure — hence the moniker *vapor compression refrigeration cycle.*

Examining the four processes in a vapor-compression refrigerator

The vapor-compression refrigeration cycle provides cooling using two heat exchangers, a compressor, and a throttling valve. Figure 13-3 shows a schematic of the cycle. The cold heat exchanger is called the *evaporator* because liquid refrigerant absorbs heat from the cool environment and evaporates into a vapor. The hot heat exchanger is called the *condenser* because refrigerant vapor is condensed back into a liquid as it rejects heat to the warm environment.

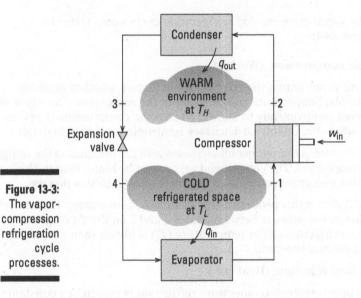

Figure 13-3: The vapor-compression refrigeration cycle processes.

You can figure out how the cycle works by drawing the path of each process on a temperature-entropy (*T-s*) diagram. The *T-s* diagram shown in Figure 13-4 illustrates how the temperature and entropy change for each process in the cycle.

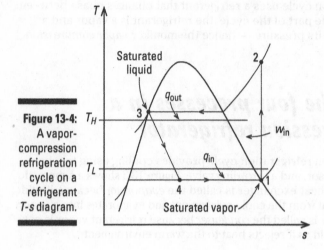

Figure 13-4:
A vapor-
compression
refrigeration
cycle on a
refrigerant
T-s diagram.

Here's how the vapor-compression refrigeration cycle works (refer to Figures 13-3 and 13-4):

1. **Isentropic compression. (Work in.)**

   Refrigerant vapor enters the compressor at State 1, which is normally defined by the temperature and pressure of the refrigerant. The vapor is compressed isentropically to State 2. (Isentropic compression is reversible and adiabatic. Chapter 8 discusses isentropic processes in detail.)

   The temperature of the refrigerant increases with the pressure as the refrigerant is compressed. The compressed vapor must be hotter than the high-temperature energy reservoir that's used in the heat rejection process.

   Figure 13-3 shows this process as the isentropic compressor. Figure 13-4 shows this as the process between States 1 and 2 on the *T-s* diagram. Notice the refrigerant outlet temperature (T_2) is higher than that of the high-temperature reservoir (T_H).

2. **Isobaric heat rejection. (Heat out.)**

   The high-pressure, high-temperature refrigerant is cooled in a constant-pressure process. A condenser located between the high-pressure side of the compressor and the expansion valve of the refrigerator (see Figure 13-3) removes the heat.

   Ambient air from a warm environment cools the hot compressed refrigerant vapor inside the condenser. This process requires the condenser temperature to be greater than the high-temperature reservoir temperature (T_H). The vapor inside the condenser is hotter than the warm environment, and the vapor changes phase into a liquid by the time it reaches the condenser outlet. In the ideal refrigeration cycle, the outlet conditions are saturated liquid.

For an air-conditioning system, the warm environment is the air outside your house; for a refrigerator, the warm environment is the kitchen. Figure 13-4 shows this process between States 2 and 3 on the *T-s* diagram.

3. **Isenthalpic expansion (throttling).**

The high-pressure refrigerant liquid passes through an expansion valve, shown in Figure 13-3 between States 3 and 4, to reduce its pressure. As the liquid decreases in pressure, its temperature also decreases. The main purpose of throttling the refrigerant is to make it colder than the low-temperature reservoir. The *enthalpy* of the refrigerant remains constant during the process, but the entropy increases, as shown on the *T-s* diagram in Figure 13-4.

A turbine could be used to decrease the pressure between States 3 and 4 (which make this cycle a reverse Rankine cycle), but extracting work from a saturated liquid while reducing its pressure isn't practical for several reasons. As a saturated liquid expands, it becomes a liquid-vapor mixture; turbines are efficient handling only a liquid or a vapor. Secondly, little work can be extracted from such a turbine compared to the turbine in the reverse Brayton cycle. Lastly, a turbine adds cost and another possible failure mode.

A *throttling valve* can be any device that provides a resistance to flowing liquid, usually by forcing the liquid through a small-diameter tube called a *capillary tube*. The resistance may also come from forcing the liquid to go through a small-diameter hole. Either method causes the fluid pressure to decrease as it squeezes through a tight passage.

No work comes out of a vapor-compression refrigeration cycle; there is only a work input.

4. **Isobaric heat addition. (Heat in.)**

The low-pressure, low-temperature refrigerant liquid is heated in a constant-pressure process in the evaporator shown in Figure 13-3. During the heating process, the liquid evaporates into a vapor. In an ideal refrigeration cycle, the vapor is saturated at the exit of the evaporator. Ambient air from a cold environment is used to warm the cold liquid inside the evaporator. The liquid inside the evaporator is colder than the cold environment.

When the refrigeration cycle is used in your house, the cold environment is the air inside the house; for your refrigerator, the cold environment is inside the fridge. Figure 13-4 shows this process between States 4 and 1 on the *T-s* diagram.

Analyzing the cycle with refrigerant property tables

You can analyze a simple vapor-compression refrigeration cycle using refrigerant R-134a, given the following information and referring to Figures 13-3 and 13-4:

- The evaporator pressure entering the compressor at State 1 (P_1) is 250 kilopascals.

- The condenser pressure leaving the compressor at State 2 (P_2) is 1,000 kilopascals.

- The temperature of the warm environment (T_H) is 310 Kelvin.

- The temperature of the cold environment (T_L) is 294 Kelvin.

Follow these steps to analyze a vapor-compression refrigeration cycle:

1. **Find the enthalpy (h_1) and entropy (s_1) at the compressor inlet.**

   Use the compressor inlet pressure P_1 of 250 kilopascals for saturated vapor ($x_1 = 1$) to find h_1 and s_1 from Table A-7 in the appendix: $h_1 = 396.1$ kJ/kg, and $s_1 = 1.73$ kJ/kg · K.

   (You can also see from Table A-7 that the saturation temperature is −4.2 degrees Celsius or 269 Kelvin, which is quite a bit colder than the low-temperature reservoir. This is basically the evaporator operating temperature.)

2. **Find the compressor outlet enthalpy (h_2).**

   The ideal compressor is isentropic, so the entropy at State 2 equals that at State 1. The refrigerant is a superheated vapor in this state. Use the compressor outlet pressure (P_2) of 1,000 kilopascals and the entropy (s_2) of 1.73 kilojoules per kilogram-Kelvin to find the enthalpy (h_2) from Table A-8 in the appendix: $h_2 = 425.1$ kJ/kg.

   (You can also use Table A-8 to find that the vapor temperature is 44.6 degrees Celsius or 317.6 Kelvin, which is quite a bit warmer than the high-temperature reservoir.)

3. **Find the condenser outlet enthalpy (h_3) and entropy (s_3).**

   The condenser pressure (P_3) is 1,000 kilopascals, and the refrigerant leaves the condenser as a saturated liquid. Use Table A-7 for these conditions to find the enthalpy and entropy: $h_3 = 255.5$ kJ/kg, and $s_3 = 1.188$ kJ/kg · K.

(You can see in Table A-7 that the saturation temperature at 1,000 kilopascals is 39.4 degrees Celsius or 312.4 Kelvin, so even at the condenser outlet, the temperature is warmer than the high-temperature reservoir. This means the condenser has a maximum temperature of 317.6 Kelvin and a minimum temperature of 312.4 Kelvin.)

The enthalpy remains constant in a throttling valve, so $h_4 = h_3$.

4. **Find the enthalpy and entropy of the saturated liquid and saturated vapor at the throttling valve exit.**

   The pressure at the exit of the throttling valve (P_4) is 250 kilopascals. Use Table A-7 in the appendix at this pressure to find the enthalpy and the entropy at both the saturated liquid and saturated vapor states: $h_f = 194.3$ kJ/kg and $h_g = 396.1$ kJ/kg; $s_f = 0.979$ kJ/kg · K and $s_g = 1.730$ kJ/kg · K.

5. **Find the refrigerant quality (x_4) at the throttling valve exit.**

   Quality is a measure of how much vapor mass is present in a liquid-vapor mixture. Use the throttling valve exit enthalpy (h_4), the saturated liquid enthalpy (h_f), and the saturated vapor enthalpy (h_g) in the following equation to determine the quality of the refrigerant at State 4:

   $$x_4 = \frac{h_4 - h_f}{h_g - h_f} = \frac{(255.5 - 194.3) \text{ kJ/kg}}{(396.1 - 194.3) \text{ kJ/kg}} = 0.303$$

6. **Find the refrigerant entropy (s_4) at the throttling valve exit.**

   Use the throttle exit quality (x_4), the saturated liquid entropy (s_f), and the saturated vapor entropy (s_g) in the following equation to determine the entropy of the refrigerant at State 4:

   $$s_4 = s_f + x_4(s_g - s_f) = 0.979 \text{ kJ/kg} \cdot \text{K} + 0.303(1.73 - 0.979) \text{ kJ/kg} \cdot \text{K}$$
   $$s_4 = 1.207 \text{ kJ/kg} \cdot \text{K}$$

As you work your analysis of a cycle, it's a good idea to determine entropy at each state. This lets you calculate irreversibility of the processes.

7. **Find the amount of energy exchanged in each process using the energy equations:**

   $$q_{in} = (h_1 - h_4) = (396.1 - 255.5) \text{ kJ/kg} = 140.6 \text{ kJ/kg}$$
   $$q_{out} = (h_2 - h_3) = (425.1 - 255.5) \text{ kJ/kg} = 169.6 \text{ kJ/kg}$$
   $$w_{in} = (h_2 - h_1) = (425.1 - 396.1) \text{ kJ/kg} = 29 \text{ kJ/kg}$$
   $$w_{out} = 0$$

8. **Check your calculations to make sure the net heat transfer to the cycle equals the net work by the cycle:**

$$q_{out} - q_{in} = w_{in}$$

$$(169.6 - 140.6) \text{ kJ/kg} = 29 \text{ kJ/kg}$$

The amount of heat the refrigerator rejects to the hot reservoir equals the heat absorbed from the cold reservoir plus the net work input required by the compressor.

Calculating the vapor-compression refrigerator coefficient of performance

I explain in detail what the coefficient of performance (COP) is in the earlier section "Calculating the reverse Brayton cycle coefficient of performance." The COP is basically a measure of how much heat a refrigeration cycle moves per unit of work input. The calculation of the coefficient of performance for the vapor-compression refrigeration cycle is similar to that of the reverse Brayton cycle. The only difference between the two calculations is that the vapor-compression cycle has no work output to include in determining the net work input. The calculation of the COP for the vapor-compression refrigeration cycle looks like this:

$$\text{COP} = \frac{q_{in}}{w_{in}} = \frac{140.6 \text{ kJ/kg}}{29 \text{ kJ/kg}} = 4.85$$

You can calculate the coefficient of performance for a Carnot refrigeration cycle operating between the two thermal energy reservoirs given in this example. The Carnot COP tells you the maximum amount of heat that can be moved between two thermal energy reservoirs per unit of work input. This knowledge allows you to compare how effective a vapor-compression refrigeration cycle is compared to the ideal refrigeration cycle.

You calculate the Carnot COP for this cycle as follows:

$$\text{COP}_{\text{Carnot, ref}} = \frac{T_L}{T_H - T_L} = \frac{294 \text{ K}}{(310 - 294) \text{ K}} = 18.4$$

The Carnot COP is greater than the vapor-compression refrigeration COP. Calculating the irreversibilities of the cycle in the following section shows you why the difference between the ideal Carnot refrigerator and the vapor-compression refrigeration cycle is so great.

Calculating vapor-compression refrigerator irreversibility

I discuss calculating irreversibility for refrigeration cycles in detail in the section "Calculating reverse Brayton cycle irreversibility." Basically, when you calculate the irreversibility of a process or cycle, you're quantifying the change in entropy of the system. You calculate the total irreversibility of the cycle by first calculating the irreversibility of each process; then you add them together to get the total irreversibility. This process allows you to compare the effectiveness of different systems to see which one gives the better performance.

Isentropic processes are reversible, so the irreversibility, or change in entropy $(s_{final} - s_{initial})$, for the compression of the refrigerant vapor in the compressor is zero. The irreversibility of the heat rejected to the high-temperature reservoir uses absolute temperatures for T_0 and T_H. The irreversibility of the heat absorbed from the low-temperature reservoir uses absolute temperature for T_0 and T_L. T_0 isn't necessarily the lowest temperature in the cycle; it's the lowest naturally occurring temperature in the environment, so T_0 equals T_H.

You can use the entropy values and the heat transfer values you calculated for the vapor compression cycle in the earlier section "Analyzing the cycle with refrigerant property tables" to find the irreversibility of each process by following these steps:

1. **Compute the irreversibility of the heat absorption process in the evaporator.**

   Use the entropy values you found for s_1 and s_4 and the heat absorbed from the cold reservoir (q_{in}) in the following equation:

   $$i_{41} = T_0\left[(s_1 - s_4) - \frac{q_{in}}{T_L}\right] = (310 \text{ K})\left[(1.730 - 1.207)\frac{\text{kJ}}{\text{kg} \cdot \text{K}} - \frac{140.6 \text{ kJ/kg}}{294 \text{ K}}\right] = 14.0 \text{ kJ/kg}$$

2. **Find the irreversibility of the heat rejection process in the condenser.**

   Use the entropy values you found for s_2 and s_3 and the heat rejected to the warm reservoir (q_{out}) in the following equation:

   $$i_{23} = T_0\left[(s_3 - s_2) + \frac{q_{out}}{T_H}\right] = (310 \text{ K})\left[(1.188 - 1.730)\frac{\text{kJ}}{\text{kg} \cdot \text{K}} + \frac{169.6 \text{ kJ/kg}}{310 \text{ K}}\right] = 1.6 \text{ kJ/kg}$$

3. **Calculate the irreversibility of the throttling process.**

Use the entropy values you found for s_3 and s_1 in the following equation. Note that no heat transfer is associated with the throttling process.

$$i_{34} = T_0 (s_4 - s_3) = (310 \text{ K})(1.207 - 1.188) \text{ kJ/kg} \cdot \text{K} = 5.9 \text{ kJ/kg}$$

The irreversibility in the evaporator and the condenser comes from transferring heat with a temperature difference between the refrigerant and the thermal energy reservoirs. The temperature difference (and the irreversibility) can be made smaller by using larger heat exchangers. But larger heat exchangers increase manufacturing costs and make the equipment larger. Consumers don't like bulky and expensive products; therefore, manufacturers make a trade-off to provide a reasonably sized heat exchanger for both the evaporator and the condenser. Irreversibility in the throttling valve is due to fluid friction as the fluid flows through the valve.

Warming Up with Heat Pumps

A heat pump is an air conditioner that can pump heat out of the house in the summer (cooling mode) and pump heat into the house during the winter (heating mode), as shown in Figure 13-5. The heat pump switches the operation of the evaporator and the condenser when changing from cooling mode to heating mode. Switching modes means a change in the flow direction of the refrigerant, using a reversing valve. In the heating mode, the evaporator is the outdoor heat exchanger, and the condenser is the indoor heat exchanger. I discuss the evaporator and the condenser in the previous section on the vapor-compression refrigeration cycle.

The heat source for a heat pump is usually the outdoor air. However, using the outdoor air is limited to regions where winter weather generally stays above freezing. When outdoor conditions are too cold, ice can form on the evaporator and prevent additional heat from being absorbed by the system, resulting in a cold house. This problem can be avoided by using the ground or water (such as a pond, lake, or river) as a heat source, as long as the heat exchanger is buried below the frost line.

Examining the four processes in a heat pump

A heat pump has the same four thermodynamic processes used in the vapor-compression refrigeration cycle (refer to the previous section). In this section, I briefly describe the processes as they apply to a heat pump operating in the heating mode. The starting point in the cycle is the inlet to the compressor. Here's how the heat pump works (refer to the states shown in Figure 13-5):

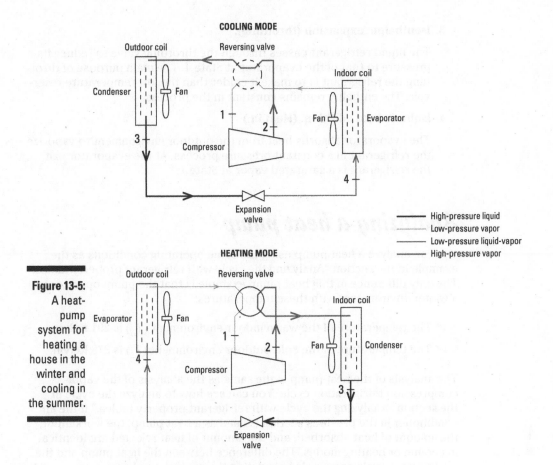

COOLING MODE

Outdoor coil   Reversing valve

Condenser   Fan   Indoor coil

Fan   Evaporator

1

2

3

Compressor

4

Expansion
valve

— High-pressure liquid
— Low-pressure vapor
— Low-pressure liquid-vapor
— High-pressure vapor

HEATING MODE

Outdoor coil   Reversing valve

Evaporator   Fan   Indoor coil

Fan   Condenser

1

2

4

Compressor

3

Expansion
valve

Figure 13-5:
A heat-
pump
system for
heating a
house in the
winter and
cooling in
the summer.

1. Isentropic compression. (Work in.)

In the ideal heat pump, saturated refrigerant vapor comes from the
evaporator and enters the compressor at State 1. The compressor isentropically compresses the vapor to the condenser operating pressure at
State 2. The vapor is superheated in this state.

2. Isobaric heat rejection. (Heat out.)

The vapor passes through the condenser in a constant-pressure process and gives up heat to the air inside the house. The warm air then is
distributed throughout the house. At the condenser exit, the vapor is a
saturated liquid at State 3.

3. **Isenthalpic expansion (throttling).**

The liquid refrigerant passes through the throttling valve to reduce its pressure to that of the evaporator at State 4. The main purpose of throttling the refrigerant is to make it colder than the low-temperature reservoir. The enthalpy remains constant in the process.

4. **Isobaric heat addition. (Heat in.)**

The evaporator absorbs heat from the outdoor environment to vaporize the refrigerant in a constant-pressure process. At the evaporator exit, the refrigerant is a saturated vapor at State 1.

Analyzing a heat pump

You can analyze a heat pump using the same operating conditions as the example in the section "Analyzing the cycle with refrigerant property tables." The only difference in this heat pump example is that the pump operates in a winter environment with these temperatures:

- ✔ The temperature of the warm indoor environment (T_H) is 291 Kelvin.
- ✔ The temperature of the cold outdoor environment (T_L) is 278 Kelvin.

The analysis of the heat pump is the same as the analysis of the vapor-compression refrigeration cycle. You can see how to analyze the cycle in the section "Analyzing the cycle with refrigerant property tables." If the air conditioner in the previous section is used as a heat pump, the work input, the amount of heat absorbed, and the amount of heat rejected are identical in cooling or heating modes. The difference between the heat pump and the refrigeration cycle appears in calculating the coefficient of performance and the irreversibility of the processes.

Calculating the heat pump coefficient of performance

Remember that any measure of performance compares what you want to what you provide. For a heat pump, you want the heat of the cycle that enters the house, which is q_{out}. In an air conditioner, you want the heat of the cycle that's leaving your house, which is q_{in}. You calculate the coefficient of performance for a heat pump by using the heat transfer from the condenser instead of the evaporator. The condenser heat transfer is of interest because it's heating your house. You calculate the coefficient of performance using this formula:

$$\text{COP}_{HP} = \frac{q_{out}}{w_{in}} = \frac{169.6 \text{ kJ/kg}}{29 \text{ kJ/kg}} = 5.85$$

If you compare the coefficient of performance of the heat pump to the COP of the vapor-compression refrigerator (air conditioner), you see that it's higher for the heating mode than the cooling mode. This difference exists because the desired quantity (q_{out}) is higher than that of the air-conditioning refrigeration cycle, where the desired quantity is q_{in}.

The Carnot heat pump has a different coefficient of performance than the Carnot refrigeration cycle for two reasons: They're calculated from different equations, and they have different thermal energy reservoir temperatures from summer to winter. You can find the Carnot heat pump COP using this formula:

$$\text{COP}_{\text{Carnot, HP}} = \frac{T_H}{T_H - T_L} = \frac{291 \text{ K}}{(291 - 278) \text{ K}} = 22.4$$

Calculating heat pump irreversibility

I discuss calculating irreversibility in detail in the earlier section "Calculating vapor-compression refrigerator irreversibility." The following steps summarize calculating irreversibility for a heat pump.

Isentropic processes are reversible, so the irreversibility, or change in entropy ($s_{final} - s_{initial}$), for the compression of the refrigerant vapor in the compressor is zero. The irreversibility of the heat rejected to the high-temperature reservoir uses absolute temperatures for T_0 and T_H. The irreversibility of the heat absorbed from the low-temperature reservoir uses absolute temperature for T_0 and T_L. T_0 isn't necessarily the lowest temperature in the cycle; it's the lowest naturally occurring temperature in the environment; so, for a heat pump operating in the heating mode, T_0 equals T_L.

You can use the entropy values and the heat transfer values you calculated for the vapor compression cycle in the section "Analyzing the cycle with refrigerant property tables" to find the irreversibility of each process. Follow these steps:

1. **Compute the irreversibility of the heat absorption process in the evaporator.**

   Use the entropy values you found for s_1 and s_4 and the heat transfer from the cold reservoir (q_{in}) in the following equation:

$$i_{41} = T_0\left[(s_1 - s_4) - \frac{q_{in}}{T_L}\right] = (278 \text{ K})\left[(1.730 - 1.207)\frac{\text{kJ}}{\text{kg} \cdot \text{K}} - \frac{140.6 \text{ kJ/kg}}{278 \text{ K}}\right] = 4.8 \text{ kJ/kg}$$

2. **Find the irreversibility of the heat rejection process in the condenser.**

   Use the entropy values you found for s_2 and s_3 and the heat transfer to the warm reservoir (q_{out}) in the following equation:

   $$i_{23} = T_0\left[(s_3 - s_2) + \frac{q_{out}}{T_H}\right] = (278\ \text{K})\left[(1.188 - 1.730)\frac{\text{kJ}}{\text{kg}\cdot\text{K}} + \frac{169.6\ \text{kJ/kg}}{291\ \text{K}}\right]$$

   $$= 11.3\ \text{kJ/kg}$$

3. **Calculate the irreversibility of the throttling process.**

   The irreversibility of the throttling valve is the same for the heat pump, whether it's operating in the heating mode or the cooling mode. In the cooling mode, you found that the irreversibility is as follows:

   $$i_{34} = 5.9\ \text{kJ/kg}$$

Compare the irreversibility of the evaporator and the condenser between the cooling mode and the heating mode. In the heating mode, the irreversibility of the evaporator (4.8 kilojoules per kilogram) is less than in the cooling mode (14.0 kilojoules per kilogram). This difference exists because the evaporator temperature (269 Kelvin) is closer to the temperature of the cold thermal energy reservoir in the winter (278 Kelvin) than in the summer (294 Kelvin).

The irreversibility of the condenser behaves in an opposite way. In the heating mode, the irreversibility of the condenser (11.3 kilojoules per kilogram) is greater than in the cooling mode (1.6 kilojoules per kilogram). This difference exists because the condenser temperature (317.6 to 312.4 Kelvin) is closer to the temperature of the warm thermal energy reservoir in the summer (310 Kelvin) than in the winter (291 Kelvin).

Part IV
Handling Thermodynamic Relationships, Reactions, and Mixtures

In this part . . .

I let you play with fire from the comfort of your own room. Here, you use the first law of thermodynamics on combustion reactions. Don't worry about the chemistry; it's very easy because the outcome is just water vapor and carbon dioxide. If fire is too dangerous for you, you can work on air-conditioning systems using that ubiquitous, nonreacting gas mixture: air plus water vapor. If you're tired of that goody-two-shoes "the ideal gas," then here's your chance to work with that bad boy, "the real gas." Whip out your calculator and use real-gas equations to figure out how much pressure you can really handle.

Chapter 14

Understanding the Behavior of Real Gases

*W*hen you're attracted to someone, you want to be close to that person. These feelings you may have aren't unique to humans. Even molecules have them: Attractive forces bind molecules together in solids and liquids. In gases, attractive forces come into effect when the molecules are packed close together, which occurs under high pressure or at a low temperature.

Molecular attractive forces change the way a gas responds to energy. When the density of a gas is low, the gas follows the ideal-gas law (refer to Chapter 3 for more on the ideal-gas law). When the density of the gas is high, the ideal-gas law is inadequate, and the gas is referred to as a *real gas* because it doesn't obey the ideal-gas law.

In this chapter, I discuss the shortcomings of the ideal-gas law and how to account for the effects of molecular attractive forces. One fun and easy way to predict the pressure, temperature, or specific volume of a gas under the influence of molecular attractive forces is to use the *generalized compressibility chart.* I show you how to use this chart to get good estimates of the pressure, temperature, or specific volume of any gas.

You can also determine the pressure of a gas using equations that modify the ideal-gas law. Using these equations may not be as fun as using the compressibility chart, but they can help you find more accurate values for pressure than the chart can. These equations are called *equations of state* because they allow you to calculate the state or condition of a gas. In this chapter, I tell you about one particular equation of state called the *van der Waals equation,* which allows you to find the pressure of a gas for a given temperature and a specific volume.

Deviating from Ideal-Gas Behavior: Real-Gas Behavior

In some thermodynamic systems, such as the Brayton cycle in Chapter 10 and the Otto cycle in Chapter 11, you assume the gas (air) used in the cycle behaves as an ideal gas. But in other thermodynamic systems, like the Rankine cycle in Chapter 12 or the refrigeration cycles in Chapter 13, you can't assume the steam or the refrigerant behaves as an ideal gas. How do you know whether a gas is behaving as an ideal gas?

Look at Figure 14-1, a generic temperature-entropy (T-s) diagram of a substance showing regions of liquid, superheated vapor, and liquid-vapor mixture. (I discuss T-s diagrams in more detail in Chapter 8.) The line in the diagram is called the *liquid-vapor dome*. At the top of the dome is the *critical point*. The critical point is the temperature (T_{cr}) and pressure (P_{cr}) where a phase change between liquid and vapor ceases to exist. Above the critical point a liquid doesn't change into a vapor; instead, it becomes a *supercritical fluid*.

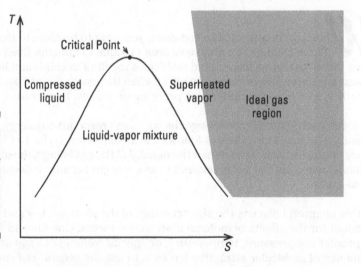

Figure 14-1:
The ideal-gas law is valid in the shaded region of the T-s diagram.

If the temperature and pressure of a vapor are far enough away from the critical point, the vapor behaves as an ideal gas, as shown in the shaded region in Figure 14-1. Air is usually far enough away from its critical point (133 Kelvin and 3.77 megapascals pressure) to behave like an ideal gas in many common thermodynamic processes. However, use caution in assuming that air behaves as an ideal gas for all processes.

Water vapor in the air is considered an ideal gas for thermodynamic analysis of air-conditioning applications such as heating and cooling. I discuss the properties of moist air for air-conditioning processes in Chapter 15. Water vapor in the air behaves as an ideal gas because it has a low pressure (usually less than 7 kilopascals) compared to its critical point pressure of 22.1 megapascals. When water is at high pressure, such as when it's used as steam in a Rankine cycle, it has a high density and must be treated as a real gas. The thermodynamic properties of real gases are usually tabulated, like those found in Tables A-5 and A-8 for water and R-134a, respectively, in the appendix.

The ideal-gas region in a *T-s* diagram is about the same for any gas relative to the critical point. If you know what the critical point is for a substance, you can figure out at what temperatures and pressures it behaves as a real gas instead of an ideal gas. Table 14-1 shows the critical point temperature and pressure for a number of gases. It also shows the molecular mass and the gas constant for those gases. (Table 14-1 is used in the example problems throughout this chapter.)

Table 14-1	Critical Point Properties of Selected Fluids			
	Molecular Mass	Gas Constant	Critical-Point Properties	
Substance	kg/kmol	R, kJ/kg · K	Temp, K	Press. MPa
Air	28.97	0.2870	132.5	3.77
Ammonia	17.03	0.4882	405.5	11.28
Carbon dioxide	44.01	0.1889	304.2	7.39
Ethanol	46.07	0.1805	516.0	6.38
Methanol	32.04	0.2595	513.2	7.95
Propane	44.10	0.1885	370.0	4.26
R-134a	102.03	0.0815	374.2	4.06
Water	18.02	0.4615	647.1	22.06

To see why the ideal-gas law isn't the right equation of state for all gases, try using the ideal-gas law to find the specific volume of superheated steam at 500 degrees Celsius and 10 megapascals pressure. Then compare it to the actual value from Table A-5 in the appendix. Follow these steps:

1. **Rearrange the ideal-gas law to solve for specific volume.**

$$v = \frac{RT}{P}$$

2. **Convert the temperature to absolute units in Kelvin.**

   $500°C + 273 = 773$ K

3. **Calculate the specific volume from the ideal-gas law.**

   Use the gas constant R for water, from Table 14-1. Convert megapascals to kilopascals.

   $$v = \frac{(0.462 \text{ kJ/kg} \cdot \text{K})(773 \text{ K})}{10,000 \text{ kPa}} = 0.03571 \text{ m}^3/\text{kg}$$

4. **Look up the specific volume of the steam in Table A-5 of the appendix.**

   $v = 0.03279$ m³/kg

5. **Determine the error of using the ideal-gas law to calculate the specific volume, as follows:**

   $$\frac{(0.03571 - 0.03279) \text{ m}^3/\text{kg}}{0.03279 \text{ m}^3/\text{kg}} = 9\%$$

You can see that the error in using the ideal-gas law to predict the behavior of a non-ideal (real) gas is significant.

Determining Properties with the Compressibility Factor

An equation of state describes the pressure-volume-temperature relationship of a substance. The simplest equation of state for a gas is the ideal-gas law:

$Pv = RT$

The ideal-gas law assumes the forces between gas molecules are insignificant. Under ordinary conditions, air and water vapor behave as ideal gases. But at high pressure or at low temperature, the molecules are close together, and attractive forces bring the molecules even closer together, which increases their density.

One way to account for the density change at high pressures and low temperatures is to use the *compressibility factor* (Z). The compressibility factor is inserted into the ideal-gas law equation as follows:

$Pv = ZRT$

When density increases due to molecular attraction, the compressibility factor is less than 1.0, because specific volume is the reciprocal of density. If density increases, specific volume decreases. So the product of Pv is less than that predicted by the ideal-gas law.

In some conditions, the density is less than that predicted by the ideal-gas law, and the compressibility factor is greater than 1.0. This situation occurs at extremely high pressures or temperatures. At extremely high pressures, repulsive forces are at work, and at extremely high temperatures, molecules have so much energy that they don't like being near each other.

Figure 14-2 shows a chart of the compressibility factor (Z). The compressibility factor is determined by calculating two quantities called the *reduced temperature* and the *reduced pressure*. The reduced temperature (T_R) is the ratio of the gas temperature (T) to the critical temperature (T_{cr}), as shown in this equation.

$$T_R = \frac{T}{T_{cr}}$$

Figure 14-2: The generalized compressibility chart is used to determine the compressibility factor for real-gas behavior.

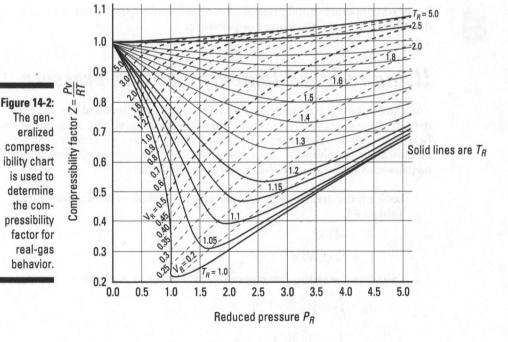

The reduced temperature tells you how close you are to the critical temperature. If you're well above the critical temperature, where T_R is greater than 2.0, a gas behaves more like an ideal gas. Under these conditions, the compressibility factor lies between 0.95 and 1.05, as shown in Figure 14-2, so the error in the product (Pv) is less than 5 percent.

The reduced pressure (P_R) is the ratio of the gas pressure (P) to the critical pressure (P_{cr}), as shown in this equation:

$$P_R = \frac{P}{P_{cr}}$$

The reduced pressure tells you how close you are to the critical pressure. In Figure 14-2, you can see that at low pressures, such as P_R less than 0.3, the compressibility factor is greater than 0.9. This means the ideal-gas law is within 10 percent of predicting the actual pressure of a gas.

At the critical point, T_R and P_R both equal 1.0. Figure 14-2 shows that the compressibility factor at the critical point is 0.2. This deviation from ideal-gas behavior is the most significant one on the chart.

Reduced temperature and pressure are dimensionless quantities, and they don't have units of temperature or pressure.

Using reduced temperature and pressure

You can see how the compressibility factor improves the ideal-gas law prediction of the specific volume in the earlier section "Deviating from Ideal-Gas Behavior: Real-Gas Behavior." The following steps show you how to calculate the specific volume of superheated steam at 500 degrees Celsius and 10 megapascals pressure with the compressibility factor:

1. **Look up the critical temperature and critical pressure for water in Table 14-1.**

   $T_{cr} = 647 \text{ K}$

   $P_{cr} = 22.06 \text{ MPa}$

2. **Calculate the reduced temperature and reduced pressure for the steam.**

   $$T_R = \frac{T}{T_{cr}} = \frac{773 \text{ K}}{647 \text{ K}} = 1.19$$

   $$P_R = \frac{P}{P_{cr}} = \frac{10 \text{ MPa}}{22.1 \text{ MPa}} = 0.45$$

3. **Look up the compressibility factor (Z) in Figure 14-2, using T_R and P_R.**

   $Z \approx 0.91$

4. **Calculate the specific volume, using the compressibility factor with the volume calculated using the ideal-gas law.**

   $$v = Z\left(\frac{RT}{P}\right) = Zv_{\text{ideal}} = 0.91\left(0.03571 \text{ m}^3/\text{kg}\right) = 0.03250 \text{ m}^3/\text{kg}$$

5. **Calculate the error in the specific volume, using the compressibility factor and the actual specific volume of the steam.**

   $$\frac{\left(0.03279 - 0.03250\right) \text{ m}^3/\text{kg}}{0.03279 \text{ m}^3/\text{kg}} = 1\%$$

You see that the error has been reduced significantly using the compressibility factor compared to just using the ideal-gas law to calculate the specific volume.

Using pseudo-reduced volume

If you have specific volume instead of temperature or pressure, you can use the *pseudo-reduced specific volume* (v_R) to find the compressibility factor. The pseudo-reduced specific volume is called "pseudo" because it's defined a bit differently than reduced pressure or reduced temperature. Instead of calculating a specific volume ratio relative to the specific volume at the critical state, v_R is calculated using the critical pressure and temperature properties in the following equation:

$$v_R = \frac{vP_{cr}}{RT_{cr}}$$

Figure 14-2 shows a set of dashed lines for v_R. You can use any two of the three properties (T_R, P_R, or v_R) to find the compressibility factor.

Here's an example that shows you how to determine the pressure of a system using the pseudo-reduced specific volume and the reduced temperature. Suppose you have steam at 400 degrees Celsius and a specific volume of 0.0165 m³/kg. You can find the steam pressure by following these steps:

1. **Calculate the reduced temperature using the critical temperature of steam from Table 14-1. Temperature must use absolute units.**

   $$T_R = \frac{T}{T_{cr}} = \frac{673 \text{ K}}{647 \text{ K}} = 1.04$$

2. **Calculate the pseudo-reduced specific volume, using the critical temperature and pressure of steam from Table 14-1.**

$$v_R = \frac{vP_{cr}}{RT_{cr}} = \frac{(0.0165 \text{ m}^3/\text{kg})(22{,}100 \text{ kPa})}{(0.462 \text{ kJ/kg} \cdot \text{K})(647 \text{ K})} = 1.22$$

3. **Look up the reduced pressure directly from Figure 14-2, using the intersection of T_R and v_R on the compressibility factor chart.**

$$P_R = 0.65$$

4. **Calculate the steam pressure by rearranging the equation defining reduced pressure.**

$$P = P_R \cdot P_{cr} = (0.65)(22 \text{ MPa}) = 14.3 \text{ MPa}$$

5. **Use the ideal-gas law to compute the pressure to see how far off it is from the actual value.**

$$P = \frac{(0.462 \text{ kJ/kg} \cdot \text{K})(673 \text{ K})}{0.0165 \text{ m}^3/\text{kg}}\left(\frac{1 \text{ MPa}}{1{,}000 \text{ kPa}}\right) = 18.8 \text{ MPa}$$

6. **Extrapolate the data from Table A-5 in the appendix to find the actual value for the pressure.**

$$P = 14.44 \text{ MPa}$$

The compressibility factor gives a reasonable estimate for the pressure. The error of the ideal-gas law prediction for pressure is 30 percent.

Finding Pressure with van der Waals

Over the years, many attempts have been made to develop an equation of state that predicts the temperature-pressure-volume relationship of gases. In this section, I show you how to use one of the most prominent equations, called the *van der Waals equation of state*. This equation isn't a magic bullet; it has its own limitations.

You can see how the van der Waals equation of state can predict the pressure of a gas and compare that prediction to an ideal-gas law prediction. Suppose you have water at 500 degrees Celsius with a specific volume of 0.03279 cubic meters per kilogram. Use the ideal-gas law to predict the gas pressure. The gas constant (R) for water is in Table 14-1. Remember to convert temperature to absolute units.

$$P = \frac{RT}{v} = \frac{(0.4615 \text{ kJ/kg} \cdot \text{K})(773 \text{ K})}{0.3279 \text{ m}^3/\text{kg}} = 10{,}880 \text{ kPa}$$

The ideal-gas law doesn't account for the effects of molecular forces, so it makes sense to include these effects for a more accurate equation of state. Johannes van der Waals improved the ideal-gas equation by adding a term for intermolecular attraction forces and another term for the physical volume of molecules in a gas. He did such a good job that they named the equation after him. The van der Waals equation of state is written like this:

$$\left(P + \frac{a}{v^2}\right)(v - b) = RT$$

The two constants in the equation are relative to the critical temperature and pressure of a gas. The first constant is given by the following equation:

$$a = \frac{27R^2 T_{cr}^2}{64 P_{cr}}$$

This constant is used in the a/v^2 term, which accounts for the intermolecular forces. The second constant accounts for the volume of the molecules and is given by this equation:

$$b = \frac{RT_{cr}}{8 P_{cr}}$$

You can follow these steps to use the van der Waals equation of state to predict the pressure of water at 500 degrees Celsius with a specific volume of 0.03279 cubic meters per kilogram.

1. **Rewrite the van der Waals equation to solve for pressure.**

   $$P = \frac{RT}{v - b} - \frac{a}{v^2}$$

2. **Look up the critical pressure and critical temperature of water in Table 14-1.**

   $P_{cr} = 22,060 \text{ kPa}$

   $T_{cr} = 647.1 \text{ K}$

3. **Calculate the value of the constant a in the van der Waals equation.**

   $$a = \frac{27(0.4615 \text{ kJ/kg} \cdot \text{K})^2 (647.1 \text{ K})^2}{64(22,060 \text{ kPa})} = 1.706 \text{ m}^6 \text{ kPa/kg}^2$$

4. **Calculate the value of the constant b in the van der Waals equation.**

   $$b = \frac{(0.4615 \text{ kJ/kg} \cdot \text{K})(647.1 \text{ K})}{8(22,060 \text{ kPa})} = 1.69 \times 10^{-3} \text{ m}^3/\text{kg}$$

5. Substitute the values for the constants a and b into the van der Waals equation and calculate the pressure of the carbon dioxide.

$$P = \frac{(0.4615 \text{ kJ/kg} \cdot \text{K})(773 \text{ K})}{(0.03279 - 1.69 \times 10^{-3}) \text{ m}^3/\text{kg}} - \frac{1.706 \text{ m}^6 \text{ kPa/kg}^2}{(0.03279 \text{ m}^3/\text{kg})^2} = 9,884 \text{ kPa}$$

You can find that the actual pressure of the water vapor is 10,000 kilopascals at 500 degrees Celsius and specific volume of 0.03279 cubic meters per kilogram in Table A-5 in the appendix. You can see that the van der Waals equation provided a better value of the water vapor pressure than the ideal-gas law.

Although this example shows the van der Waals equation is better than the ideal-gas law, sometimes the van der Waals equation is inadequate, especially around the critical point. The van der Waals equation of state is of interest because it's a simple model that attempts to capture real-gas behavior. Because of its limitations, the van der Waals equation of state isn't used in computing thermodynamic properties. Much more sophisticated equations have been developed to determine properties, but these equations are beyond the scope of introductory thermodynamics.

Chapter 15

Mixing Gases That Don't React with Each Other

. . . .

In This Chapter

▶ Calculating thermodynamic properties for a mixture of nonreacting gases

▶ Estimating pressure, volume, and temperature for a mixture of nonreacting gases

▶ Using compressibility factors for mixtures of gases

▶ Figuring out thermodynamic properties of moist air

▶ Using the first law to analyze heating and cooling processes for air

. . . .

As a long, hard winter draws to a close, you look forward to that first warm day when you can open the windows and let some fresh air into the house. People are like thermodynamic engines — they breathe air to support chemical reactions that enable them to live, play, and work. The heat engines I describe in Chapters 10–12 also breathe air, in a manner of speaking, to produce work, but they don't live and play as people do.

So what exactly is air? There's no such thing as an air molecule. Air is a gas made up of a mixture of different molecules: nitrogen, oxygen, carbon dioxide, argon, water, and traces of many other gas molecules. For each gas, the thermodynamic properties — such as enthalpy, internal energy, density, and entropy — differ from one another. If you were to analyze a heat engine using the thermodynamic properties of each individual gas in air, you'd have a very complicated problem on your hands. Therefore, air is treated as if it were a pure substance for most thermodynamic analyses. When you analyze heat engines with the first law of thermodynamics, you use the thermodynamic properties of air that are tabulated for you in the appendix. This greatly simplifies the analysis of heat engines.

Many processes involve a mixture of several different gases, and you may get inaccurate results if your thermodynamic analysis assumes that the process uses a pure substance. This chapter discusses how to work with processes involving a mixture of gases that don't react with each other during a thermo-dynamic process. (Gases that react with each other form new compounds, as in combustion processes. I discuss the thermodynamic analysis of combus-tion reactions in Chapter 16.)

The most important nonreacting gas mixture in thermodynamics is moist air. In simple terms, the moist air gas mixture is made of dry air and water vapor. Thermodynamic processes involving moist air focus on heating and cooling air for comfort. This chapter looks at what happens to the air as it interacts with an air-conditioning cycle (refer to Chapter 13 for a discussion of the thermodynamic analysis of air-conditioning cycles).

Determining Thermodynamic Properties for a Mixture of Gases

Many thermodynamic processes involve a mixture of gases. In the heat engine cycles in Chapters 10–12, you treat the air in the heat transfer and work processes as if it were a pure ideal gas in the cycle to simplify the thermodynamic analysis. In reality, the combustion process in a heat engine produces a mix of gases different from air. The exhaust gases can contain carbon dioxide, carbon monoxide, and water vapor in addition to oxygen and nitrogen. The thermodynamic properties of the exhaust gases are a little bit different from those of the air you use in the analysis of the ideal heat engine cycles. This difference means that heat transfer and work processes with an exhaust gas mixture may result in slightly different temperatures or pres-sures than you get when you use air in the analysis.

This section shows you how to calculate thermodynamic properties — such as internal energy, enthalpy, entropy, and specific heat — for gas mixtures. Using the properties of gas mixtures can give you more accurate results in your thermodynamic analysis of a process than assuming the process uses only air throughout.

Using mass and molar fractions for gas mixtures

When you analyze a mixture of gases in a thermodynamic process, you need to figure out how much each gas contributes to the overall process. The contribution of each gas depends on the amount present, which is described using either a molar basis or a mass basis. You may need to convert between a molar and a mass basis for your analysis. Converting from one to another is quite simple, and I show you how in this section. I also show you how to determine the relative amount of each gas in a mixture using a mass fraction and a molar fraction. You use the mass fractions or the molar fractions to determine the thermodynamic properties of a mixture of gases.

For example, suppose you need to do a thermodynamic analysis using the combustion products from a gasoline engine. I describe how the Otto cycle represents a gasoline engine in Chapter 11. The combustion gas mixture can have the following proportions of gases: 8 moles of CO_2, 9 moles of H_2O, and 47 moles of N_2. (See Chapter 16 for more information about combustion reactions.)

TIP

The number of moles of each gas in a mixture also describes the mixture on a volumetric basis.

To find the molar fractions and the mass fractions of a gas mixture, follow these steps:

1. **Calculate the total number of moles in the mixture to figure out the molar fractions of each gas.**

   You calculate the total number of moles using this equation:

   $$n_{tot} = n_1 + n_2 + n_3 + \ldots$$

   The numbers in the subscripts stand for the name of each gas in a mixture (for example, $1 = CO_2$, $2 = H_2O$, and so on). You can add as many gases to this equation as you need. For this example, the total number of moles is

   $$n_{tot} = (8 + 9 + 47) \text{ moles} = 64 \text{ moles of exhaust}$$

2. **Calculate the mole fraction (y_i) for each gas in a mixture using this equation:**

   $$y_i = \frac{n_i}{n_{tot}}$$

   The molar fractions (in decimal form) for the gases in the combustion gas mixture are as follows:

   $$y_{CO_2} = \frac{8 \text{ mol}}{64 \text{ mol}} = 0.125$$

   $$y_{H_2O} = \frac{9 \text{ mol}}{64 \text{ mol}} = 0.141$$

   $$y_{N_2} = \frac{47 \text{ mol}}{64 \text{ mol}} = 0.734$$

3. **Convert from a molar basis to a mass basis using the molar mass (M_i) of each gas.**

   The molar masses of a few gases found in combustion reactions are listed in Table A-11 in the appendix. The mass (m_i) of each gas is related to the number of moles (n_i) of each gas, as shown in this equation:

   $$m_i = n_i \cdot M_i$$

   The subscript "i" stands for the name of the gas you're using in your calculations.

4. **Calculate the mass (m_i) of each gas in the combustion gas mixture, as follows:**

$$m_{CO_2} = (44 \text{ g/mole})(8 \text{ moles}) = 352 \text{ grams } CO_2$$

$$m_{H_2O} = (18 \text{ g/mole})(9 \text{ moles}) = 162 \text{ grams } H_2O$$

$$m_{N_2} = (28 \text{ g/mole})(47 \text{ moles}) = 1,316 \text{ grams } N_2$$

5. **Calculate the total mass of the mixture, using this equation:**

$$m_{tot} = m_1 + m_2 + m_3 + \ldots$$

For this example, the total mass is as follows:

$$m_{tot} = (352 + 162 + 1,316) \text{ grams} = 1,830 \text{ grams of combustion gases}$$

6. **Calculate the mass fraction (mf_i) for each gas in a mixture, using this equation:**

$$mf_i = \frac{m_i}{m_{tot}}$$

Therefore, the mass fractions for the gases in the combustion gas mixture are as follows:

$$mf_{CO_2} = \frac{352 \text{ grams}}{1,830 \text{ grams}} = 0.192$$

$$mf_{H_2O} = \frac{162 \text{ grams}}{1,830 \text{ grams}} = 0.089$$

$$mf_{N_2} = \frac{1,316 \text{ grams}}{1,830 \text{ grams}} = 0.719$$

Finding properties of a gas mixture

You can use the mass fractions (mf_i) to determine the thermodynamic properties of a mixture, as shown by these equations:

- ✔ **Internal energy:** $u_m = mf_1 u_1 + mf_2 u_2 + mf_3 u_3 + \ldots$
- ✔ **Enthalpy:** $h_m = mf_1 h_1 + mf_2 h_2 + mf_3 h_3 + \ldots$
- ✔ **Entropy:** $s_m = mf_1 s_1 + mf_2 s_2 + mf_3 s_3 + \ldots$
- ✔ **Constant volume specific heat:** $c_{v,m} = mf_1 c_{v1} + mf_2 c_{v2} + mf_3 c_{v3} + \ldots$
- ✔ **Constant pressure specific heat:** $c_{p,m} = mf_1 c_{p1} + mf_2 c_{p2} + mf_3 c_{p3} + \ldots$

The numbers in the subscripts represent the names of the gases in the mixture.

When you need to use molar versions of thermodynamic properties, you simply replace the mass fraction terms (mf_1, mf_2, mf_3 ...) with molar fraction terms (y_1, y_2, y_3 ...) in the molar equations for enthalpy, entropy, internal energy, and specific heat.

Here's an example that shows you how to calculate the internal energy of a gas mixture using the combustion gas mixture from the preceding example. Suppose the combustion gas mixture temperature is 1,843 degrees Celsius. You calculate the internal energy of the gas mixture by following these steps:

1. **Find the internal energy of each gas at 1,843 degrees Celsius in Table A-9 in the appendix.**

   You need to interpolate the table to find the properties. I show you how to do table interpolation in Chapter 3. Note that the properties in Table A-9 are molar properties.

   Because Table A-9 lists only molar enthalpy, you can find the molar internal energy using this equation: $\bar{u} = \bar{h} - \bar{R}T$. So for carbon dioxide, the molar internal energy is 98,500 kJ/kmol by interpolating Table A-9; then the internal energy is

   $$\bar{u} = 98{,}500 \text{ kJ/kmol} - (8.314 \text{ kJ/kmol-K})(1{,}843°C + 273)\,K = 80{,}900 \text{ kJ/kmol}$$

   Then you have to convert the molar internal energy to specific internal energy using the molar mass of each gas, as shown here:

   $u_{CO_2} = (80{,}880 \text{ kJ/kmol})/(44\text{kg/kmol}) = 1{,}839 \text{ kJ/kg}$

   $u_{H_2O} = (61{,}210 \text{ kJ/kmol})/(18 \text{ kg/kmol}) = 3{,}400 \text{ kJ/kg}$

   $u_{N_2} = (42{,}740 \text{ kJ/kmol})/(28 \text{ kg/kmol}) = 1{,}526 \text{ kJ/kg}$

2. **Multiply the internal energy (u_i) of each gas by its mass fraction (mf_i) and add the results together to determine the internal energy of the gas mixture:**

   $$u_m = [0.192(1{,}839) + 0.089(3{,}400) + 0.719(1{,}526)] \text{ kJ/kg} = 1{,}753 \text{ kJ/kg}$$

You can compare the internal energy of this gas mixture to the internal energy of the air used in the Otto cycle example in Chapter 11. This example uses the same gas temperature as the Otto cycle example. In the Otto cycle example, the internal energy of the air is 1,790 kilojoules per kilogram at 1,843 degrees Celsius. You can see a difference in the internal energy, assuming the gas mixture is air rather than an actual combustion gas mixture.

Every gas mixture has an apparent molar mass (M_m). The molar mass is used to compute the apparent ideal-gas constant (R_m) for a mixture. For example, the apparent molar mass (M_{air}) for air is 28.97 kilograms per kilomole, and the ideal-gas constant (R_{air}) for air is 0.287 kilojoules per kilogram-Kelvin.

You can calculate the apparent molar mass (M_m) for the exhaust mixture by using this equation:

$$M_m = \frac{m_{tot}}{n_{tot}} = \frac{1,830 \text{ kg}}{64 \text{ kmol}} = 28.59 \text{ kg/kmol}$$

Then you can calculate the apparent ideal-gas constant (R_m) for the exhaust mixture using the universal gas constant ($\bar{R}$) and the apparent molar mass (M_m), as shown in this equation:

$$R_m = \frac{\bar{R}}{M_m} = \frac{8.314 \text{ kJ/kmol} \cdot \text{K}}{28.59 \text{ kJ/kmol}} = 0.291 \text{ kJ/kmol} \cdot \text{K}$$

Getting the Compressibility Factor for Real-Gas Mixtures

In an ideal gas, the molecules are spaced far enough apart that they don't influence each other's behavior; thus, you can predict the pressure, specific volume, and temperature of the gas using the ideal-gas law discussed in Chapter 3:

$$Pv = RT$$

When gases are at high pressure or low temperature, the molecules are packed closer together, and intermolecular forces can affect the pressure, specific volume, or temperature of the gas. A gas that doesn't quite follow the ideal-gas law predictions due to intermolecular interactions is called a *real gas*. To predict the pressure, temperature, or specific volume of a real gas, a *compressibility factor* (Z) is included in the ideal-gas law to account for the effects of molecular interaction, as shown in this equation:

$$Pv = ZRT$$

I discuss the compressibility factor for real gases (as single components, not as a mixture) in more detail in Chapter 14. You can find the compressibility factor for a gas mixture (Z_m) using several different assumptions. In the following sections, I discuss three different sets of assumptions used by Amagat's law, Dalton's law, and Kay's rule for determining compressibility factors of real-gas mixtures.

Making assumptions for mixture compressibility factors

The compressibility factor for a mixture (Z_m) is weighted by the amount of each gas present in the mixture. You calculate Z_m by using the mole fraction and the compressibility factor for each gas, as shown in the following equation. The value of the individual compressibility factors depends upon the method you use, because each method uses different parameters for calculating compressibility factors. (The subscript numbers represent the names of the gases.)

$$Z_m = y_1 Z_1 + y_2 Z_2 + y_3 Z_3 + \dots$$

You determine the compressibility factor of each component using the following parameters for each of the three methods:

- **Amagat's law:** Mixture pressure and temperature
- **Dalton's law:** Mixture volume and temperature
- **Kay's rule:** Pseudocritical temperature and pressure

I cover the details of these methods in the following sections.

Amagat's law is more accurate at high pressures because it uses the mixture pressure instead of the component pressure. In so doing, this law accounts for the influence of forces between molecules at high pressure. Dalton's law is more accurate at low pressures, where the influence of intermolecular forces is negligible. Overall, Kay's rule is accurate over a wide range of temperatures and pressures.

You can find the compressibility factor for each gas in a mixture by using the generalized compressibility chart in Figure 14-2 in Chapter 14. The same compressibility chart is used for each of the three different methods.

To use the chart, you first need to calculate any two of the following properties for each gas: reduced pressure (P_R), reduced temperature (T_R), or pseudo-reduced volume (v_R). I discuss the compressibility chart, reduced pressure, reduced temperature, and pseudo-reduced volume for a real gas in detail in Chapter 14.

The reduced pressure (P_R) is the ratio of the gas pressure (P) in the mixture to the critical pressure (P_{cr}) of the gas. You calculate (P_R) by using this equation:

$$P_R = \frac{P}{P_{cr}}$$

The reduced temperature (T_R) is the ratio of the gas temperature (T) in the mixture to the critical temperature (T_{cr}) of the gas. You calculate T_R with this equation, using absolute temperature:

$$T_R = \frac{T}{T_{cr}}$$

The pseudo-reduced volume (v_R) is calculated using the critical pressure and temperature properties in the following equation:

$$v_R = \frac{vP_{cr}}{RT_{cr}}$$

Table A-11 in the appendix lists the molar mass (M), gas constant (R), critical pressure (P_{cr}), and critical temperature (T_{cr}) for several fluids.

You can determine the pressure (P_m) and specific volume (v_m) of a real-gas mixture by using the compressibility factor of the mixture (Z_m) to modify the ideal-gas law, as shown in these equations:

$$P_m = \frac{Z_m R_m T_m}{v_m}$$

$$v_m = \frac{Z_m R_m T_m}{P_m}$$

Finding compressibility factors with Amagat's law

Amagat's law states that the volume of a gas mixture equals the sum of the volumes each gas would occupy if it existed alone at the mixture temperature and pressure. The compressibility factor of each component in a real-gas mixture is evaluated using the temperature and pressure of the mixture. To find the compressibility factor using Amagat's law, you calculate the reduced temperature and the reduced pressure for each component and look up the corresponding compressibility factor using the compressibility chart in Chapter 14 (see Figure 14-2).

 If you don't know both the mixture temperature and pressure in a process involving real gases, you have to use an iterative procedure to find the compressibility factor with Amagat's law. A good first estimate for the mixture temperature or pressure can be calculated using the ideal-gas law.

The following example shows you how to calculate the compressibility factor using Amagat's law for a real-gas mixture. Suppose the combustion gas mixture of CO_2, H_2O, and N_2 from the earlier section "Using mass and molar fractions for gas mixtures" is at 10 megapascals pressure and 580 Kelvin. You can determine the compressibility factor of the mixture, as shown in these steps:

1. **Calculate the reduced pressure (P_R) and temperature (T_R) for each of the three gases in the mixture.**

   Use the critical pressure and critical temperature values for the gases from Table A-11 in the appendix. In calculating the reduced temperature, remember to use absolute temperature values.

   - CO_2: $P_{R, CO_2} = \dfrac{P_m}{P_{cr}} = \dfrac{10 \text{ MPa}}{7.39 \text{ MPa}} = 1.4$ and $T_{R, CO_2} = \dfrac{T_m}{T_{cr}} = \dfrac{580 \text{ K}}{304.2 \text{ K}} = 1.9$

   - H_2O: $P_{R, H_2O} = \dfrac{P_m}{P_{cr}} = \dfrac{10 \text{ MPa}}{22.06 \text{ MPa}} = 0.5$ and $T_{R, H_2O} = \dfrac{T_m}{T_{cr}} = \dfrac{580 \text{ K}}{647.1 \text{ K}} = 0.9$

   - N_2: $P_{R, N_2} = \dfrac{P_m}{P_{cr}} = \dfrac{10 \text{ MPa}}{3.39 \text{ MPa}} = 3.0$ and $T_{R, N_2} = \dfrac{T_m}{T_{cr}} = \dfrac{580 \text{ K}}{126.0 \text{ K}} = 4.6$

2. **Use the values from Step 1 for T_R and P_R on the generalized compressibility chart in Figure 14-2 in Chapter 14 to find the compressibility factor (Z) for each gas.**

   You find the following:

   $Z_{CO_2} = 0.95$

   $Z_{H_2O} = 0.70$

   $Z_{N_2} = 1.025$

3. **Calculate the compressibility factor for the gas mixture (Z_m) using the molar fraction for each gas.**

   $$Z_m = y_{CO_2} Z_{CO_2} + y_{H_2O} Z_{H_2O} + y_{N_2} Z_{N_2}$$

   $Z_m = (0.125)(0.95) + (0.141)(0.70) + (0.734)(1.025) = 0.97$

The combustion gas mixture in this example is very nearly an ideal gas under these conditions, because Z_m is very close to 1.0. The water vapor shows a significant departure from ideal-gas behavior under these conditions, but it only makes a small contribution to the overall behavior.

Finding compressibility factors with Dalton's law

Another way to look at a mixture of gases is to assume all the gases occupy the same volume but have different pressures. This is the viewpoint of *Dalton's law*, which says that the total pressure of a mixture equals the sum of the pressures exerted by each gas, as if each gas occupied the same volume at the same temperature of the mixture. To find the compressibility factor using Dalton's law, you calculate the reduced temperature and the pseudo-reduced volume for each component and look up the corresponding compressibility factor using Figure 14-2 in Chapter 14.

TIP

If you don't know both the mixture temperature and volume in a process involving real gases, you have to use an iterative procedure to find the compressibility factor with Dalton's law. A good first estimate for the mixture temperature or volume can be calculated using the ideal-gas law.

The following example shows you how to find the compressibility factor using Dalton's law for a combustion gas mixture. The gas mixture has 8 moles of CO_2, 9 moles of H_2O, and 47 moles of N_2 at 10 megapascals pressure and 580 Kelvin. (These are the same values used with Amagat's law in the preceding section.) You use the mixture temperature and specific volume to find the compressibility factor with Dalton's law, but the mixture specific volume is unknown at this point, so an iterative solution process is needed, as shown in these steps:

1. **Estimate the mixture specific volume (v_m) using the ideal-gas law.**

$$v_m = \frac{R_m T_m}{P_m} = \frac{(0.291 \text{ kJ/kmol} \cdot \text{K})(580 \text{ K})}{10,000 \text{ kPa}} = 0.017 \text{ m}^3/\text{kg}$$

2. **Calculate the pseudo-reduced volume (v_R) for each gas in the mixture.**

   The reduced temperature (T_R) for each gas is the same as in the example for Amagat's law.

   • CO_2: $v_{R,CO_2} = \frac{vP_{cr}}{RT_{cr}} = \frac{(0.017 \text{ m}^3/\text{kg})(7,390 \text{ kPa})}{(0.1889 \text{ kJ/kg} \cdot \text{K})(304.2 \text{ K})} = 2.2$ and $T_{R,CO_2} = 1.9$

   • H_2O: $v_{R,H_2O} = \frac{vP_{cr}}{RT_{cr}} = \frac{(0.017 \text{ m}^3/\text{kg})(22,060 \text{ kPa})}{(0.4615 \text{ kJ/kg} \cdot \text{K})(647.1 \text{ K})} = 1.3$ and $T_{R,H_2O} = 0.9$

   • N_2: $v_{R,N_2} = \frac{vP_{cr}}{RT_{cr}} = \frac{(0.017 \text{ m}^3/\text{kg})(3,390 \text{ kPa})}{(0.2968 \text{ kJ/kg} \cdot \text{K})(126.0 \text{ K})} = 1.5$ and $T_{R,N_2} = 4.6$

3. **Look up the compressibility factor for each gas with the values for T_R and v_R from Step 2, using Figure 14-2 in Chapter 14.**

   $Z_{CO_2} = 0.96$

   $Z_{H_2O} = 0.76$

   $Z_{N_2} = 1.02$

4. **Calculate the compressibility factor for the gas mixture (Z_m) using the molar fraction for each gas.**

   $Z_m = y_{CO_2} Z_{CO_2} + y_{H_2O} Z_{H_2O} + y_{N_2} Z_{N_2}$

   $Z_m = (0.125)(0.96) + (0.141)(0.76) + (0.734)(1.02) = 0.98$

5. **Calculate the specific volume (v_m) of the real-gas mixture using the compressibility factor.**

   $v_m = Z_m v_{ideal} = (0.98)(0.017 \text{ m}^3/\text{kg}) = 0.0166 \text{ m}^3/\text{kg}$

The gas mixture is nearly an ideal gas in this case. Even though the water vapor shows a significant departure from ideal-gas behavior, the nitrogen component compensates for it. Because the specific volume of the gas mixture is nearly identical to that of the ideal-gas volume assumed as an initial estimate, you don't need to do another iteration to refine the solution.

If the calculated specific volume (v_m) of the real gas mixture is significantly different than your initial estimate, you need to do another iteration by taking the calculated value for v_m and using it in Step 2 to find v_R for each gas in the mixture. Then repeat Steps 3 through 5. You can stop iterating when the value of v_m found in Step 5 doesn't change very much between successive iterations.

Calculating compressibility factors with Kay's Rule

Another method for determining the compressibility factor for a mixture is to use Kay's rule. This method is quite simple and is generally accurate within 10 percent for a wide range of temperatures and pressures. To use Kay's rule, you calculate the pseudocritical temperature (T'_c) and the pseudocritical pressure (P'_c) of a gas mixture.

The *pseudocritical temperature* is an imaginary critical temperature of a gas mixture. Table A-11 shows an example for air. Because air isn't a pure gas, the critical temperature listed (132.5 Kelvin) is the pseudocritical temperature. In a similar fashion, the *pseudocritical pressure* of a gas mixture is an imaginary critical pressure. For air, this pressure is 3.77 megapascals.

You use these two pseudocritical properties to determine the reduced temperature and the reduced pressure of the mixture directly. Then you can determine the compressibility factor from Figure 14-2 in Chapter 14.

You calculate the pseudocritical temperature and pressure by using the molar fraction of each gas and the critical temperature and pressure of each gas, as the following equations show:

$$T'_{cr} = y_1 T_{cr,1} + y_2 T_{cr,2} + y_3 T_{cr,3} + \ldots$$
$$P'_{cr} = y_1 P_{cr,1} + y_2 P_{cr,2} + y_3 P_{cr,3} + \ldots$$

Calculate the reduced temperature and the reduced pressure properties with these equations:

$$T_R = \frac{T_m}{T'_{cr}}$$

$$P_R = \frac{P_m}{P'_{cr}}$$

You can calculate the compressibility factor for the combustion gas mixture used in the earlier Amagat's- and Dalton's-law examples using Kay's rule by following these steps:

1. **Calculate the pseudocritical temperature (T'_{cr}) of the mixture using the volume fractions and the critical temperature of each gas.**

$$T'_{cr} = y_{CO_2}T_{crCO_2} + y_{H_2O}T_{crH_2O} + y_{N_2}T_{crN_2}$$

$$T'_{cr} = (0.125)(304 \text{ K}) + (0.141)(647 \text{ K}) + (0.734)(126 \text{ K}) = 222 \text{ K}$$

2. **Calculate the pseudocritical pressure (P'_{cr}) of the mixture using the volume fractions and the critical pressure of each gas.**

$$P'_{cr} = y_{CO_2}P_{crCO_2} + y_{H_2O}P_{crH_2O} + y_{N_2}P_{crN_2}$$

$$P'_{cr} = (0.125)(7.39 \text{ MPa}) + (0.141)(22.06 \text{ MPa}) + (0.734)(3.39 \text{ MPa})$$

$$P'_{cr} = 6.5 \text{ MPa}$$

3. **Calculate the reduced temperature and the reduced pressure of the mixture, using the pseudocritical properties.**

$$T_R = \frac{T_m}{T'_{cr}} = \frac{580 \text{ K}}{222 \text{ K}} = 2.6$$

$$P_R = \frac{P_m}{P'_{cr}} = \frac{10 \text{ MPa}}{6.5 \text{ MPa}} = 1.5$$

4. **Use Figure 14-2 in Chapter 14 to look up the values for T_R and P_R that you found in Step 3 to find the mixture compressibility factor (Z_m).**

You find that $Z_m = 1.00$.

The combustion gas mixture in this example is an ideal gas under these conditions because Z_m equals 1.0.

Working with Psychrometrics: Air and Water Vapor Mixtures

The most commonly analyzed gas mixture in thermodynamics is moist air — that is, dry air mixed with water vapor. *Dry air* is defined as the mixture of nitrogen, oxygen, carbon dioxide, and argon, among other trace gases, excluding water vapor. The analysis of moist air processes is called *psychrometrics*. The moist air gas mixture deserves special attention because it serves a variety of applications related to air conditioning. In this section, I discuss heating and humidifying air and cooling and dehumidifying air. Furthermore, I introduce a few new thermodynamic properties here that don't appear in ordinary gas mixtures.

Finding the wet-bulb temperature with a sling psychrometer

When working with a moist air gas mixture, you need to be able to figure out how much water vapor is in the air. Then you can have fun using the conservation of energy and conservation of mass equations to figure out basic heating, ventilating, and air-conditioning problems.

One way to figure out how much moisture is in the air is to use a sling psychrometer, as shown in Figure 15-1. The instrument has two thermometers. One thermometer has a small, wet sock wrapped around the bulb at the bottom. This thermometer measures the *wet-bulb temperature*. The other thermometer is an ordinary thermometer, and it measures the *dry-bulb temperature*.

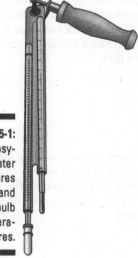

Figure 15-1:
A sling psy-
chrometer
measures
dry-bulb and
wet-bulb
tempera-
tures.

You swing the sling psychrometer around in circles for a few minutes to measure both the wet- and dry-bulb temperatures. The air flowing over the wet sock makes the water evaporate from it. As the water evaporates, it cools the sock.

When the humidity is high, the water doesn't evaporate very well, so the wet-bulb temperature is close to the dry-bulb temperature. When the humidity is low, the water evaporates more freely, and the wet-bulb temperature can be much lower than the dry-bulb temperature. The wet-bulb temperature decreases because the water in the wick gives up energy to evaporate. The wet-bulb and dry-bulb temperatures are used to determine the amount of moisture in the air. I show you how to calculate the moisture content in the next section.

It's muggy out there: Calculating specific and relative humidity

You can describe how much moisture is in the air in two ways (besides complaining about how sticky it feels outside on a hot muggy afternoon). The total amount of moisture in the air on a mass basis is called the *specific humidity*. The specific humidity (ω) is also called the *humidity ratio* because it's a measure of the mass of water vapor (m_v) in the air compared to the mass of dry air (m_a). Mathematically, the specific humidity is written as follows:

$$\omega = \frac{m_v}{m_a}$$

Another way to describe how much moisture is in the air is to use relative humidity. *Relative humidity* (ϕ) is the amount or mass of moisture in the air (m_v) compared to the amount of moisture mass (m_g) the air can hold if it's saturated. Air is saturated when the dry-bulb and the wet-bulb temperatures are the same. The air can't absorb any more moisture, so no moisture evaporates from the wet-bulb sock. Relative humidity is the term you hear most often in weather reports describing how much moisture the air contains. Mathematically, relative humidity is written like this:

$$\phi = \frac{m_v}{m_g}$$

Discovering humidity depends on pressure

In order to perform thermodynamic analysis of processes involving moist air, you must relate the specific humidity (ω) and the relative humidity (ϕ) to the vapor pressure (P_v) in the air. The total pressure of air consists of the partial pressure of the dry air (P_a) and the water vapor pressure (P_v). You calculate the total pressure in the air (P) with this equation:

$$P = P_a + P_v$$

When the air is saturated with moisture, the vapor pressure of the water equals the saturation pressure (P_g) of the water at a given temperature. The relative humidity (ϕ) can be calculated using the vapor pressure (P_v) and the saturation pressure (P_g) as follows:

$$\phi = \frac{P_v}{P_g}$$

You can use the ideal-gas law to determine the moisture mass (m_v) and the dry-air mass (m_a) for calculating the specific humidity (ω). You can see that some ideal-gas terms cancel out, leaving you with only the molar mass, partial pressure of air, and water vapor pressure for calculating the specific humidity, as shown in this equation:

$$\omega = \frac{m_v}{m_a} = \frac{M_v P_v V / \bar{R} T}{M_a P_a V / \bar{R} T} = \frac{M_v P_v}{M_a P_a}$$

Because the molar masses of the water vapor and dry air are known, you can calculate the specific humidity (ω) using only the vapor pressure (P_v) and the partial pressure (P_a) of the dry air, as follows:

$$\omega = \frac{M_v P_v}{M_a P_a} = \frac{(18.02 \text{ kg/kmol}) P_v}{(28.97 \text{ kg/kmol}) P_a} = 0.622 \, \frac{P_v}{P_a}$$

Converting between relative and specific humidity

You can convert between relative humidity (ϕ) and specific humidity (ω) by substituting terms from the three preceding equations for ϕ, ω, and P. Doing the math gives you the following equations between relative humidity (ϕ) and specific humidity (ω) in terms of the total pressure of the air (P) and the saturation pressure of the water in the air (P_g):

$$\phi = \frac{\omega P}{(0.622 + \omega) P_g} \ldots \text{or} \ldots \omega = \frac{0.622 \phi P_g}{P - \phi P_g}$$

No thermodynamic instruments directly measure the vapor pressure or the saturation pressure in the air so that you can calculate the relative humidity or the specific humidity. But you can write a set of mass and energy balance equations on the evaporation process from the wet bulb of a sling psychrometer to calculate the specific humidity.

The mass balance determines the evaporation rate of water from a wet bulb. The evaporation rate of water from the wet bulb is calculated by multiplying the air mass flow rate by the difference in specific humidity of saturated air (ω_2) and the specific humidity of the ambient air (ω_1):

$$\dot{m}_f = \dot{m}_a (\omega_2 - \omega_1)$$

In this equation, any property associated with the dry-bulb temperature is represented by the numerical subscript "1," and any property associated with the wet-bulb temperature is represented by the numerical subscript "2."

The energy balance on the evaporation process equates the enthalpy of the ambient air (h_1) plus the enthalpy of the evaporated water to the enthalpy of the saturated air (h_2), as shown in this equation:

$$h_1 + (\omega_2 - \omega_1) h_{f2} = h_2$$

In this equation, h_{f2} is the enthalpy of saturated liquid water at the wet-bulb temperature.

Or, substituting specific heat of air for enthalpy, the energy balance is

$$(c_P T_1 + \omega_1 h_{g1}) + (\omega_2 - \omega_1)h_{f2} = (c_P T_2 + \omega_2 h_{g2})$$

The terms in this equation are defined as follows:

✔ c_p is the constant-pressure specific heat of air, 1.005 kJ/kg · K.

✔ T_1 is the dry-bulb temperature.

✔ T_2 is the wet-bulb temperature.

✔ h_{g1} is the enthalpy of saturated water vapor at the dry-bulb temperature.

✔ h_{g2} is the enthalpy of saturated water vapor at the wet-bulb temperature.

The specific humidity (ω) is calculated using the dry- and wet-bulb temperatures (T_1 and T_2, respectively) in the following equation:

$$\omega_1 = \frac{c_p(T_2 - T_1) + \omega_2\left(h_{g_2} - h_{f_2}\right)}{h_{g_1} - h_{f_2}}$$

You calculate the specific humidity at the wet-bulb temperature (ω_2) using the following equation:

$$\omega_2 = \frac{0.622 P_{g_2}}{P_2 - P_{g_2}}$$

In this equation, P_{g2} is the saturation vapor pressure at the wet-bulb temperature, and P_2 is the total pressure of the air.

My glasses are fogging up: Defining the dew point

You may have seen a person's glasses fog up when they came indoors on a cold day. Maybe it has even happened to you. Why do glasses that are clear outside fog up when their wearer comes inside? The answer lies in the dew-point temperature of the air.

The *dew point* is the temperature at which moisture from the air begins to condense on a surface. The dew-point temperature (T_{dp}) of the air depends on the humidity of the air, which is related to the water vapor pressure in the air. The dew-point temperature is the saturation temperature (T_{sat}) that corresponds to the partial pressure (P_v) of the water vapor in the air. You can find the saturation temperature (T_{sat}) that corresponds to the partial pressure of the water vapor (P_v) by using Table A-4 in the appendix.

When a person wearing glasses walks indoors on a cold day, the glasses are cold relative to the room. If the temperature of the glasses is lower than the dew-point temperature of the room, moisture condenses on the lenses.

Working out problems with temperature and humidity

Imagine receiving a sling psychrometer and a barometer for your birthday and going outside to try them out. You measure the wet-bulb temperature at 16 degrees Celsius and the dry-bulb temperature at 21 degrees Celsius. The barometer indicates the air pressure is 100 kilopascals. You can calculate the specific humidity and the relative humidity and find the dew point by following these steps:

1. **Find the saturation pressure at the wet-bulb temperature (P_{g2}) by looking up the vapor pressure of water at 16 degrees Celsius in Table A-3 of the appendix.**

   You'll need to do some interpolation in the table. I show you how to interpolate a table in Chapter 3.

   $P_{g2} = 1.818$ kPa

2. **Calculate the specific humidity at the wet-bulb temperature (ω_2) for saturated conditions at the wet-bulb temperature.**

   $$\omega_2 = \frac{0.622 P_{g2}}{P_2 - P_{g2}} = \frac{0.622(1.818 \text{ kPa})}{(100 - 1.818 \text{ kPa})} = 0.0115 \frac{\text{kg H}_2\text{O}}{\text{kg dry air}}$$

3. **Find the enthalpy of saturated water vapor (h_{g2}) at the wet-bulb temperature of 16 degrees Celsius in Table A-3 of the appendix.**

   You'll need to do some interpolation in the table.

   $h_{g2} = 2{,}531$ kJ/kg

4. **Find the enthalpy of saturated liquid water (h_{f2}) at the wet-bulb temperature of 16 degrees Celsius in Table A-3 of the appendix.**

   $h_{f2} = 67$ kJ/kg

5. **Find the enthalpy of saturated water vapor (h_{g1}) at the dry-bulb temperature of 21 degrees Celsius in Table A-3 of the appendix.**

   You'll need to do some interpolation in the table.

   $$h_{g2} = 2{,}540 \text{ kJ/kg}$$

6. **Calculate the specific humidity at the dry-bulb temperature (ω_1) by substituting the values determined in Steps 2–5 into the specific humidity equation, as follows:**

$$\omega_1 = \frac{(1.005 \text{ kJ/kg} \cdot \text{K})(16-21)^\circ C + (0.0115)(2{,}531-67) \text{ kJ/kg}}{(2{,}540-67) \text{ kJ/kg}}$$

$$= 0.00943 \frac{\text{kg water}}{\text{kg dry air}}$$

7. **Find the saturation pressure at the dry-bulb temperature (P_{g1}) by looking up the vapor pressure of water at 21 degrees Celsius in Table A-3 of the appendix.**

   $$P_{g1} = 2.487 \text{ kPa}$$

8. **Calculate the relative humidity (ω) by substituting the values determined in Steps 6 and 7 into the relative humidity equation, as follows:**

$$\phi = \frac{\omega P}{(0.622+\omega)P_g} = \frac{(0.00943)(100 \text{ kPa})}{(0.622+0.00943)(2.487 \text{ kPa})} = 0.60 \text{ or } 60\%$$

9. **Calculate the vapor pressure (P_v) of the moisture in the air using the equation for the relative humidity:**

   $$P_v = \phi P_{g1} = (0.60)(2.487 \text{ kPa}) = 1.5 \text{ kPa}$$

10. **Find the saturation temperature (T_{sat}) for the vapor pressure of 1.5 kilopascals in Table A-4 of the appendix.**

    This is the dew-point temperature (T_{dp}).

    $$T_{dp} = T_{sat@Pv} = 13^\circ C$$

After making all these calculations, you're on your way to becoming a weatherman!

Using the psychrometric chart

Surely the weatherman doesn't do all those calculations in the preceding section every day, right? There's an easy way to find the humidity values from the wet-bulb and dry-bulb temperatures. The *psychrometric chart* in Figure 15-2 has been developed to help you do thermodynamic analysis of air-conditioning processes.

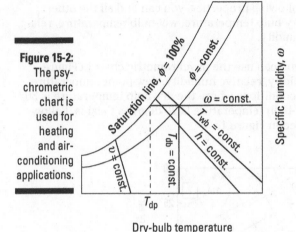

Dry-bulb temperature

The psychrometric chart is packed full of information:

- ✔ **Dry-bulb temperatures:** The dry-bulb temperatures of the air appear along the horizontal axis at the bottom. A vertical line drawn on the chart represents a line of constant dry-bulb temperature. The dry-bulb temperature increases going from left to right on the chart.

- ✔ **Wet-bulb temperatures:** The wet-bulb temperatures appear as lines slanting downward to the right. The wet-bulb temperature increases going from the lower-left corner to the upper-right corner.

- ✔ **Enthalpy of moist air:** The enthalpy of moist air appears as slanted lines on the chart. These slanted lines are almost parallel to the wet-bulb lines. You use the enthalpy lines in thermodynamic analysis of air-conditioning processes. The enthalpy increases going from the lower-left corner to the upper-right corner.

- ✔ **Specific humidity:** The specific humidity is shown on the vertical axis on the right side of the chart. A horizontal line drawn on the chart represents a line of constant specific humidity. The humidity increases from the bottom to the top.

- ✔ **Relative humidity:** The relative humidity appears as a line that curves upward to the right on the chart. The relative humidity increases from zero along the bottom horizontal axis to 100 percent along the saturation line shown on the upper-left portion of the chart.

- ✔ **Constant specific volume:** Some psychrometric charts show lines of constant specific volume. These lines are slanted like the wet-bulb and enthalpy lines, but they have a distinctly different slope. The specific volume increases from the lower left to the upper right in the chart.

If you know any two of the following properties, you can find all the other properties from the chart: dry-bulb temperature, wet-bulb temperature, relative humidity, or specific humidity.

Here is an example of how you can use the psychrometric chart in Figure 15-3 to find the specific humidity, relative humidity, dew-point temperature, and enthalpy from ambient conditions where the dry-bulb temperature is 30 degrees Celsius and the wet-bulb temperature is 20 degrees Celsius. Follow these steps, which are labeled in Figure 15-3:

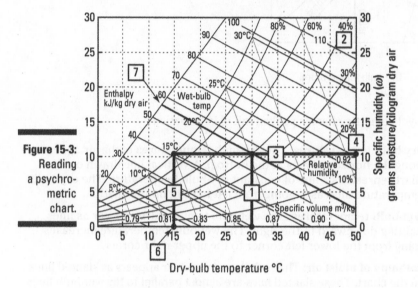

Figure 15-3: Reading a psychrometric chart.

1. **Draw a vertical line from the dry-bulb temperature until it intersects with the wet-bulb temperature.**

2. **Read the relative humidity from the curved line that passes through the intersection of the dry- and wet-bulb lines.**

   You find that the relative humidity is 40 percent.

3. **Draw a horizontal line through the intersection of the dry- and wet-bulb temperature lines.**

   The line should extend from the 100-percent saturation line on the left to the vertical axis on the right. This line is a constant specific humidity line.

4. **Read the specific humidity where the horizontal line intersects the right-hand axis.**

   You find that the specific humidity is about 11 grams of moisture per kilogram of dry air.

5. **Draw a vertical line down from the intersection of the horizontal line you drew in Step 3 and the saturation line. Extend this line to the horizontal axis.**

6. **Read the dew-point temperature from the intersection of this vertical line with the dry-bulb temperature scale.**

   You find the dew-point is 15 degrees Celsius.

7. **Read the enthalpy of the air from the constant enthalpy line that runs through the intersection of the dry- and wet-bulb temperature lines.**

   You find that the enthalpy is about 56 kilojoules per kilogram of dry air.

The next section on air conditioning uses the psychrometric chart to perform thermodynamic analysis of heating and cooling applications.

Making Life Comfortable with Air Conditioning

A person's comfort zone is fairly small. Most people like the temperature to be between 20 and 26 degrees Celsius (68 to 78 degrees Fahrenheit). Humidity also plays a role in comfort, and a relative humidity in the range of 40 to 60 percent is preferable. Air movement is a third aspect of comfort. On a hot summer day, a gentle breeze makes you feel cool. But in the winter, the wind can make you feel very cold. Everyone has their own level of comfort, which makes people battle over thermostat settings and blankets on the bed. In this section, I show you how to do a thermodynamic analysis of basic air-conditioning processes — unfortunately, it won't help you win any thermostat battles!

Analysis of air-conditioning processes is easy with the psychrometric chart. You can draw a heating or cooling process on the chart and quickly find out how much energy you need to heat or cool a space. The analysis I cover in this section focuses only on heating and cooling systems that blow air over a heat exchanger to warm or cool the air. I don't cover systems that use hot water or steam radiators to heat the air. Those kinds of systems heat the space using natural convection.

Heating and humidifying the air

Most forced-air heating systems use a furnace, a heat pump, or an electric resistance heater to warm the air. A fan pulls cool air into the furnace, where it picks up heat. The fan then sends that heat throughout the house via a series of ducts.

If you look at a simple heating process on a psychrometric diagram, like the one in Figure 15-4, from States 1 to 2 you see the process follows a straight line. The total moisture in the air and the room temperature, described by the dry-bulb temperature, stay the same. Consequently, you see in Figure 15-4 that the relative humidity decreases as the air warms up from State 1 to State 2. This process can make a person feel uncomfortable and cause dry skin if the relative humidity gets too low.

Sometimes a humidifier is added to the heating system to make the air more comfortable. You see a humidifier in the process from State 2 to State 3 in Figure 15-4, where warm water increases the temperature and specific humidity of the air.

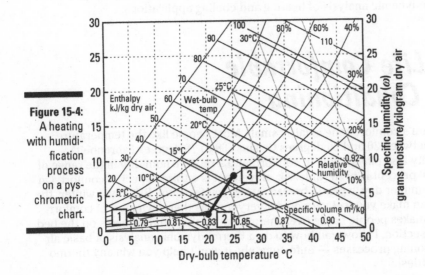

Figure 15-4:
A heating with humidification process on a pyschrometric chart.

In a simple heating process, you can calculate the amount of energy added to the dry air ($\dot{Q}_{in}$) by using the mass flow rate of the dry air ($\dot{m}_{air}$) and the change in enthalpy of the air ($h_2 - h_1$), with the following equation. The enthalpy in this equation includes the enthalpy of the water vapor in the moist air.

$$\dot{Q}_{in} = \dot{m}_{air}(h_2 - h_1)$$

For a humidification process like the one in Figure 15-4, from States 2 to 3 you can calculate the amount of water added to the air ($\dot{m}_w$) using the mass flow rate of the dry air ($\dot{m}_{air}$) and the change in specific humidity ($\omega_3 - \omega_2$) of the air, with this equation:

$$\dot{m}_w = \dot{m}_{air}(\omega_3 - \omega_2)$$

Suppose you have a heating system that draws 30 cubic meters of air per minute from outdoors. The air is at 5 degrees Celsius and 40 percent relative humidity. (Refer to State 1 in Figure 15-4.) The air is heated to 20 degrees Celsius in the furnace (State 2 in Figure 15-4). Then warm water is sprayed into the air to humidify and further heat the air until it reaches 25 degrees Celsius and 40 percent relative humidity. In a thermodynamic analysis of this heating and humidification process, you can find out how much energy and water are required for the process using the psychrometric chart. Just follow these steps:

1. **Estimate the enthalpy at State 1, identified by the point shown in Figure 15-4 at 5 degrees Celsius and 40 percent relative humidity.**

   You find that $h_1 \approx 11$ kJ/kg dry air.

2. **Draw a straight horizontal line from State 1 until it reaches State 2 at 20 degrees Celsius.**

   This line has already been drawn for you in Figure 15-4.

3. **Estimate the enthalpy at State 2.**

   You find that $h_2 \approx 26$ kJ/kg dry air.

4. **Estimate the specific volume of the air at State 1.**

   You find that $v_1 \approx 0.785$ m³/kg dry air.

5. **Calculate the mass flow rate of the dry air blowing through the furnace by using the volume flow rate and the specific volume at State 1, as shown in this equation:**

$$\dot{m}_{air} = \frac{\dot{V}_1}{v_1} = \frac{30 \text{ m}^3/\text{min}}{0.785 \text{ m}^3/\text{kg dry air}} = 38.2 \text{ kg dry air/min}$$

6. **Calculate the heat transferred to the air in the furnace using the following equations:**

$$\dot{Q}_{in} = \dot{m}_{air}(h_2 - h_1)$$

$$Q_{in} = (38.2 \text{ kg/min})(26 - 11) \text{ kJ/kg}\left(\frac{1 \text{ min}}{60 \text{ sec}}\right)\left(\frac{1 \text{ kW}}{1 \text{ kJ/s}}\right) = 9.6 \text{ kW}$$

7. **Estimate the specific humidity at State 2 from the chart. Note the units are grams of water per kilogram of dry air.**

   You find that $\omega_2 \approx 2$ g H_2O/kg of dry air.

8. **Estimate the specific humidity at State 3 from the chart.**

   You find that $\omega_3 \approx 8$ g H_2O/kg of dry air.

9. **Calculate the mass flow rate of water added to the air, using the mass flow rate of dry air and the change in specific humidity from States 2 to 3.**

$$\dot{m}_w = 38.2 \text{ kg/min}(8 - 2) \text{ g } H_2O/\text{kg dry air} = 229 \text{ g } H_2O/\text{min}$$

You can use the heat transfer rate and the water flow rate to determine what size furnace you need to heat the space and what size humidifier you need.

Cooling and dehumidifying the air

In a typical air-conditioning system, a fan blows warm air over a heat exchanger, where it loses heat and moisture. The heat exchanger is the evaporator section of an air-conditioning cycle that I discuss in Chapter 13. If you draw a cooling process on a psychrometric diagram, like the one in Figure 15-5, you see that the relative humidity increases until the air becomes saturated as it cools from State 1 to State 2. Additional cooling of the air follows the saturation line until the air leaves the heat exchanger at State 3. Moisture is removed from the air from State 2 to State 3. The moisture condenses into liquid water and leaves the evaporator through a drain. You can see that the specific humidity decreases in the process.

In a simple cooling and dehumidification process like the one in Figure 15-5, the amount of energy removed from the air ($\dot{Q}_{out}$) can be calculated using the mass flow rate of the dry air ($\dot{m}_{air}$), the change in enthalpy of the moist air ($h_3 - h_1$), and the energy removed by the water condensation ($\dot{m}_w h_w$), with this equation:

$$\dot{Q}_{out} = \dot{m}_{air}(h_1 - h_3) - \dot{m}_w h_w$$

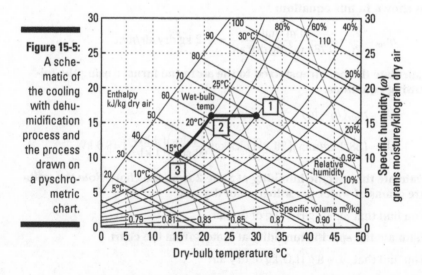

Figure 15-5: A schematic of the cooling with dehumidification process and the process drawn on a pyschrometric chart.

In the dehumidification process in Figure 15-5, from States 2 to 3 the amount of water removed from the air ($\dot{m}_w$) is calculated using the mass flow rate of the dry air ($\dot{m}_{air}$) and the change in specific humidity ($\omega_3 - \omega_2$) of the air with this equation:

$$\dot{m}_w = \dot{m}_{air}(\omega_3 - \omega_2)$$

Suppose you have a cooling system that draws 50 cubic meters of air per minute from outdoors. The air is at 33 degrees Celsius and 60 percent relative humidity. This condition is shown as State 1 in Figure 15-5. The air is cooled to 15 degrees Celsius at saturated conditions, as shown by State 3. You can calculate how much energy and water are removed from the air in this process by using either the psychrometric chart — as discussed in the preceding section, "Heating and humidifying the air" — or by using energy equations. In this section, I show you how to do the calculations with energy equations.

You can calculate the heat transfer rate and the condensate removal rate for this air-conditioning process by following these steps:

1. **Find the saturation pressure of water (P_{g1}) at State 1 at 33 degrees Celsius in Table A-4 of the appendix.**

   Interpolate the table to find

   P_{g1} = 5.035 kPa

2. **Calculate the vapor pressure (P_{v1}) at 60 percent relative humidity, using this equation:**

   $P_{v1} = \phi P_{g1} = 0.60(5.035 \text{ kPa}) = 3.021 \text{ kPa}$

3. **Calculate the partial pressure of the dry air (P_a); assume the total ambient pressure is 100 kPa.**

   $P_a = P - P_{v1} = (100 - 3.021) \text{ kPa} = 96.98 \text{ kPa}$

4. **Calculate the specific volume of the air (v_1) at State 1, using the ideal-gas law.**

   Convert the air temperature of 33 degrees Celsius to 306 Kelvin.

   $$v_1 = \frac{RT_1}{P_a} = \frac{(0.287 \text{ kPa} \cdot \text{m}^3/\text{kg} \cdot \text{K})(306 \text{ K})}{(96.98 \text{ kPa})} = 0.906 \text{ m}^3/\text{kg dry air}$$

5. **Calculate the air mass flow rate using the volumetric flow rate and the specific volume at State 1, with this equation:**

   $$\dot{m}_a = \frac{\dot{V}_1}{v_1} = \frac{50 \text{ m}^3/\text{min}}{0.906 \text{ m}^3/\text{kg}} = 55.2 \text{ kg dry air/min}$$

6. **Calculate the specific humidity at State 1, using the following equation:**

   $$\omega_1 = \frac{0.622 \, P_{v_1}}{P - P_{v_1}} = \frac{(0.622)(3.021 \text{ kPa})}{(100 - 3.021) \text{ kPa}} = 0.019 \frac{\text{kg water}}{\text{kg dry air}}$$

7. **Find the saturation pressure of water at State 3 at 15 degrees Celsius in Table A-4 of the appendix.**

   P_{g3} = 1.705 kPa

8. **Calculate the specific humidity at State 3, using the following equation:**

   Because the air is saturated at this state, the relative humidity, ϕ_3, is 100 percent.

   $$\omega_3 = \frac{0.622\phi_3 P_{g_3}}{P - \phi_3 P_{g_3}} = \frac{(0.622)(1.705 \text{ kPa})}{(100 - 1.705) \text{ kPa}} = 0.011 \frac{\text{kg water}}{\text{kg dry air}}$$

9. **Calculate the mass flow rate of the water condensing from the air with this equation:**

   $$m_w = m_a(\omega_1 - \omega_2) = (55.2 \text{ kg/min})(0.019 - 0.011)\frac{\text{kg water}}{\text{kg dry air}} = 0.44 \text{ kg/min}$$

10. **Find the enthalpy of saturated water vapor at State 1 (h_{g1}) for 33 degrees Celsius, using Table A-4 of the appendix.**

    $h_{g1} = 2,562 \text{ kJ/kg}$

11. **Find the enthalpy of saturated water vapor at State 3 (h_{g3}) for 15 degrees Celsius, using Table A-4 of the appendix.**

    $h_{g2} = 2,529 \text{ kJ/kg}$

12. **Calculate the enthalpy of the dry air at State 1, using the following equation.**

    Use temperature in degrees Celsius.

    $h_1 = c_p T_1 + \omega_1 h_{g_1}$

    $h_1 = (1.005 \text{ kJ/kg°C})(33°C) + (0.019)(2,562 \text{ kJ/kg}) = 81.8 \text{ kJ/kg dry air}$

13. **Calculate the enthalpy of the dry air at State 3, using the following equation.**

    $h_3 = c_p T_3 + \omega_3 h_{g_3}$

    $h_3 = (1.005 \text{ kJ/kg°C})(15°C) + (0.011)(2,529 \text{ kJ/kg}) = 42.9 \text{ kJ/kg dry air}$

14. **Calculate the heat transfer rate from the air, using the following form of the energy equation.**

    You can find h_w, the enthalpy of saturated water at 15°C by interpolating Table A-3 in the appendix. Convert the result from kilojoules per minute into kilowatts.

    $$\dot{Q}_{out} = m_{air}(h_1 - h_3) - m_w h_w$$

    $$\dot{Q}_{out} = \left[55.2 \frac{\text{kg}}{\text{min}}(81.8 - 42.9) \frac{\text{kJ}}{\text{kg}} - \left(0.44 \frac{\text{kg}}{\text{min}}\right)\left(63.0 \frac{\text{kJ}}{\text{kg}}\right) \right]\left(\frac{1 \text{ min}}{60 \text{ sec}}\right)\left(\frac{1 \text{ kW}}{1 \text{ kJ/sec}}\right)$$

    $\dot{Q}_{out} = 35.3 \text{ kW}$

Chapter 16

Burning Up with Combustion

• •

In This Chapter

▶ Exploring systems whose chemical composition changes during combustion

▶ Balancing combustion reaction equations

▶ Defining thermodynamic properties related to combustion

▶ Using the first law to analyze steady-flow and closed combustion systems

▶ Calculating the adiabatic flame temperature for a combustion reaction

• •

hemical reactions are everywhere in your everyday life. Did you power on your laptop today? The battery uses a chemical reaction to make electricity. Have you ever baked a cake? Heat causes baking powder to produce carbon dioxide gas; the reaction makes the cake rise so it becomes fluffier. Ever seen a rusty nail? The iron oxidizes in moist air or water to form iron oxide. What about driving your car? The burning of gasoline is a chemical reaction that generates heat, which can produce work. Chemical reactions are important not only in life, but also in thermodynamics.

Combustion chemical reactions are the heat source that drives the heat engines I discuss in Chapters 11–13. Engines use fuels to create power. But what actually occurs when fuel is ignited in your car? How much energy can you get from fuel? Why do you get exhaust, and where does it come from?

In this chapter, I focus on the chemical reaction of combustion in detail. First, I show you how to set up and balance chemical equations (yes, it's time for a trip back to high school chemistry) for fuels, specifically hydrocarbons such as gasoline. I define some new terms that are useful when dealing with chemical equations. I show you how to use the first law of thermodynamics to qualitatively analyze the energy (heat) transfer of both open and closed systems undergoing combustion. Finally, I show you how to calculate the temperature of a flame. The flame temperature can be used in the design of a combustion chamber for a heat engine. A materials engineer can then select materials that can survive that temperature.

Forming Combustion Reaction Equations

A *combustion reaction* takes place when a fuel is burned with oxygen. The most common fuels are hydrocarbons, which are made from hydrogen and carbon atoms. Combustion usually takes place in air (a mixture of nitrogen, oxygen, and several other gases), although for some special processes (like welding), it takes place with oxygen alone.

Although actual combustion processes are complicated due to intermediate reactions, writing the chemical reaction for a hydrocarbon fuel combustion process is easy. The hydrogen atoms in the fuel bond with oxygen to form water, and the carbon atoms react with oxygen to form carbon dioxide. Every combustion reaction consists of reactants and products, as shown in Figure 16-1. The mass of the reactants is equal to the mass of the products because mass like energy can't be created or destroyed but can change form.

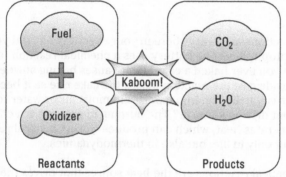

Figure 16-1: Combustion starts with reactants and ends with products.

You write chemical reactions in molar quantities. However, sometimes you need to convert molar quantities into mass quantities in thermodynamic analysis. The number of moles of a substance, N, is related to the mass of a substance, m, using the molar mass (molecular weight), M, with the following equation:

$$M = \frac{m}{N}$$

Table A-11 in the appendix lists relevant molar mass values for several compounds found in combustion reactions.

To analyze combustion reactions, you need to know a little bit about how air is handled in a reaction. Air is composed of about 21 percent oxygen and 79 percent nitrogen by volume or moles, meaning that each mole of oxygen corresponds to 79/21 = 3.76 moles of nitrogen. A mole of a substance corresponds to 6.02×10^{23} molecules (Avogadro's number).

In the following sections, I show you how to figure out how much air is needed for the reaction, as well as how to account for excess air to write out a combustion reaction equation.

Figuring out how much air you need: Writing stoichiometric reaction equations

Writing out a combustion reaction equation starts with figuring out exactly how much air you need to completely burn the fuel. The amount of air that supplies just enough oxygen for complete combustion of fuel is called *theoretical air*. Combustion reactions completed with theoretical air are called *stoichiometric reactions*. You can write the stoichiometric reaction for any simple hydrocarbon fuel combustion process using the *general combustion reaction equation*:

$$C_x H_y + \left(x + \frac{y}{4} \right)(O_2 + 3.76N_2) \rightarrow xCO_2 + \frac{y}{2}H_2O + 3.76\left(x + \frac{y}{4} \right)N_2$$

The reactants are on the left side, and the products are on the right side. This format is used for all hydrocarbon reactions. The general combustion reaction equation uses the molar ratio for oxygen and nitrogen in air.

Even though chemical reaction equations are written on a molar basis, the number of moles of reactants doesn't necessarily equal the number of moles of products. But the mass of reactants equals the mass of products. Conservation of mass applies here. There is no such law as the conservation of moles or volume.

When writing a chemical reaction equation, keeping track of individual elements on both sides of the equation is helpful. Here's one way to do it for burning a simple hydrocarbon fuel in air using the following format:

Element: Reactants: (# moles)(# atoms) = Products: (# moles)(# atoms)

C: $(1)(x) = (x)(1)$

H: $(1)(y) = (y/2)(2)$

O: $(x + y/4)(2) = (x)(2) + (y/2)(1)$

N: $(3.76)(x + y/4)(2) = (3.76)(x + y/4)(2)$

A simple hydrocarbon fuel has only carbon and hydrogen atoms like methane (CH_4) and octane (C_8H_{18}). For more complex fuels like ethanol (CH_3OH) and nitromethane (CH_3NO_2), you can't use the general combustion reaction equation; you need to write out the balanced reaction equation by balancing all the atoms in the reactants with those in the products.

If the amount of air supplied for a combustion reaction is less than theoretical air, the reaction is incomplete. When incomplete combustion occurs, carbon monoxide is formed along with carbon dioxide, and soot (carbon particles) may also appear in the exhaust.

Accounting for excess air in combustion

In actual combustion systems, excess air is usually available to ensure the reaction is complete. The amount of excess air above the theoretical air is given as a percentage. For example, 50 percent excess air is the same as 150 percent theoretical air. You can write the balanced chemical equation for octane (C_8H_{18}) with 20 percent excess air as follows:

1. **Write out the values for x and $y/2$ in the general combustion reaction equation.**

   You have 8 carbon atoms and 18 hydrogen atoms in octane:

   $$x = 8$$

   $$\frac{y}{2} = \frac{18}{2} = 9$$

2. **Find the number of moles of air required to burn one mole of octane, using this equation:**

   $$x + \frac{y}{4} = 8 + \frac{9}{2} = 12.5$$

3. **Calculate the number of moles of nitrogen in the air used during combustion with this equation:**

   $$3.76\left(x + \frac{y}{4}\right) = 3.76 \times 12.5 = 47.0$$

4. **Place these results into the general combustion reaction equation for 100-percent theoretical air to get the following:**

   $$C_8H_{18} + 12.5(O_2 + 3.76N_2) \rightarrow 8CO_2 + 9H_2O + 47.0N_2$$

5. **Multiply the amount of air in the reaction by 120 percent, and the combustion equation becomes this:**

   $$C_8H_{18} + (1.2)12.5O_2 + (1.2)(12.5)3.76N_2 \rightarrow 8CO_2 + 9H_2O +$$
   $$(0.2)(12.5)O_2 + 56.4N_2$$

6. **Write the balanced chemical equation for burning octane in 20-percent excess air:**

$$C_8H_{18} + 15O_2 + 56.4N_2 \rightarrow 8CO_2 + 9H_2O + 2.5O_2 + 56.4N_2$$

The following summary of the elements in the reaction shows that the total mass or number of moles of each element is the same for both the reactants and products.

Element: Reactants: (# moles)(# atoms) = Products: (# moles)(# atoms)

C: $(1)(8) = (8)(1)$

H: $(1)(18) = (9)(2)$

O: $(15)(2) = (8)(2) + (9)(1) + (2.5)(2)$

N: $(56.4)(2) = (56.4)(2)$

Note: The coefficient for oxygen in the product is equal to the "extra" amount of oxygen being supplied. In this case, the extra oxygen supplied is 20 percent more than the theoretical amount, making the coefficient 20 percent of 12.5, which equals 2.5 moles of oxygen per mole of octane.

Octane is a relatively simple hydrocarbon and is thermodynamically equivalent to gasoline. Gasoline is a mixture of many hydrocarbons, and to account for each individual hydrocarbon would make analysis complicated.

Defining Combustion-Related Thermodynamic Properties

In all the other chapters of this book, all the substances you work with maintain their chemical composition despite changes in physical properties. In chemical reactions, molecular bonds of the reactants break and new bonds form in the products of combustion. The chemical energy related to these bonds changes when the molecules change chemical form. Therefore, you have to account for the changes in chemical energy when evaluating the heat transfer of a chemical reaction. More importantly, you must account for the chemical energy when attempting an energy balance. So, when balancing the energy of a combustion equation, remember that the change in energy of the system is due to the change in state *and* the change in chemical composition.

When you have a change in chemical composition, you need a *reference state* that you can use to compare the energy associated with different reactions. For the purpose of evaluating combustion-related energy, you normally assume a reference state temperature of 25 degrees Celsius and a pressure of 100 kilopascals, and you assume ideal-gas conditions for gaseous substances.

In the following sections, I discuss thermodynamic properties that allow you to determine the change in energy of combustion reactions.

Enthalpy of formation

All compounds in a combustion reaction — such as CH_4, H_2O, and CO_2 — are made from elements like C, H_2, and O_2. When simple elements combine in a chemical reaction to form a compound, energy is either released to the environment in an *exothermic* reaction or absorbed from the environment in an *endothermic* reaction. The amount of energy released by or absorbed in the reaction to form one mole of a compound at 25 degrees Celsius and 100 kilopascals is called the *enthalpy of formation* ($\bar{h}_f^{\circ}$). For this symbol, the bar over the *h* is the molar-based enthalpy and has the units of kilojoules per kilomole (kJ/kmol).

For an exothermic reaction, the enthalpy of formation is negative because heat is transferred *from the system*. In an endothermic reaction, the enthalpy of formation is positive because heat is transferred *to the system*. You can see the direction of heat transfer for both exothermic and endothermic formation reactions in Figure 16-2.

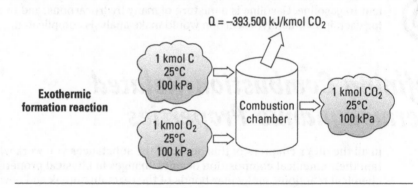

$$Q = -393,500 \text{ kJ/kmol } CO_2$$

Exothermic formation reaction

1 kmol C
25°C
100 kPa

1 kmol O_2
25°C
100 kPa

Combustion chamber

1 kmol CO_2
25°C
100 kPa

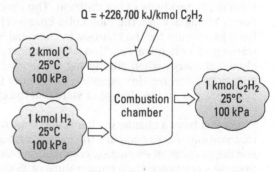

$$Q = +226,700 \text{ kJ/kmol } C_2H_2$$

Figure 16-2: Exothermic and endothermic enthalpy of formation reactions.

Endothermic formation reaction

2 kmol C
25°C
100 kPa

1 kmol H_2
25°C
100 kPa

Combustion chamber

1 kmol C_2H_2
25°C
100 kPa

The enthalpy of formation for a compound is measured by the amount of heat absorbed or released while allowing the compound to either cool down or warm up to 25 degrees Celsius after the reaction is complete. This is the standard reference temperature for comparing the enthalpy of formation for all compounds. All stable elements have an enthalpy of formation of zero. A stable element is defined as the chemically stable form of the element at 25 degrees Celsius. So C, N_2, and O_2 are stable, whereas N and O are not.

Table 16-1 gives the enthalpy of formation for several compounds found in many common combustion reactions. Note the enthalpy of formation for CO_2 is negative, which means that the enthalpy of one kilomole of C and one kilomole of O_2 is more than the enthalpy of one kilomole of CO_2. Physically, energy is released when CO_2 is formed.

Table 16-1 contains two enthalpy values for water and for octane because each of these substances can be in the liquid phase or the vapor phase at 25 degrees Celsius. The difference between the values is the enthalpy of vaporization, which, for water, is about 44,000 kilojoules per kilomole at 25 degrees Celsius. The *enthalpy of vaporization* is the amount of energy it takes to change a saturated liquid into a saturated vapor.

Table 16-1	Enthalpy of Formation of Some Compounds at 25 Degrees Celsius and 100 Kilopascals	
Substance	**Formula**	**Enthalpy of Formation** $\bar{h}_f^\circ$, **kJ/kmol**
Acetylene	C_2H_2	+226,700
Benzene	C_6H_6	+82,900
Carbon dioxide	CO_2	−393,500
Carbon monoxide	CO	−110,500
Diesel	$C_{14.4}H_{24.9}$	−174,000
Methane	CH_4	−74,900
Octane liquid	C_8H_{18}	−250,100
Octane vapor	C_8H_{18}	−208,600
Propane	C_3H_8	−103,900
Propylene	C_3H_6	+20,400
Water liquid	H_2O	−285,800
Water vapor	H_2O	−241,800

Enthalpy of combustion

When you completely burn 1 kilomole (or 1 kilogram on a mass basis) of fuel in a steady-flow combustion process, the amount of energy released is called the *enthalpy of combustion* ($\bar{h}_c$). This symbol is the molar-based enthalpy and has the units of kilojoules per kilomole. You can calculate the enthalpy of combustion by finding the difference between the total enthalpy of the products and the reactants, using this equation:

$$\bar{h}_C = H_P - H_R$$

This equation is a bit more complicated than it appears. The total enthalpy of the products (H_P) and the reactants (H_R) is the sum of the enthalpy of formation of all the products and the reactants. The enthalpy of the products and reactants are normally written on a molar basis. Often you will see H_R and H_P written without the over-bar that usually signifies a molar quantity. You rewrite the enthalpy of combustion equation into this form:

$$\bar{h}_C = \sum N_P \bar{h}_{f,P}^{\circ} - \sum N_R \bar{h}_{f,R}^{\circ}$$

The summation terms in this equation simply mean that you multiply the number of moles, N, of each compound by its enthalpy of formation, $\bar{h}_f^{\circ}$.

You can calculate the enthalpy of combustion on a mass basis also; then you don't have the bar over the enthalpy terms. The superscript ($^{\circ}$) indicates the value is at the reference state of 25 degrees Celsius and 100 kilopascals. The subscripts f, P, and R are reminders for *formation*, *products*, and *reactants*.

The enthalpy of combustion is sometimes called the *enthalpy of reaction*. You can find enthalpy of combustion values for several compounds in Table 16-2 at 25 degrees Celsius and 100 kilopascals. The enthalpy of combustion varies with temperature and pressure, so the values in the table will change if the temperature of the reactants and products isn't at 25 degrees Celsius.

The enthalpy of combustion values given in Table 16-2 assume the water in the products is in the liquid state at 25 degrees Celsius. In some textbooks, you find similar tables that give enthalpy of combustion values for water in the products in both the liquid and vapor states. I omitted the vapor state data in Table 16-2 for the sake of brevity. Table 16-2 gives enthalpy of combustion values for the fuel in both the liquid and vapor forms if the fuel can exist in either state at 25 degrees Celsius.

Table 16-2	**Enthalpy of Combustion at 25 Degrees Celsius**			
		Molar Mass	*Liquid H₂O Products*	
Fuel	*Formula*	*kg/kmol*	*Liq. Fuel kJ/kmol*	*Vapor Fuel kJ/kmol*
Methane	CH_4	16		−55,496
Propane	C_3H_8	44	−49,973	−50,343
Octane	C_8H_{18}	114	−47,893	−48,256
Gasoline	C_7H_{17}	101	−48,201	−48,582
Diesel	$C_{14.4}H_{24.9}$	198	−45,700	−46,074
Ethanol	C_2H_5OH	46	−29,676	−30,596
Nitromethane	CH_3NO_2	61	−11,618	−12,247

You can estimate the enthalpy of combustion for the gasoline you put in your car. Because gasoline is a mix of many different hydrocarbons, you may approximate gasoline as octane for this analysis. Use the enthalpy of formation of every compound in the combustion reaction to determine the enthalpy of combustion for a fuel. The chemical reaction for octane is

$$C_8H_{18} + 12.5(O_2 + 3.76N_2) \rightarrow 8CO_2 + 9H_2O + 47.0N_2$$

To calculate the enthalpy of combustion, assume both the reactants and products are at the same state of 25 degrees Celsius and 100 kilopascals. This means that before the reaction takes place, the air and the octane are at 25 degrees Celsius; after the combustion reaction, all the gases that are produced are cooled down to 25 degrees Celsius. Recall from the section "Enthalpy of formation" that nitrogen and oxygen are stable elements with an enthalpy of formation equal to zero. The enthalpy of combustion is

$$\bar{h}_C = \left(N\bar{h}_f^\circ\right)_{CO_2} + \left(N\bar{h}_f^\circ\right)_{H_2O} - \left(N\bar{h}_f^\circ\right)_{C_8H_{18}}$$

Water and gasoline exist in both the liquid and vapor phase at 25 degrees Celsius, but you use the enthalpy of formation in the liquid phase when you calculate the enthalphy of combustion. You can find the enthalpy of formation of the compounds of this reaction in Table 16-1. Calculate the molar enthalpy of combustion by multiplying the number of moles of each compound by its corresponding enthalpy of formation, like this:

$$\overline{h}_C = (8 \text{ kmol})(-393{,}500 \text{ kJ/kmol}) + (9 \text{ kmol})(-285{,}800 \text{ kJ/kmol})$$
$$- (1 \text{ kmol})(-250{,}100 \text{ kJ/kmol})$$

Completing the math, you find that the enthalpy of combustion for octane is

$$\overline{h}_C = -5{,}470{,}000 \text{ kJ/kmol of fuel}$$

To convert this molar quantity into a mass quantity, you divide it by the molar mass of octane (114 kilograms per kilomole), as shown in Table 16-2.

$$h_C = \frac{\overline{h}_C}{M_{\text{fuel}}} = \frac{5{,}470{,}000 \text{ kJ/kmol}}{114 \text{kg/kmol}} = -47{,}980 \text{ kJ/kg of fuel}$$

The enthalpy of combustion you just calculated is very close to the value for liquid octane shown in Table 16-2. Because most fuels don't have uniform composition, their enthalpies of combustion are measured experimentally.

Heating values

The *heating value* of a fuel is the amount of heat released in a steady flow process when the fuel is completely consumed and the products revert to the same state as the reactants. This term is basically synonymous with the enthalpy of combustion, but its value is always positive, making it equal to the absolute value of the enthalpy of combustion, as shown in this equation:

$$\text{heating value} = |h_C|$$

The heating value depends on the state or phase of water in the products. The *higher heating value* (HHV) has liquid water in the products. The *lower heating value* (LHV) has water vapor in the products. The HHV and LHV are related as follows:

$$HHV = LHV + \left(N\overline{h}_{fg} \right)_{H_2O}$$

Here, N is the number of moles of water in the products per mole of fuel, and h_{fg} is the molar enthalpy value associated with the vaporization of water.

Air-fuel ratio

Imagine that you've just spent a day at the drag races and were quite impressed with the top-fuel dragsters, which burned nitromethane. But you've been studying your thermodynamics, and you noticed in Table 16-2 that the enthalpy of combustion for nitromethane is only –11,600 kilojoules per do top-fuel dragsters burn nitromethane rather than gasoline? Do they know something that you don't know about combustion reactions? Yes, they do know something you don't know about — yet! They know about the air-fuel ratio.

The *air-fuel ratio*, AFR, is the ratio of the mass of air to the mass of fuel required for a combustion reaction. The minimum AFR is based on burning fuel with theoretical air proportions. You calculate the air-fuel ratio using the following equation:

$$AFR = \frac{\dot{m}_{air}}{\dot{m}_{fuel}}$$

Many combustion analyses are done on a molar basis, but when you're working with AFR, you usually use a mass basis.

You start with the combustion reaction to determine the AFR. For nitromethane, the balanced reaction with 100-percent theoretical air is

$$CH_3NO_2 + 0.75(O_2 + 3.76N_2) \rightarrow 1CO_2 + 1.5H_2O + 3.32N_2$$

You can determine the mass of the air and the mass of the fuel consumed in the reaction by multiplying the number of moles of each component by its respective molar mass. You can find the molar mass values for air in Table A-11 of the appendix, and you can find the molar mass values for nitromethane in Table 16-2. So the mass quantities for air and fuel in the reaction are as follows:

$$m_{air} = (N \cdot M)_{air} = 0.75(1 + 3.76)(28.97 \text{ kg/kmol}) = 103.4 \text{ kg air}$$

$$m_{fuel} = (N \cdot M)_{fuel} = (1)(61 \text{ kg/kmol}) = 61 \text{ kg fuel}$$

Use these values to find the AFR for nitromethane, like this:

$$AFR = \frac{103.4 \text{ kg air}}{61 \text{ kg fuel}} = 1.7 \text{ kg air/kg fuel}$$

You can calculate the AFR for gasoline by the same method; you'll find the AFR is 14.6 for 100-percent theoretical air. In order to figure out which fuel is better, you need to find the higher heating value for each fuel on a mass basis because the AFR is on a mass basis. Take the absolute value of the molar enthalpy of combustion for each fuel from Table 16-2 and divide it by the molar mass. Here's the result:

$$HHV_{nitro} = \frac{|\bar{h}_c|}{M} = \frac{11{,}618 \text{ kJ/kmol}}{61 \text{ kg/kmol}} = 190 \text{ kJ/kg}$$

$$HHV_{gasoline} = \frac{|\bar{h}_c|}{M} = \frac{48{,}201 \text{ kJ/kmol}}{101 \text{ kg/kmol}} = 477 \text{ kJ/kg}$$

Gasoline has more than twice the energy content on a mass basis than nitromethane. But to compare their performance in an engine, consider two identical engines operating at the same rpm. The mass (or volume) of air consumed by both engines is identical. This is a key point! So, you need to figure out how much energy is burned per unit mass of air. The AFR comes into play here. You determine the combustion energy per unit mass by dividing the higher heating value by the AFR, like this:

$$E_{nitro} = \frac{HHV}{AFR} = \frac{190 \text{ kJ/kg fuel}}{1.7 \text{ kg air/kg fuel}} = 112 \text{ kJ/kg air}$$

$$E_{gasoline} = \frac{HHV}{AFR} = \frac{477 \text{ kJ/kg fuel}}{14.6 \text{ kg air/kg fuel}} = 31 \text{ kJ/kg air}$$

This calculation turns the table upside-down and shows you that an engine running on nitromethane has over three times the energy of a gasoline engine.

Using the First Law of Thermodynamics on Steady-Flow Combustion Systems

Combustion reactions take place in either an *open system,* in which air and fuel continuously flow into the combustion chamber to produce heat, or in a *closed system,* where the combustion reaction takes place once (per cycle, usually) within the system. Examples of open systems are the Brayton cycle (Chapter 10) and the Rankine cycle (Chapter 12). The Otto and diesel cycles are examples of closed combustion systems because the fuel ignites and releases heat within a closed volume.

In Chapter 5 you find out about the first law of thermodynamics and its significance in terms of energy balance. In this chapter, you examine chemically reacting systems that see a change in chemical energy. You can still apply the energy balance relations, but you must rewrite the equation to explicitly show the components of chemical energy change.

When solving for energy balance, the enthalpy of formation is essential for the analysis of a reacting system. But enthalpy of formation alone doesn't fully convey the total enthalpy of a reactant or product. Therefore, your expression for enthalpy must include the sensible enthalpy relative to the reference state. Combining these two notions, the enthalpy on a molar basis ($\bar{h}$) of a chemical component is

$$\bar{h} = \bar{h}_f^\circ + \left(\bar{h} - \bar{h}^\circ \right) = \bar{h}_f^\circ + \Delta\bar{h}$$

where $\bar{h}_f^\circ$ is the enthalpy of formation at the reference state, and the term in the parentheses is the difference in enthalpy between any specified state ($\bar{h}$) and the enthalpy at the reference state ($\bar{h}^\circ$).

The enthalpy for ideal gases at the reference state temperature of 25 degrees Celsius is defined as zero in many thermodynamics textbooks. However, some books define the enthalpy as zero at absolute zero temperature, so $\bar{h}^\circ$ has a non-zero value at the reference state. The values in Table A-9 in the appendix aren't zero because 25 degrees Celsius is actually 298.15 K.

Analyzing an Example Steady-Flow System

The first law of thermodynamics isn't just for analyzing engines; you can use it on your barbeque. A gas grill is an open-system, *steady-flow* combustion process, as shown in Figure 16-3. Fuel and oxidizer enter the system and combustion products leave the system. The energy equation for this process can be written as follows, as long as the combustion process isn't doing any work (and usually there isn't any work in a steady-flow combustion process).

$$Q_{out} = \sum N_R \left(\bar{h}_f^\circ + \Delta\bar{h} \right)_R - \sum N_P \left(\bar{h}_f^\circ + \Delta\bar{h} \right)_P$$

The heat you get out of the combustion process is equal to the difference between the energy of the reactants and the energy of the products.

If your gas grill runs on propane, C_3H_8, the balanced chemical reaction with 100-percent theoretical air is

$$C_3H_8 + 5\left(O_2 + 3.76N_2 \right) \rightarrow 3CO_2 + 4H_2O + 18.8N_2$$

You can imagine that inside your gas grill you may have excess air, but for now you don't have to worry about that; focus on just the energy equation for a balanced chemical reaction. Suppose the propane and the air prior to combustion are at 25 degrees Celsius (298 Kelvin), and after combustion the products are at 1,400 Kelvin.

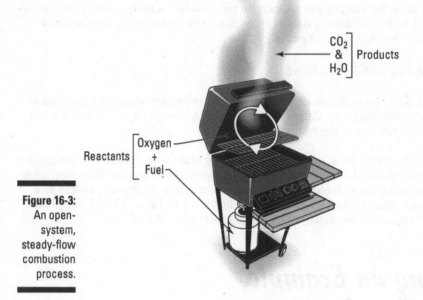

Reactants $\begin{bmatrix} \text{Oxygen} \\ + \\ \text{Fuel} \end{bmatrix}$

$\begin{bmatrix} CO_2 \\ \& \\ H_2O \end{bmatrix}$ Products

Figure 16-3:
An open-
system,
steady-flow
combustion
process.

You can find the heat of the combustion process by following these steps:

1. **Find the enthalpy of formation for propane, carbon dioxide, and water vapor by referring to Table 16-1.**

   At 1,400 Kelvin, water is in the vapor phase, so you use the enthalpy of formation for vapor instead of liquid water in this case. Table 16-3 summarizes the enthalpy of formation terms for each substance in the energy equation.

2. **Look up the molar enthalpy of the product gases CO_2, H_2O, and N_2 at 1,400 Kelvin.**

   You find them in Table A-9 of the appendix. (I show you how to interpolate data tables in Chapter 3.) Table 16-3 summarizes the enthalpy terms in the energy equation.

Table 16-3	A Summary of Enthalpy Terms Used in the Open-Flow Combustion Process Example	
Substance	Enthalpy of Formation, $\bar{h}_f^\circ$ kJ/kmol	Enthalpy of Products, $\bar{h}$, at 1,400 K kJ/kmol
C_3H_8	−103,900	Not present
CO_2	−393,500	55,900
H_2O	−241,800	43,500
O_2	0	Not present
N_2	0	34,940

3. **Insert the enthalpy terms summarized in Table 16-3 into the energy equation, like this:**

$$Q_{out} = (1 \text{ kmolC}_3\text{H}_8)(-103,900 \text{ kJ/kmol})$$
$$-(3 \text{ kmol CO}_2)(-393,500 + 55,900 \text{ kJ/kmol})$$
$$-(4 \text{ kmol H}_2\text{O})(-241,800 + 43,500 \text{ kJ/kmol})$$
$$-(18.8 \text{ kmol N}_2)(34,940 \text{ kJ/kmol})$$

4. **Complete the algebra to get the following results.**

   You can divide Q_{out} by the molar mass of propane (44 kilograms per kilomole) and by 1,000 to get the answer in megajoules per kilogram of fuel.

$$Q_{out} = 1,045,000 \text{ kJ/kmolC}_3\text{H}_8 = 23.8 \text{ MJ/kg fuel}$$

If you look at the specifications on your grill, you can find its heat output. For example, if your grill puts out 46,000 British thermal units per hour of heat, that's about equal to 48 megajoules per hour. You can calculate the fuel consumption rate of your gas grill using Q_{out} of the fuel:

$$\dot{m}_{fuel} = \frac{48 \text{ MJ/hr}}{23.8 \text{ MJ/kg}} = 2 \text{ kg/hr}$$

And now you can impress your friends by predicting how long you can grill before you run out of propane!

Using the First Law of Thermodynamics on Closed Combustion Systems

Not all combustion reactions take place in a steady flow device. Some combustion reactions take place in a confined volume, like those in the Otto and diesel cycles. You can find out more about these cycles in Chapter 11. In that chapter, I give you the amount of heat input for both the Otto and diesel cycles as part of the problem statement. Where does the heat input value come from? Here, you find out how to determine the heat input to an engine from a combustion process.

The energy equation for an open-flow combustion system is expressed in terms of enthalpy (h). For a closed system, you use the internal energy (u) of the reaction in the energy equation. Remember that for an ideal gas, enthalpy is a combined property of internal energy and the product of pressure times the volume of a system, that is, $h = u + Pv$.

The ideal-gas law states that $Pv = RT$. So, you can use these two equations and solve for internal energy such that $u = h - RT$ for ideal gases. The general form of the energy equation for a combustion process in a closed system is

$$Q - W = \sum N_R \left(\bar{h}_f^\circ + \Delta \bar{h} - \bar{R}T \right)_R - \sum N_P \left(\bar{h}_f^\circ + \Delta \bar{h} - \bar{R}T \right)_P$$

This equation differs in a few ways from the one for the open-flow system energy equation. In an open-flow system, work output or input usually isn't associated with the combustion process, so the work term is zero. Usually, in an open-system heat engine, a turbine or a compressor is used for work while a combustor is used for heat input to the engine. In a closed system, work may be extracted during the combustion process as it is in the diesel cycle, where the piston moves during combustion to provide a constant-pressure process. In the Otto cycle, the volume remains constant during the combustion process, so the work term is zero in this case.

Analyzing an Example Closed System

Here's an example of a closed-system combustion process. Suppose you have a mixture of 1 kilomole of diesel fuel with 110 percent theoretical air starting at 298 Kelvin and burning completely in a closed, rigid container that serves as a bomb calorimeter. (A *bomb calorimeter* is used for measuring the energy content in reactions.) The combustion temperature reaches 1,600 Kelvin. The process is shown in Figure 16-4.

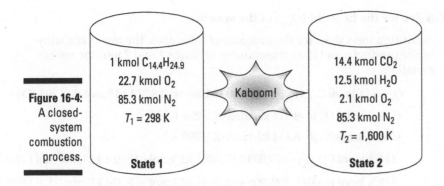

Figure 16-4:
A closed-system combustion process.

State 1:
1 kmol $C_{14.4}H_{24.9}$
22.7 kmol O_2
85.3 kmol N_2
$T_1 = 298$ K

Kaboom!

State 2:
14.4 kmol CO_2
12.5 kmol H_2O
2.1 kmol O_2
85.3 kmol N_2
$T_2 = 1,600$ K

You can determine the heat transfer (Q_{out}) from the closed-system combustion process by following these steps:

1. **Write the balanced chemical reaction for diesel fuel with 10-percent excess air.**

   When you have excess air, you need to account for the extra oxygen and nitrogen in the products.

   $$C_{14.4}H_{24.9} + 22.7(O_2 + 3.76N_2) \rightarrow 14.4CO_2 + 12.5H_2O + 2.1O_2 + 85.3N_2$$

2. **Write the energy balance equation for the combustion reaction.**

   $$Q_{out} = 1\left(\bar{h}_f^{\circ} - \bar{R}T_1\right)_{C_{14.4}H_{24.9}} + 22.7\left(-\bar{R}T_1\right)_{O_2} + 85.3\left(-\bar{R}T_1\right)_{N_2} - 14.4\left(\bar{h}_f^{\circ} + \Delta\bar{h} - \bar{R}T_2\right)_{CO_2}$$
   $$-12.5\left(\bar{h}_f^{\circ} + \Delta\bar{h} - \bar{R}T_2\right)_{H_2O} - 2.1\left(\Delta\bar{h} - \bar{R}T_2\right)_{O_2} - 85.3\left(\Delta\bar{h} - \bar{R}T_2\right)_{N_2}$$

3. **Find the enthalpy of formation ($\bar{h}_f^{\circ}$) for diesel fuel, carbon dioxide, and water vapor from Table 16-1.**

   The enthalpy of formation for the stable elements is zero.

   • $C_{14.4}H_{24.9}$: $\bar{h}_f^{\circ} = -174,000$ kJ/kmol
   • CO_2: $\bar{h}_f^{\circ} = -393,500$ kJ/kmol
   • H_2O: $\bar{h}_f^{\circ} = -241,800$ kJ/kmol

4. **Find the molar enthalpy, $\Delta\bar{h}$, of the products at 1,600 Kelvin using Table A-9 of the appendix.**

   • CO_2: $\Delta\bar{h} = 67,580$ kJ/kmol
   • H_2O: $\Delta\bar{h} = 52,910$ kJ/kmol
   • O_2: $\Delta\bar{h} = 44,270$ kJ/kmol
   • N_2: $\Delta\bar{h} = 41,910$ kJ/kmol

5. Solve for the heat out (Q_{out}) of the system.

Substitute the values for the enthalpy of formation, the molar enthalpy of the products, and the temperatures at States 1 and 2 into the energy equation.

$$Q_{out} = (1 \text{ kmol } C_{14.4}H_{24.9})(-174,000 \text{ kJ/kmol} - (8.314 \text{ kJ/kmol} \cdot K)(298 \text{ K}))$$
$$+ (22.7 \text{ kmol } O_2)(-8.314 \text{ kJ/kmol} \cdot K)(298 \text{ K})$$
$$+ (85.3 \text{ kmol } N_2)(-8.314 \text{ kJ/kmol} \cdot K)(298 \text{ K})$$
$$- (14.4 \text{ kmol } CO_2)((-393,500 + 67,580) \text{ kJ/kmol} - (8.314 \text{ kJ/kmol} \cdot K)(1,600 \text{ K}))$$
$$- (12.5 \text{ kmol } H_2O)((-241,800 + 52,910) \text{ kJ/kmol} - (8.314 \text{ kJ/kmol} \cdot K)(1,600 \text{ K}))$$
$$- (2.1 \text{ kmol } O_2)(44,270 \text{ kJ/kmol} - (8.314 \text{ kJ/kmol} \cdot K)(1,600 \text{ K}))$$
$$- (85.3 \text{ kmol } N_2)(41,910 \text{ kJ/kmol} - (8.314 \text{ kJ/kmol} \cdot K)(1,600 \text{ K}))$$

Completing the math gives this result for the heat output:

$$Q_{out} = 4,460,000 \text{ kilojoules per kilomole of fuel}$$

In a first-law analysis of a combustion process, the energy released by the fuel is based on a unit mass or unit mole of fuel. When you analyze a heat engine cycle, such as the Otto or diesel cycle, the heat input is based on a unit mass of air. You can use the air-fuel ratio and the molar mass of the fuel to convert the heat out from the combustion process from a unit mole of fuel to a unit mass of air, as shown in this equation:

$$Q_{out,air} = \frac{Q_{out,fuel}}{(M_{fuel})(AFR)}$$

The air-fuel ratio (AFR) for this reaction is

$$AFR = \frac{(22.7)(4.76 \text{ kmol air})(28.97 \text{ kg/kmol})}{198 \text{ kg/kmol fuel}} = 15.8 \text{ kg air/kg fuel}$$

The molar mass of diesel fuel from Table 16-2 is 198 kilograms per kilomole of fuel. Substitute in the heat out from the combustion process, the molar mass of the fuel, and the air-fuel ratio to calculate the heat out based on the air mass, as follows:

$$Q_{out,air} = \frac{4,460,000 \text{ kJ/kmol fuel}}{(198 \text{ kg/kmol fuel})(15.8 \text{ kg air/kg fuel})} = 1,426 \text{ kJ/kg air}$$

This heat output from the fuel combustion reaction can be used as the heat input to a diesel cycle engine analysis.

Ouch! That's Hot: Determining the Adiabatic Flame Temperature

The flames in that cozy fireplace of yours aren't the same temperature throughout. The color of the flames gives you a good indication of temperature. A red-orange colored flame is colder than a bright yellow flame. The flames are hottest where the combustion reaction takes place; this is known as the reaction zone. You can do some quick calculations to figure out how hot those flames are in the reaction zone.

If you look at the energy equation for a combustion process, you see that the difference between the enthalpy of the reactants and the products is heat because most steady-flow combustion processes don't produce work.

Suppose for a moment that you don't allow any heat to escape from the combustion process. In that case, all the energy of combustion goes into heating the combustion gases. When no heat transfer occurs between the combustion process and the environment, the process is adiabatic; it gives the maximum possible combustion temperature, which is known as the *adiabatic flame temperature*. The energy equation for the adiabatic flame temperature is

$$\sum N_R \left(\bar{h}_f^\circ + \Delta \bar{h} \right)_R = \sum N_P \left(\bar{h}_f^\circ + \Delta \bar{h} \right)_P$$

The enthalpy of the products is equal to the enthalpy of the reactants because the heat transfer (Q) and the work (W) terms of the energy equation are zero. The chemical energy of the fuel is converted into sensible enthalpy of the combustion products. Finding the adiabatic flame temperature is a bit tedious, because you have to make an educated guess at the temperature and then do some iterating to find the result.

The adiabatic flame temperature isn't fixed for a particular fuel. It depends on the following factors:

- ✔ The amount of air used in the combustion process
- ✔ The starting temperature of the reactants
- ✔ How complete the reaction is

Excess air in a combustion process reduces the adiabatic flame temperature because the energy released must heat up an additional amount of air. The maximum possible flame temperature occurs when you have complete combustion with 100-percent theoretical air.

Figuring Out an Example Adiabatic Flame Temperature

Take a peek into your gas furnace when it's on, and you'll see nice rows of big blue flames. If the furnace burns natural gas, you're burning methane (CH_4). You can calculate the adiabatic flame temperature of your furnace — the temperature at the hottest part of the flame. The temperature of the air above the flames isn't colder because the combustion products mix with a lot of excess air after combustion is complete. You can find the adiabatic flame temperature of a combustion reaction by following these steps:

1. **Write the balanced equation for methane combustion in 100-percent theoretical air.**

$$CH_4 + 2(O_2 + 3.76N_2) \rightarrow CO_2 + 2H_2O + 7.5N_2$$

2. **Write the energy equation for the combustion reaction.**

For simplicity, assume the fuel and the air prior to combustion are at the reference state of 25 degrees Celsius. Thus, no sensible enthalpy ($\Delta \bar{h}$) term is associated with the reactants, so the energy equation shows that the enthalpy of formation for methane is equal to the total enthalpy of the products. The equation looks like this:

$$N\bar{h}^{\circ}_{f,CH4} = \sum N_P \left(\bar{h}_f^{\circ} + \Delta \bar{h} \right)_P$$

3. **Look up the enthalpy of formation ($\bar{h}_f^{\circ}$) of CH_4, CO_2, and H_2O (vapor) in Table 16-1.**

   • CH_4: $\bar{h}_f^{\circ} = -74,900$ kJ/kmol

   • CO_2: $\bar{h}_f^{\circ} = -393,500$ kJ/kmol

   • H_2O: $\bar{h}_f^{\circ} = -241,800$ kJ/kmol

4. **Substitute the enthalpy of formation of the fuel, CO_2, and H_2O into the energy balance equation.**

$$(1 \text{ kmol } CH_4)(-74,900 \text{ kJ/kmol}) =$$
$$(1 \text{ kmol } CO_2)(-393,500 + \Delta \bar{h}_{CO_2}) \text{ kJ/kmol}$$
$$+(2 \text{ kmol } H_2O)(-241,800 + \Delta \bar{h}_{H_2O}) \text{ kJ/kmol}$$
$$+(7.52 \text{ kmol } N_2)(\Delta \bar{h}_{N_2} \text{ kJ/kmol})$$

5. **Simplify the energy equation so that it's in terms of the sensible enthalpy of the products CO_2, H_2O, and N_2.**

$$802,200 \text{ kJ} = \Delta \bar{h}_{CO_2} + 2\Delta \bar{h}_{H_2O} + 7.52 \, \Delta \bar{h}_{N_2}$$

The temperature of the products is unknown at this point, so you can't really solve this problem directly; it requires iteration.

6. **Find the average enthalpy of the products.**

You can make an educated guess for the adiabatic flame temperature to get the iteration off to a good start.

You can do a neat trick to get a good initial estimate for the adiabatic flame temperature. Divide the total amount of known energy (E) in the energy balance equation by the number of moles in the product (N_P) (in this case, 10.5) and the molar mass of nitrogen (M_{N_2}). This calculation gives you the average enthalpy ($\Delta \bar{h}_{ave}$) of the products:

$$\Delta \bar{h}_{ave} = \frac{E}{M_{N_2} N_P} = \frac{802,200 \text{ kJ}}{(28 \text{ kg/kmol})(10.5 \text{ kmol})} = 2,729 \text{ kJ/kg}$$

7. **Look up the temperature of nitrogen at the enthalpy value of 2,729 kJ/ kg using Table A-9 of the appendix.**

First convert the enthalpy to molar basis: $\bar{h} = (2,729 \text{ kJ/kg})(28 \text{ kg/kmol}) = 76,400 \text{ kJ/kmol}$. You need to interpolate the table.

$$T_{N_2} = 2,559 \text{ K}$$

The actual temperature of the products is a bit lower than 2,559 Kelvin because the enthalpies of CO_2 and H_2O are higher than the enthalpy of N_2.

8. **Choose an initial guess value for iteration to find the adiabatic flame temperature of the reaction.**

The value 2,400 Kelvin is a nice round number, and it's a little bit lower than 2,559 Kelvin, so I chose this value for my initial guess for the flame temperature. Also, the enthalpy values for the combustion products are listed at this temperature in Table A-9 without needing interpolation.

9. **Find the enthalpy of CO_2, H_2O, and N_2 at 2,400 Kelvin in Table A-9 of the appendix and substitute these values into the energy equation.**

$$\Delta \bar{h}_{CO_2} + 2\Delta \bar{h}_{H_2O} + 7.52 \ \Delta \bar{h}_{N_2} = \left(115,800 + 2(93,740) + 7.52(70,640)\right) \text{ kJ} = 834,500 \text{ kJ}$$

The enthalpy of the products ($\sum N_P \Delta \bar{h}_P$) is 834,500 kilojoules per kilomole at 2,400 Kelvin. This is a bit higher than the target value of 802,200 kilojoules per kilomole.

10. **Choose a second temperature value to use in your iteration to find the adiabatic flame temperature of the reaction.**

Because the enthalpy of the products is higher than the target, choose a lower temperature. The next lowest entry in Table A-9 is 2,200 Kelvin.

11. **Find the enthalpy of CO_2, H_2O, and N_2 at 2,200 Kelvin from Table A-9 and substitute them into the energy equation.**

$$\Delta \bar{h}_{CO_2} + 2\Delta \bar{h}_{H_2O} + 7.52 \ \Delta \bar{h}_{N_2} = \left(103,600 + 2(83,150) + 7.52(63,360)\right) \text{ kJ} = 746,400 \text{ kJ}$$

The enthalpy of the products ($\sum N_P \Delta \bar{h}_P$) after the second iteration is 746,400 kilojoules per kilomole. This number is lower than the target value of 802,200 kilojoules per kilomole. The actual flame temperature can be found by linear interpolation between 2,200 and 2,400 Kelvin.

12. **Estimate the adiabatic flame temperature by interpolating between 2,200 and 2,400 Kelvin so that the enthalpy of the products equals 802,200 kilojoules per kilomole.**

$$\frac{(T_P - 2,200) \text{ K}}{(2,400 - 2,200) \text{ K}} = \frac{(802,200 - 746,400) \text{ kJ/kmol}}{(834,500 - 746,400) \text{ kJ/kmol}}$$

Solving this equation for the temperature of the products (T_P) gives you an adiabatic flame temperature of 2,327 Kelvin.

Part V
The Part of Tens

Taking thermodynamics tests was particularly embarrassing for Mr. Ed because of his inability to keep his calculations to himself.

©RICHTENNANT

In this part . . .

Looking at nature and trying to figure out how it works isn't easy. But when you discover something new and it becomes a certified law of nature, you've done something great and it makes you famous. In Part V, I present to you ten of the most famous names in thermodynamics. In addition, I give you a brief overview of other thermodynamic systems that aren't covered in Part III. These systems contain several ideas for improving energy consumption. May you be inspired to develop something new and something great.

Chapter 17

Ten Famous Names in Thermodynamics

• •

In This Chapter

▶ Introducing ten of the most influential names in thermodynamics

▶ Finding out about the progression of thermodynamics through theory and engineering

• •

*I*n this chapter, I introduce you to ten famous names in thermodynamics. Many people on this list achieved fame through the need to develop more efficient engines (which at the time had an average efficiency of 2 percent). The drive to create bigger, better, more powerful machines can be seen through the groundbreaking discoveries and inventions made by these scientists and engineers. Others dedicated time to understanding the relationships between the nature of heat and its ability to produce useful work. The work of these ten thermodynamicists reflects the progression of thermodynamics not only in engineering but also in theory.

George Brayton

In 1872, George Brayton patented his two-stroke Ready Motor. The motor was similar to a steam engine, so it didn't require a spark plug to ignite fuel. Brayton employed a continuously burning flame to ignite each engine cycle. By doing so, he extended the combustion phase of the cycle and produced more power per unit of fuel. The continuous combustion process, which is now referred to as the Brayton cycle, is the basis for the gas turbine, and it's used in jet engines, rocket engines, and both internal and external combustion engines. In 1876, the Ready Motor was exhibited at the Centennial Exposition in Philadelphia. Despite being well-received, the engine became outdated within a few years. You can still see one of Brayton's engines at the Smithsonian in the American History Museum.

Nicolas Léonard Sadi Carnot

Often called the "Father of Thermodynamics," Sadi Carnot's most important contribution to the field of thermodynamics was his idea of an idealized heat engine (the Carnot cycle), generalizing the critical aspects of the steam engine. His thoughts were publicized in 1824, where he laid the groundwork for the second law of thermodynamics. Later, he showed that the efficiency (known as the Carnot efficiency) of such a system depends only on the temperatures of the two reservoirs between which the engine operates. The Carnot cycle is the most efficient engine possible, because it assumes that there is no friction or heat transfer using a temperature difference.

Anders Celsius

In 1742, Anders Celsius proposed an international temperature scale, now known as the Celsius scale, which much of the world uses today. His proposal was the result of experiments that were meant to define a temperature scale based on science rather than forms of measurement. The scale, measured with a 100-step thermometer (hence the name centigrade), was established to be 100 degrees at the boiling point of water and 0 at the freezing point. In some of his other experiments, Celsius was also able to show (with remarkable accuracy) that the boiling point of water varies with atmospheric pressure. He later wrote an expression on how to establish the boiling point of water should the barometric pressure deviate from ambient conditions.

Rudolf Diesel

Rudolf Diesel designed many heat engines, one of which was a solar-powered steam engine. But Diesel is most known as the first engineer to prove that fuel can be ignited without a flame or a spark, which he accomplished in the 1890s. In the engine he developed, fuel is injected at the end of compression when the temperature of the cylinder is above the fuel flash point (the temperature at which the fuel spontaneously combusts). Because of their high compression ratio, diesel engines have the highest thermal efficiency of any regular internal combustion engine. Originally, diesel engines were used as stationary engines; eventually, they were used in various modes of transportation, first with submarines and ships (replacing coal), then in locomotives and trucks.

Daniel Gabriel Fahrenheit

Daniel Fahrenheit, as you can tell by his name, is best known for his experiments and research in temperature scales. In the early 1700s, Fahrenheit invented the alcohol and mercury thermometers. The temperature scale that was the result of his research is built on the basis of three predetermined temperatures. The first was found by placing a thermometer into a mixture of ice water and salt and recording its lowest stable temperature, 0 degrees Fahrenheit. The next temperature was taken at the freezing point of water, 32 degrees Fahrenheit. The final reading was taken from a thermometer at body heat, around 96 degrees Fahrenheit.

James Prescott Joule

A physicist and brewer, James Joule was interested in demonstrating the link between heat and mechanical energy in the 1840s. Historically, heat was measured in thermal units (calories) while work was measured mechanically in force times distance, for which the unit is joules (named after Joule). Using a paddle powered by a falling weight, Joule found that measurements of work and heat were proportional, and he was able to estimate a heat equivalent of the work being done by the weight. This work (and evidence) led to the establishment and acceptance of the theory of conservation of energy, explaining why heat can do work. Joule also worked with Lord Kelvin to develop an absolute temperature scale. Their experiments led to the estimate that 819 foot-pounds of force are required to raise a pound of water by 1 degree Fahrenheit (the actual value is 778.17 foot-pounds of force).

Nikolaus August Otto

In the 1860s, most engines were noisy, shaky, and incredibly inefficient. Nikolaus Otto sought to resolve these issues (and thus increase efficiency and power) by finding a better way to combust fuel. He did this by using only one piston and extending the intake, compression, combustion, power, and exhaust processes from one stroke to four. Critics believed that such a machine would be inefficient because it wouldn't generate power at every rotation. However, Otto was the first to create a practical multistroke engine. His working model was the first internal combustion, four-stroke engine to efficiently burn fuel in a piston. What made Otto's design even more progressive was the fact that it used a spark plug to ignite the fuel.

William John Macquorn Rankine

William Rankine, a Scottish engineer, generated a comprehensive theory of the steam engine and other heat engines, starting in 1840. He is well-known for discovering relationships between saturation pressure and temperature, and the density of gases. His findings led to expressions for the latent heat of evaporation of a liquid. Rankine calculated the efficiency of heat engines. He based his findings on his own theory and was able to deduce, like Carnot, that the maximum efficiency of a heat engine depends only on the two separate temperatures at which it functions. The Rankine cycle, which analyzes a heat engine with a condenser and uses a condensable substance (like water) as a working fluid, is the basic cycle seen in steam power generation plants.

William Thomson or Lord Kelvin

Lord Kelvin is credited with developing the concept and basis of absolute zero. Through laboratory experiments, he confirmed his prediction that the melting point of ice would drop in the case of decreased pressure. Dissatisfied with the gas thermometer's dependence on the properties of a substance, Kelvin proposed an absolute temperature scale in 1848. This scale would be independent of the physical properties associated with any substance. This implied the idea of absolute zero, the point where no heat can be transferred. In 1851, Kelvin independently proposed the second law of thermodynamics, despite its first appearance the year before.

James Watt

After realizing that most engine designs wasted a large amount of heat through the cooling and heating of the cylinder, James Watt set out to improve existing engines in the 1760s. After a lot of experimenting, Watt demonstrated that most of the steam was used just to heat the cylinder because the steam was condensed by cold water. Watt's contribution was to add a separate condensation chamber that avoided significant heat loss. This addition improved efficiency by maintaining the cylinder at the injected-steam temperature. Watt also measured the power of a steam engine he built. He determined that one "horsepower" is about 550 foot-pounds/second. Later, the unit for power in the metric system (watt) was named after him.

Chapter 18

Ten More Cycles of Note

*I*n Chapters 10–13, I discuss many major thermodynamic cycles, such as the Carnot, Brayton, Otto, and diesel cycles. But keep in mind that heat transfer and heat engines can yield work in many other ways, too. Engineers are still trying to make better, more efficient machines. In this chapter, I introduce you to ten more thermodynamic cycles that are lesser-known, but still have an impact on thermodynamics. Many of these cycles are designed to improve the efficiency and performance of the standard cycles.

Two-Stroke Engines

The Otto and diesel cycles presented in Chapter 11 need four strokes to complete a cycle. But these cycles can be completed in only two strokes by modifying the piston-cylinder arrangement. A two-stroke engine, as the name suggests, completes the thermodynamic cycle (compression, combustion, expansion, exhaust, and intake) in only two strokes. A two-stroke engine uses the beginning of the compression stroke and the latter part of the power stroke to perform the intake and exhaust processes at the same time, a technique known as *scavenging*. You can see the intake and exhaust ports in Figure 18-1. The air and fuel are then compressed during the compression stroke and ignited. Because the exhaust gases are not completely expelled and some of the fuel tends to be removed along with the exhaust, two-stroke cycles are less efficient than four-stroke ones. But due to their high power-to-weight ratios and simplicity, they're widely used in functions requiring small weight, such as chainsaws and motorbikes.

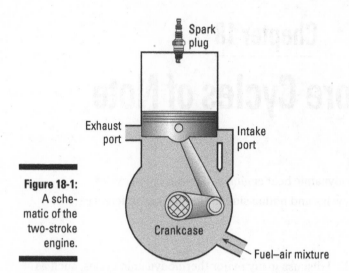

Figure 18-1:
A schematic of the two-stroke engine.

Wankel Engines

A Wankel engine is basically an Otto cycle that uses rotary motion rather than reciprocating motion to complete the four basic thermodynamic processes of a heat engine. Wankel engines have a unique and confusing design, so look carefully at Figure 18-2. Instead of pistons, the Wankel motor has a rotary design that uses pressure to create a rotating motion. A basic, single-rotor Wankel engine is made of a triangular shaped rotor, an oval-like housing, and a drive shaft.

As the drive shaft rotates, the rotor both rotates and revolves around a gear. Each rotation of the rotor creates a new cavity between the housing and one of the three rotor edges. As the cycle progresses, each cavity becomes smaller, thus raising the pressure in the cavity. After combustion, the cavity is at its greatest pressure, creating the expansion stroke.

The Wankel engine completes intake, compression, combustion, expansion, and exhaust like an Otto engine. However, it achieves this in only one revolution, making its power output higher than that of an Otto engine. Because the smooth, circular motion minimizes stress on the engine parts, Wankel engines can operate at incredibly high revolutions per minute, making their design optimal for speed. Unfortunately, some racing series have banned its use altogether.

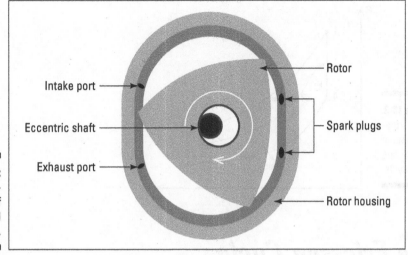

Figure 18-2:
A sche-
matic of
the Wankel
engine.

Intake port

Eccentric shaft

Exhaust port

Rotor

Spark plugs

Rotor housing

The Stirling Cycle

The Stirling cycle uses a gas working fluid that's confined to the engine. The gas is heated by an external source as opposed to an internal heat source where fuel is mixed with air in a combustion process. During the heating process from States 1 to 2, as shown in Figure 18-3, the piston expands, creating a nearly isothermal expansion, to get work out of the engine. At the end of the expansion stroke, a constant-volume heat exchange removes excess heat from the gas from States 2 to 3. The excess heat is stored in a device called a regenerator. Basically, a regenerator uses energy from the working fluid during one part of the cycle and then returns it to the fluid in another. The piston compresses the gas while rejecting heat from States 3 to 4 in a near isothermal manner. Then heat from the regenerator is added to the gas at the end of the piston stroke, from States 4 to 1.

The heat source for a Stirling engine can operate continuously; it doesn't need to be periodic as it does in an Otto or diesel cycle. This continuous operation makes the Stirling cycle a good candidate for harvesting concentrated solar energy.

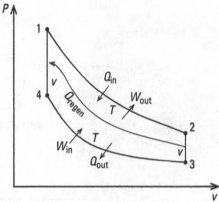

Figure 18-3:
A *P-v* dia-
gram of
the Stirling
cycle.

The Ericsson Cycle

The Ericsson cycle is very much like the Stirling cycle. It uses an external heating source on the gas contained within the engine. Figure 18-4 shows the *P-v* diagram of the cycle. The gas is compressed by a piston from States 1 to 2 while heat is removed in a nearly isothermal process. The gas expands while heat is added to the gas from the regenerator in a constant-pressure process from States 2 to 3. Additional heat is provided by an external heat source from States 3 to 4, and work is extracted at the same time, making this a nearly isothermal process. Finally, excess heat is removed to the regenerator at the end of the piston expansion stroke in a constant-pressure compression process from States 4 to 1.

The main difference between the Ericsson and Stirling cycles is the heat exchange with the regenerator. The Ericsson cycle uses a constant-pressure process, whereas the Stirling cycle uses a constant-volume process.

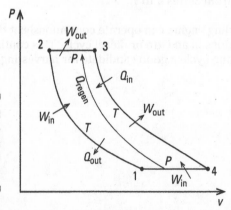

Figure 18-4:
A *P-v* dia-
gram of the
Ericsson
cycle.

The Atkinson Cycle

Although the Atkinson cycle was invented in 1882, its unique features of an extended power stroke and a short compression stroke are used in many modern hybrid automobile engines. By making the power stroke longer, the power output occurs at a lower rate, which makes the efficiency higher than that of most four-stroke cycles. The cycle incorporates all four thermodynamic engine processes into one turn of the crankshaft. The ideal Atkinson cycle consists of a six-step operation, as shown in the P-v diagram in Figure 18-5. Isentropic compression occurs from States 1 to 2, isochoric heating from States 2 to 3, isobaric heating during expansion from States 3 to 4, isentropic expansion from States 4 to 5, isochoric cooling from States 5 to 6, and isobaric cooling during compression from States 6 to 1.

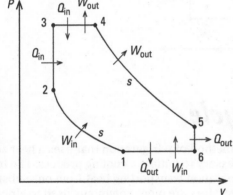

Figure 18-5:
A P-v diagram of the Atkinson cycle.

The Miller Cycle

Another cycle that four-stroke combustion engines employ is the Miller cycle. The Miller cycle is known for its "fifth stroke," a process that basically splits the compression stroke into two different steps. Figure 18-6 shows the P-v diagram of the Miller cycle. During compression from States 5 to 1, the intake valve is left open longer than in typical four-stroke cycles like the Otto cycle. After the intake valve closes, the isentropic compression stroke continues from States 1 to 2.

In order to account for the air loss created by the open valve, most Miller cycles use a supercharger to boost the amount of air inside the cylinder. The supercharger, along with the reduced compression stroke, produces higher cylinder pressure at a lower temperature, which means that the efficiency is greater than that of typical gasoline engines. The rest of the cycle is similar to the Otto cycle, with isochoric heat addition from States 2 to 3, isentropic expansion and power out from States 3 to 4, and heat out during the isochoric exhaust stroke from States 4 to 5.

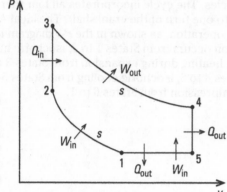

Figure 18-6:
A P-v diagram of the Miller cycle.

The Absorption Cycle

The absorption cycle is a method of refrigeration that uses a heat source to provide the energy necessary to initiate a cooling process. The basis of this process relies on evaporation transporting heat from one substance to another. Absorption refrigerators are more commonly used to air-condition commercial buildings using heat waste from turbines, water heaters, and other industrial processes.

Figure 18-7 shows an example of an absorption cycle. The absorber dissolves ammonia in water while rejecting heat. A liquid pump raises the pressure of the ammonia/water mixture, and a generator drives off the ammonia refrigerant vapor from the high-pressure liquid when heat is put into the system. The high-pressure water is returned from the generator to the absorber through a throttling valve. The hot ammonia vapor goes to a condenser, where it loses heat. The condensed liquid ammonia is throttled to a low pressure, where it absorbs heat from the environment in the evaporator. The warm ammonia vapor then goes back into the absorber to complete the cycle.

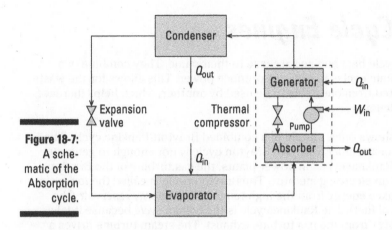

Figure 18-7:
A schematic of the Absorption cycle.

The Einstein Cycle

The Einstein cycle is another example of an absorption refrigerator, but unlike most designs, it uses no moving or mechanical parts. The cycle consists of a single-pressure absorption refrigerator that requires only a heat source to function. Figure 18-8 shows an Einstein cycle that uses ammonia, butane, and water as working fluids. The water pumps ammonia while the ammonia pumps butane, and butane serves as a refrigerant. The evaporation of butane absorbs heat at lower temperatures. A bubble pump moves the fluids through heat exchangers. Heat input to the generator creates bubbles that force fluids to circulate through the system.

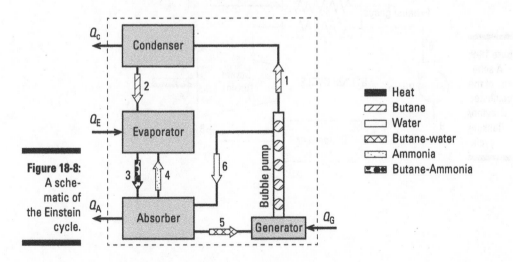

Figure 18-8:
A schematic of the Einstein cycle.

Combined-Cycle Engines

Combined-cycle heat engines are true to their name. They combine two thermodynamic cycles in order to produce power. This allows for the waste heat generated from one cycle to be used by another, which helps increase overall efficiency.

Figure 18-9 shows one example of a combined Brayton/Rankine cycle. The turbine exhaust temperature of a Brayton cycle is hot enough to generate steam for a Rankine cycle. In power plants, the gas turbine of the Brayton cycle drives an electric generator. The Brayton cycle is called the *top cycle* because it takes energy from the high-temperature energy reservoir (the combustion of fuel). The Rankine cycle is the *bottom cycle* because it takes energy left over from the gas turbine exhaust. The steam turbine drives a separate electric generator to produce power.

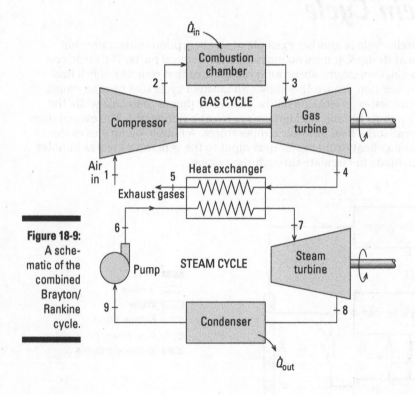

Figure 18-9:
A schematic of the combined Brayton/Rankine cycle.

Binary Vapor Cycles

Binary vapor cycles are similar to combined cycles in that two cycles are used together. Figure 18-10 shows one configuration of a binary cycle. Heat is added to a Rankine cycle with mercury as the working fluid. The mercury cycle operates at higher temperatures than the water cycle. The waste heat from the mercury turbine is used to generate saturated steam in a second Rankine cycle. The saturated steam passes through a superheater to gain additional energy before entering the steam turbine.

The binary vapor cycle is also used in geothermal power generation. Pumps bring hot water from a geothermal well through a heat exchanger to heat a secondary fluid — such as pentane — with a lower boiling point than water. This process causes the pentane to flash vapor, which can drive turbines. The cooled water is then sent back to the underground reservoir. The pentane is condensed and pumped back up to a high pressure to complete the cycle.

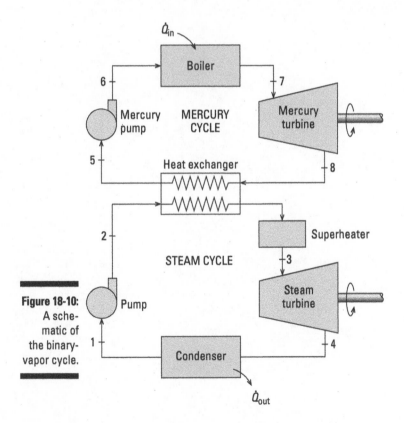

Figure 18-10: A schematic of the binary-vapor cycle.

Appendix

Thermodynamic Property Tables

Table A-1			Ideal-Gas Properties of Air			
T	u	h	s°	P_r	v_r	c_p
K	kJ/kg	kJ/kg	kJ/kg-K			kJ/kg-K
200	142.9	200.3	6.463	0.271	492.5	1.005
220	157.2	220.4	6.559	0.378	388.5	1.003
240	171.5	240.4	6.646	0.512	312.8	1.003
260	185.8	260.5	6.726	0.677	256.2	1.003
273.15	195.1	273.5	6.775	0.803	226.9	1.001
280	200.1	280.5	6.801	0.877	213.0	1.003
298	213.0	298.6	6.863	1.091	182.3	1.005
320	228.8	320.7	6.935	1.400	152.5	1.007
340	243.2	340.9	6.996	1.731	131.1	1.008
360	257.6	361.0	7.053	2.115	113.5	1.008
380	272.1	381.2	7.108	2.558	99.1	1.010
400	286.6	401.5	7.160	3.065	87.1	1.013
500	360.0	503.5	7.387	6.774	49.2	1.030
600	435.2	607.5	7.577	13.10	30.5	1.051
700	512.8	713.7	7.740	23.18	20.1	1.075
800	592.7	822.4	7.885	38.42	13.9	1.099
900	674.9	933.3	8.016	60.56	9.91	1.121
1,000	759.3	1,046	8.135	91.71	7.27	1.142
1,100	845.6	1,161	8.245	134.3	5.46	1.160
1,200	933.5	1,278	8.346	191.3	4.18	1.175
1,300	1,023	1,396	8.441	265.8	3.26	1.185
1,400	1,114	1,515	8.529	361.8	2.58	1.200
1,500	1,205	1,636	8.612	483.4	2.07	1.210
2,000	1,678	2,252	8.966	1,659	0.80	1.249
2,500	2,166	2,883	9.248	4,427	0.38	1.274
3,000	2,664	3,526	9.482	10,010	0.20	1.291

Table A-2				Compressed Liquid Water Properties			
T	**v**	**u**	**h**	**s**	**v**	**u**	**h**	**s**
°C	**m³/kg**	**kJ/kg**	**kJ/kg**	**kJ/kg-K**	**m³/kg**	**kJ/kg**	**kJ/kg**	**kJ/kg-K**
	P = 0.01 MPa (45.81°C)				P = 0.1 MPa (99.61°C)			
Sat	14.67	2,438	2,585	8.15	0.001043	417.3	417.4	1.302
0.01	0.001	0.0001397	0.01014	5.115e-7	0.001	0.001478	0.1015	5.403e-6
5	0.001	20.97	20.98	0.07609	0.001	20.97	21.07	0.07609
10	0.001	41.99	42.00	0.1510	0.001	41.99	42.09	0.0151
15	0.001001	62.98	62.99	0.2245	0.001001	63.98	63.08	0.2244
20	0.001002	83.94	83.95	0.2966	0.001002	83.94	84.03	0.2965
30	0.001004	125.8	125.8	0.4369	0.001004	125.8	125.9	0.4368
40	0.001008	167.5	167.5	0.5724	0.001008	167.5	167.6	0.5724
50					0.001012	209.3	209.4	0.7037
60					0.001017	251.1	251.2	0.8310
70					0.001023	292.9	293.0	0.9548
80					0.001029	334.8	334.9	1.075
90					0.001036	376.8	376.9	1.192
	P = 1 MPa (179.88°C)				P = 10 MPa (311.00°C)			
Sat	0.001127	761.7	762.8	2.139	0.001452	1,393	1,408	3.360
0.01	0.000999	0.01478	1.015	5.32e-5	0.0009952	0.1367	10.09	0.0004
20	0.001001	83.88	84.88	0.2964	0.0009972	83.35	93.32	0.2945
40	0.001007	167.4	168.4	0.5720	0.001003	166.3	176.4	0.5685
60	0.001017	250.9	251.9	0.8305	0.001013	249.3	259.5	0.8258
80	0.001029	334.6	335.6	1.075	0.001024	332.6	342.8	1.069
100	0.001043	418.7	419.7	1.306	0.001039	416.1	426.5	1.299
125	0.001064	524.4	525.5	1.580	0.001059	521.1	531.7	1.572
150	0.001090	631.4	632.5	1.841	0.001084	627.3	638.1	1.831
175	0.001121	740.1	741.2	2.091	0.001113	734.9	746	2.079
200					0.001148	844.5	856	2.318
250					0.001240	1,073	1,085	2.778
300					0.001397	1,328	1,342	3.247

Table A-3   Saturated Water Liquid-Vapor Properties — Temperature Table

Temp.	Sat. Press.	Specific Volume		Int. Energy		Enthalpy		Entropy	
		Sat. Liq.	Sat. Vap.	Sat. Liq.	Sat. Vap.	Sat. Liq.	Sat. Vap.	Sat. Liq.	Sat. Vap.
T	P_{sat}	v_f	v_g	u_f	u_g	h_f	h_g	s_f	s_g
°C	kPa	m³/kg	m³/kg	kJ/kg	kJ/kg	kJ/kg	kJ/kg	kJ/kg-K	kJ/kg-K
5	0.87	0.001000	147.1	20.97	2,382	20.98	2,511	0.0761	9.026
10	1.23	0.001000	106.4	41.99	2,389	41.99	2,520	0.1510	8.901
20	2.34	0.001002	57.79	83.94	2,403	83.94	2,538	0.2966	8.667
30	4.25	0.001004	32.89	125.8	2,417	125.8	2,556	0.4369	8.453
40	7.38	0.001008	19.52	167.5	2,430	167.5	2,574	0.5724	8.257
50	12.35	0.001012	12.03	209.3	2,443	209.3	2,592	0.7037	8.076
60	19.94	0.001017	7.671	251.1	2,457	251.1	2,610	0.8311	7.909
70	31.19	0.001023	5.042	292.9	2,470	293.0	2,627	0.9548	7.755
80	47.39	0.001029	3.407	334.8	2,482	334.9	2,644	1.075	7.612
90	70.14	0.001036	2.361	376.8	2,495	376.9	2,660	1.192	7.479
100	101.3	0.001044	1.673	418.9	2,506	419.0	2,676	1.307	7.355
120	198.5	0.001060	0.8919	503.5	2,529	503.7	2,706	1.528	7.130
140	361.3	0.001080	0.5089	588.7	2,550	589.1	2,734	1.739	6.930
160	617.8	0.001102	0.3071	674.9	2,568	675.5	2,758	1.943	6.750
180	1,002	0.001127	0.1940	762.1	2,584	763.2	2,778	2.140	6.586
200	1,554	0.001156	0.1274	850.6	2,595	852.4	2,793	2.331	6.432
220	2,318	0.001190	0.0862	940.9	2,602	943.6	2,802	2.518	6.286
240	3,344	0.001229	0.0598	1,033	2,604	1,037	2,804	2.701	6.144
260	4,689	0.001276	0.0422	1,128	2,599	1,134	2,797	2.884	6.002
280	6,412	0.001332	0.0302	1,227	2,586	1,236	2,780	3.067	5.857
300	8,581	0.001404	0.0217	1,332	2,563	1,344	2,749	3.253	5.704
320	11,270	0.001499	0.0155	1,445	2,525	1,461	2,700	3.448	5.536
340	14,590	0.001638	0.0108	1,570	2,465	1,594	2,622	3.659	5.336
360	18,650	0.001892	0.0070	1,725	2,351	1,760	2,481	3.915	5.052

Table A-4   Saturated Water Liquid-Vapor Properties — Pressure Table

Press.	Sat. Temp.	Specific Volume		Int. Energy		Enthalpy		Entropy	
		Sat. Liq.	Sat. Vap.	Sat. Liq.	Sat. Vap.	Sat. Liq.	Sat. Vap.	Sat. Liq.	Sat. Vap.
P	T_{sat}	v_f	v_g	u_f	u_g	h_f	h_g	s_f	s_g
kPa	°C	m³/kg	m³/kg	kJ/kg	kJ/kg	kJ/kg	kJ/kg	kJ/kg-K	kJ/kg-K
1	6.977	0.001000	129.2	29.29	2,385	29.29	2,514	0.1059	8.976
3	24.08	0.001003	45.67	101.0	2,409	101.0	2,546	0.3545	8.578
5	32.88	0.001005	28.19	137.8	2,420	137.8	2,561	0.4763	8.395
10	45.81	0.001010	14.67	191.8	2,438	191.8	2,585	0.6492	8.150
15	53.97	0.001014	10.02	225.9	2,449	225.9	2,599	0.7548	8.008
25	64.97	0.001020	6.204	271.9	2,463	271.9	2,618	0.893	7.831
50	81.33	0.001030	3.24	340.4	2,484	340.5	2,646	1.091	7.594
75	91.77	0.001037	2.217	384.3	2,497	384.4	2,663	1.213	7.456
100	99.62	0.001043	1.694	417.3	2,506	417.4	2,675	1.303	7.359
200	120.2	0.001061	0.8857	504.5	2,529	504.7	2,707	1.530	7.127
300	133.5	0.001073	0.6058	561.1	2,544	561.4	2,725	1.672	6.992
400	143.6	0.001084	0.4625	604.3	2,554	604.7	2,739	1.777	6.896
500	151.9	0.001093	0.3749	639.7	2,561	640.2	2,749	1.861	6.821
600	158.9	0.001101	0.3157	669.9	2,567	670.5	2,757	1.931	6.760
800	170.4	0.001115	0.2404	720.2	2,577	721.1	2,769	2.046	6.663
1,000	179.9	0.001127	0.1944	761.7	2,584	762.8	2,778	2.139	6.586
1,500	198.3	0.001154	0.1318	843.1	2,594	844.9	2,792	2.315	6.445
2,000	212.4	0.001177	0.09963	906.4	2,600	908.8	2,800	2.447	6.341
2,500	224.0	0.001197	0.07998	959.1	2,603	962.1	2,803	2.555	6.257
5,000	264.0	0.001286	0.03944	1,148	2,597	1,154	2,794	2.920	5.973
7,500	290.6	0.001368	0.02532	1,282	2,575	1,292	2,765	3.165	5.778
10,000	311.1	0.001452	0.01803	1,393	2,544	1,408	2,725	3.360	5.614
12,500	327.9	0.001547	0.01350	1,492	2,505	1,511	2,674	3.529	5.462
15,000	342.2	0.001658	0.01034	1,586	2,455	1,610	2,610	3.685	5.310

Table A-5 Superheated Steam Properties

T	vol	u	h	s	vol	u	h	s
°C	m³/kg	kJ/kg	kJ/kg	kJ/kg-K	m³/kg	kJ/kg	kJ/kg	kJ/kg-K
	P = 0.01 MPa (45.81°C)				P = 0.1 MPa (99.61°C)			
Sat	14.67	2,438	2,585	8.150	1.694	2,506	2,675	7.359
100	17.20	2,515	2,687	8.448	1.696	2,507	2,676	7.361
200	21.83	2,661	2,880	8.904	2.172	2,658	2,875	7.834
300	26.45	2,812	3,077	9.281	2.639	2,810	3,074	8.216
400	31.06	2,969	3,280	9.608	3.103	2,968	3,278	8.543
500	35.68	3,132	3,489	9.898	3.565	3,132	3,488	8.834
600	40.29	3,302	3,705	10.16	4.028	3,302	3,705	9.097
700	44.91	3,480	3,929	10.40	4.490	3,479	3,928	9.340
800	49.53	3,664	4,159	10.63	4.952	3,664	4,159	9.565
900	54.14	3,855	4,396	10.84	5.414	3,855	4,396	9.777
1,000	58.76	4,053	4,641	11.04	5.875	4,053	4,640	9.976
1,100	63.37	4,257	4,891	11.23	6.337	4257	4,891	10.17
	P = 1 MPa (179.88°C)				P = 10 MPa (311.00°C)			
Sat	0.1944	2,584	2,778	6.586	0.018030	2,544	2,725	5.614
200	0.206	2,622	2,828	6.694	0.001148	844.5	856	2.318
300	0.2579	2,793	3,051	7.123	0.001397	1,328	1,342	3.247
400	0.3066	2,957	3,264	7.465	0.02641	2,832	3,096	6.212
500	0.3541	3,124	3,478	7.762	0.03279	3,046	3,374	6.597
600	0.4011	3,297	3,698	8.029	0.03837	3,242	3,625	6.903
700	0.4478	3,475	3,923	8.273	0.04358	3,435	3,871	7.169
800	0.4943	3,660	4,155	8.5	0.04859	3,629	4,115	7.408
900	0.5407	3,852	4,393	8.712	0.05349	3,826	4,361	7.627
1,000	0.5871	4,050	4,638	8.912	0.05832	4,028	4,611	7.831
1,100	0.6335	4,255	4,889	9.102	0.06312	4,234	4,865	8.024
1,200	0.6798	4,466	5,145	9.282	0.06789	4,445	5,124	8.205
1,300	0.7261	4,681	5,407	9.454	0.07265	4,660	5,387	8.378

Table A-6 — Saturated R-134a Liquid-Vapor Properties — Temperature Table

Temp.	Sat. Press.	Specific Volume		Int. Energy		Enthalpy		Entropy	
		Sat. Liq.	Sat. Vap.	Sat. Liq.	Sat. Vap.	Sat. Liq.	Sat. Vap.	Sat. Liq.	Sat. Vap.
T	P_{sat}	v_f	v_g	u_f	u_g	h_f	h_g	s_f	s_g
°C	kPa	m³/kg	m³/kg	kJ/kg	kJ/kg	kJ/kg	kJ/kg	kJ/kg-K	kJ/kg-K
−20	132.7	0.000736	0.14740	173.5	367	173.6	386.6	0.9002	1.741
−10	200.6	0.000754	0.09959	186.5	372.7	186.7	392.7	0.9506	1.733
0	292.8	0.000772	0.06931	199.8	378.3	200.0	398.6	1.000	1.727
10	414.6	0.000793	0.04944	213.2	383.8	213.6	404.3	1.048	1.722
20	571.7	0.000816	0.03600	227.0	389.2	227.5	409.7	1.096	1.718
30	770.2	0.000842	0.02664	241.1	394.3	241.7	414.8	1.144	1.714
40	1,017	0.000872	0.01997	255.5	399.1	256.4	419.4	1.190	1.711
50	1,318	0.000907	0.01509	270.4	403.6	271.6	423.4	1.237	1.707
60	1,682	0.000950	0.01144	285.9	407.4	287.5	426.6	1.285	1.702

Table A-7 — Saturated R-134a Liquid-Vapor Properties — Pressure Table

Press.	Sat. Temp.	Specific Volume		Int. Energy		Enthalpy		Entropy	
		Sat. Liq.	Sat. Vap.	Sat. Liq.	Sat. Vap.	Sat. Liq.	Sat. Vap.	Sat. Liq.	Sat. Vap.
P	T_{sat}	v_f	v_g	u_f	u_g	h_f	h_g	s_f	s_g
kPa	°C	m³/kg	m³/kg	kJ/kg	kJ/kg	kJ/kg	kJ/kg	kJ/kg-K	kJ/kg-K
100	−26.36	0.000726	0.19260	165.4	363.3	165.4	382.6	0.8676	1.747
150	−17.13	0.000741	0.13130	177.2	368.6	177.4	388.3	0.9148	1.739
200	−10.08	0.000753	0.09988	186.4	372.6	186.6	392.6	0.9503	1.733
250	−4.284	0.000764	0.08069	194.1	375.9	194.3	396.1	0.979	1.730
400	8.931	0.000791	0.05121	211.8	383.2	212.1	403.7	1.043	1.723
600	21.57	0.000820	0.03430	229.2	390.0	229.7	410.6	1.104	1.717
800	31.33	0.000846	0.02562	243.0	395.0	243.6	415.5	1.150	1.714
1,000	39.39	0.000870	0.02032	254.6	398.8	255.5	419.2	1.188	1.711
1,200	46.31	0.000894	0.01672	264.9	402.0	265.9	422.0	1.220	1.709

Table A-8   Superheated R-134a Properties

T	vol	u	h	s	vol	u	h	s
°C	m³/kg	kJ/kg	kJ/kg	kJ/kg-K	m³/kg	kJ/kg	kJ/kg	kJ/kg-K
P = 100 kPa (−26.37°C)					P = 250 kPa (−4.28°C)			
Sat	0.1926	363.3	382.6	1.747	0.080690	375.9	396.1	1.730
−20	0.1984	367.8	387.6	1.768	0.000736	173.5	173.7	0.9001
−10	0.2074	374.9	395.6	1.799	0.000753	186.5	186.7	0.9506
0	0.2163	382.1	403.7	1.829	0.08244	379.2	399.8	1.743
10	0.2251	389.4	412.0	1.858	0.08640	387.0	408.6	1.775
20	0.2337	396.9	420.3	1.887	0.09024	394.7	417.3	1.805
30	0.2423	404.6	428.8	1.916	0.09399	402.6	426.1	1.835
40	0.2509	412.4	437.5	1.944	0.09768	410.6	435.0	1.864
50	0.2594	420.4	446.3	1.972	0.1013	418.7	444.1	1.892
60	0.2678	428.5	455.3	1.999	0.1049	427.0	453.2	1.920
80	0.2847	445.2	473.7	2.053	0.1120	444.0	472.0	1.975
100	0.3014	462.6	492.7	2.105	0.1189	461.5	491.2	2.028
P = 500 kPa (15.71°C)					P = 1,000 kPa (39.37°C)			
Sat	0.04112	386.9	407.5	1.720	0.020320	398.8	419.2	1.711
20	0.04212	390.5	411.6	1.734	0.000815	226.7	227.5	1.095
30	0.04434	399.0	421.2	1.766	0.000841	240.9	241.7	1.143
40	0.04646	407.4	430.6	1.797	0.02041	399.5	419.9	1.713
50	0.04850	415.9	440.1	1.826	0.02180	409.1	430.9	1.748
60	0.05049	424.4	449.6	1.856	0.02307	418.5	441.5	1.781
80	0.05433	441.8	468.9	1.912	0.02540	437.0	462.4	1.841
100	0.05805	459.6	488.7	1.966	0.02755	455.7	483.2	1.899
120	0.06169	478.0	508.9	2.019	0.02959	474.6	504.2	1.954
140	0.06526	497.0	529.6	2.070	0.03155	494.0	525.6	2.007
160	0.06878	516.5	550.9	2.121	0.03346	513.8	547.3	2.058

Table A-9   Ideal-Gas Properties of Combustion Gases

Temp.	Nitrogen		Oxygen		Carbon Dioxide		Water	
	$\bar{h}$	$\bar{s}°$	$\bar{h}$	$\bar{s}°$	$\bar{h}$	$\bar{s}°$	$\bar{h}$	$\bar{s}°$
K	$\dfrac{kJ}{kmol}$	$\dfrac{kJ}{kmol\ K}$	$\dfrac{kJ}{kmol}$	$\dfrac{kJ}{kmol\ K}$	$\dfrac{kJ}{kmol}$	$\dfrac{kJ}{kmol\ K}$	$\dfrac{kJ}{kmol}$	$\dfrac{kJ}{kmol\ K}$
298	−4.355	191.6	−4.395	205.1	−5.554	213.8	−5.022	188.8
300	53.89	191.8	54.39	205.3	68.77	214.0	62.14	189.0
400	2,973	200.2	3,027	213.9	4,003	225.3	3,450	198.8
500	5,912	206.7	6,086	220.7	8,307	234.9	6,922	206.5
600	8,896	212.2	9,245	226.5	12,910	243.3	10,500	213.1
700	11,940	216.9	12,500	231.5	17,760	250.8	14,190	218.7
800	15,020	221.0	15,840	235.9	22,810	257.5	18,000	223.8
900	18,220	224.8	19,240	239.9	28,030	263.6	21,940	228.5
1,000	21,460	228.2	22,700	243.6	33,400	269.3	26,000	232.7
1,200	28,210	234.3	29,760	250.0	44,480	279.4	34,510	240.5
1,400	34,940	239.5	36,960	255.6	55,900	288.2	43,490	247.4
1,600	41,910	244.1	44,270	260.4	67,580	296.0	52,910	253.7
1,800	48,980	248.3	51,680	264.8	79,440	303.0	62,690	259.5
2,000	56,140	252.1	59,180	268.8	91,450	309.3	72,790	264.8
2,200	63,360	255.5	66,780	272.4	103,600	315.1	83,150	269.7
2,400	70,640	258.7	74,460	275.7	115,800	320.4	93,740	274.3
2,600	77,970	261.6	82,230	278.8	128,100	325.3	104,600	278.6
2,800	85,330	264.3	90,090	281.7	140,400	329.9	115,500	282.7
3,000	92,720	266.9	98,020	284.5	152,900	334.2	126,600	286.5

Note: Entropy values at 100 kPa. Also, enthalpy values are zero at 298.15 K.

Table A-10   Thermodynamics Properties of Various Materials

Material	Temperature	Density	Specific Heat	Conductivity
Gases	T, °C	ρ, kg/m^3	C_p, kJ/kg-K	k_t, W/m-K
Air	25	1.184	1.005	0.026
	100	0.946	1.010	0.031
	500	0.457	1.093	0.056
	1,000	0.277	1.184	0.082
CO_2	25	1.799	0.840	0.017
	100	1.434	0.915	0.023
	500	0.694	1.156	0.053
	1,000	0.421	1.292	0.085
Liquids				
Ammonia (sat. liq.)	−20	665	4.52	0.622
	0	639	4.60	0.559
	25	602	4.80	0.485
Ethanol	25	783	2.46	0.161
Kerosene	20	820	2.00	0.15
Methanol	25	787	2.55	0.198
Octane	20	703	2.10	0.129
R-134a (sat. liq.)	−50	1,443	1.23	0.113
	0	1,295	1.34	0.095
	25	1,207	1.43	0.083
Water	0	1,000	4.22	0.548
	25	997	4.18	0.595
	50	988	4.18	0.631
	75	975	4.19	0.653
	100	958	4.22	0.665
Solids				
Aluminum	27	2,700	0.902	237
Copper	27	8,900	0.386	401
Ice	0	921	2.11	2.0
Iron	27	7,840	0.45	80.2
Steel	27	7,830	0.50	36.0

Note: Gas pressure at 101.325 kPa

Table A-11   Critical Point Properties of Various Materials

Substance	Molecular Mass	R, Gas Constant	Critical-Point Properties	
	kg/kmol	kJ/kg-K	Temp., K	Press., MPa
Air	28.97	0.2870	132.5	3.77
Ammonia	17.03	0.4882	405.5	11.28
Carbon dioxide	44.01	0.1889	304.2	7.39
Ethanol	46.07	0.1805	516.0	6.38
Methanol	32.04	0.2595	513.2	7.95
Nitrogen	28.01	0.2968	126.0	3.39
Oxygen	32.00	0.2598	155.0	5.08
Propane	44.10	0.1885	370.0	4.26
R-134a	102.03	0.0815	374.2	4.06
Water	18.02	0.4615	647.1	22.06

Index

• S •

le & Macs

d For Dummies
-0-470-58027-1

one For Dummies,
Edition
-0-470-87870-5

cBook For Dummies, 3rd
tion
-0-470-76918-8

c OS X Snow Leopard For
mmies
-0-470-43543-4

siness

okkeeping For Dummies
-0-7645-9848-7

Interviews
Dummies,
Edition
-0-470-17748-8

sumes For Dummies,
Edition
-0-470-08037-5

rting an
line Business
Dummies,
Edition
-0-470-60210-2

ck Investing
Dummies,
Edition
-0-470-40114-9

ccessful
ne Management
Dummies
-0-470-29034-7

Computer Hardware

BlackBerry
For Dummies,
4th Edition
978-0-470-60700-8

Computers For Seniors
For Dummies,
2nd Edition
978-0-470-53483-0

PCs For Dummies,
Windows
7th Edition
978-0-470-46542-4

Laptops For Dummies,
4th Edition
978-0-470-57829-2

Cooking & Entertaining

Cooking Basics
For Dummies,
3rd Edition
978-0-7645-7206-7

Wine For Dummies,
4th Edition
978-0-470-04579-4

Diet & Nutrition

Dieting For Dummies,
2nd Edition
978-0-7645-4149-0

Nutrition For Dummies,
4th Edition
978-0-471-79868-2

Weight Training
For Dummies,
3rd Edition
978-0-471-76845-6

Digital Photography

Digital SLR Cameras &
Photography For Dummies,
3rd Edition
978-0-470-46606-3

Photoshop Elements 8
For Dummies
978-0-470-52967-6

Gardening

Gardening Basics
For Dummies
978-0-470-03749-2

Organic Gardening
For Dummies,
2nd Edition
978-0-470-43067-5

Green/Sustainable

Raising Chickens
For Dummies
978-0-470-46544-8

Green Cleaning
For Dummies
978-0-470-39106-8

Health

Diabetes For Dummies,
3rd Edition
978-0-470-27086-8

Food Allergies
For Dummies
978-0-470-09584-3

Living Gluten-Free
For Dummies,
2nd Edition
978-0-470-58589-4

Hobbies/General

Chess For Dummies,
2nd Edition
978-0-7645-8404-6

Drawing
Cartoons & Comics
For Dummies
978-0-470-42683-8

Knitting For Dummies,
2nd Edition
978-0-470-28747-7

Organizing
For Dummies
978-0-7645-5300-4

Su Doku For Dummies
978-0-470-01892-7

Home Improvement

Home Maintenance
For Dummies,
2nd Edition
978-0-470-43063-7

Home Theater
For Dummies,
3rd Edition
978-0-470-41189-6

Living the
Country Lifestyle
All-in-One
For Dummies
978-0-470-43061-3

Solar Power Your Home
For Dummies,
2nd Edition
978-0-470-59678-4

Internet

Blogging For Dummies,
3rd Edition
978-0-470-61996-4

eBay For Dummies,
6th Edition
978-0-470-49741-8

Facebook For Dummies,
3rd Edition
978-0-470-87804-0

Web Marketing
For Dummies,
2nd Edition
978-0-470-37181-7

WordPress
For Dummies,
3rd Edition
978-0-470-59274-8

Language & Foreign Language

French For Dummies
978-0-7645-5193-2

Italian Phrases
For Dummies
978-0-7645-7203-6

Spanish For Dummies,
2nd Edition
978-0-470-87855-2

Spanish
For Dummies,
Audio Set
978-0-470-09585-0

Math & Science

Algebra I
For Dummies,
2nd Edition
978-0-470-55964-2

Biology For Dummies,
2nd Edition
978-0-470-59875-7

Calculus For Dummies
978-0-7645-2498-1

Chemistry For Dummies
978-0-7645-5430-8

Microsoft Office

Excel 2010 For Dummies
978-0-470-48953-6

Office 2010 All-in-One
For Dummies
978-0-470-49748-7

Office 2010 For Dummies,
Book + DVD Bundle
978-0-470-62698-6

Word 2010 For Dummies
978-0-470-48772-3

Music

Guitar For Dummies,
2nd Edition
978-0-7645-9904-0

iPod & iTunes For
Dummies, 8th Edition
978-0-470-87871-2

Piano Exercises
For Dummies
978-0-470-38765-8

Parenting & Education

Parenting For Dummies,
2nd Edition
978-0-7645-5418-6

Type 1 Diabetes
For Dummies
978-0-470-17811-9

Pets

Cats For Dummies,
2nd Edition
978-0-7645-5275-5

Dog Training For Dummies,
3rd Edition
978-0-470-60029-0

Puppies For Dummies,
2nd Edition
978-0-470-03717-1

Religion & Inspiration

The Bible For Dummies
978-0-7645-5296-0

Catholicism For Dummies
978-0-7645-5391-2

Women in the Bible
For Dummies
978-0-7645-8475-6

Self-Help & Relationship

Anger Management
For Dummies
978-0-470-03715-7

Overcoming Anxiety
For Dummies,
2nd Edition
978-0-470-57441-6

Sports

Baseball
For Dummies,
3rd Edition
978-0-7645-7537-2

Basketball
For Dummies,
2nd Edition
978-0-7645-5248-9

Golf For Dummies,
3rd Edition
978-0-471-76871-5

Web Development

Web Design
All-in-One
For Dummies
978-0-470-41796-6

Web Sites
Do-It-Yourself
For Dummies,
2nd Edition
978-0-470-56520-9

Windows 7

Windows 7
For Dummies
978-0-470-49743-2

Windows 7
For Dummies,
Book + DVD Bundle
978-0-470-52398-8

Windows 7 All-in-One
For Dummies
978-0-470-48763-1

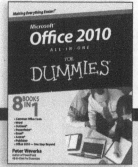

Available wherever books are sold. For more information or to order direct: U.S. customers visit www.dummies.com or call 1-877-762-29
U.K. customers visit www.wileyeurope.com or call (0) 1243 843291. Canadian customers visit www.wiley.ca or call 1-800-567-4797.